An Introduction to the Sun and Stars

Compiled by a team of experts from The Open University and the University of Bath, this textbook has been designed for elementary university courses in astronomy and astrophysics. It starts with a detailed discussion of our nearest star, the Sun, and describes how solar physicists have come to understand its internal workings. It then considers how astronomers go about studying the basic physical properties and life-cycles of more distant stars, and culminates with a discussion of the formation of exotic objects such as neutron stars and black holes. Written in an accessible style that avoids complex mathematics, and illustrated in colour throughout, this book is suitable for self-study and will appeal to amateur astronomers as well as undergraduate students. It contains numerous helpful learning features such as boxed summaries, student exercises with full solutions, and a glossary of terms. The book is also supported by a website hosting further teaching materials: **http://publishing.cambridge.org/resources/0521546222**

About the editors:

SIMON GREEN completed his PhD at the University of Leicester and is now a Senior Lecturer in the Planetary and Space Sciences Research Institute at The Open University. He has written and taught astronomy courses for 15 years and is a regular speaker at astronomical societies and public understanding of science events. His research is centred on small Solar System bodies (comets, asteroids, interplanetary dust and space debris) through ground-based observations, analysis of spacecraft data, laboratory experimentation and modelling. He has represented the scientific community on a number of ESA and UK Research Council committees and is currently a Royal Astronomical Society council member.

MARK JONES is a Staff Tutor in the Department of Physics and Astronomy at The Open University. He studied for his PhD in the X-ray astronomy group at the University of Leicester, before carrying out post-doctoral research in space-based infrared astronomy at Queen Mary and Westfield (University of London). He has a wide range of astronomical interests and has carried out research on interplanetary dust, the interstellar medium, interacting binary stars and active galaxies. He has been a Fellow of the Royal Astronomical Society since 1991.

Background image: This composite image from the Solar and Heliospheric Observatory combines EIT (Extreme ultraviolet Imaging Telescope) images from three wavelengths (17.1 nm, 19.5 nm and 28.4 nm) to reveal solar features unique to each wavelength. (Courtesy of SOHO/EIT consortium. SOHO is a project of international cooperation between ESA and NASA)

Thumbnail images: (from the left) The Pleiades; the planetary nebula IC 418; rings of material around supernova 1987A; the Trifid Nebula. (All these images are at visible wavelengths.) (Royal Observatory Edinburgh/Anglo Australian Telescope Board/David Malin; The planetary nebula IC 418, © Hubble Space Site; Rings of material around supernova 1987A, © Hubble Space Site; © NASA and Jeff Hester (Arizona State University))

An Introduction to the Sun and Stars

Edited by Simon F. Green and Mark H. Jones

Authors:

S. Jocelyn Bell Burnell

Simon F. Green

Barrie W. Jones

Mark H. Jones

Robert J. A. Lambourne

John C. Zarnecki

PUBLISHED BY THE PRESS SYNDICATE OF THE UNIVERSITY OF CAMBRIDGE

The Pitt Building, Trumpington Street, Cambridge, United Kingdom

CAMBRIDGE UNIVERSITY PRESS

The Edinburgh Building, Cambridge, CB2 2RU, UK

40 West 20th Street, New York, NY 10011–4211, USA

477 Williamstown Road, Port Melbourne, VIC 3207, Australia

Ruiz de Alarcón 13, 28014 Madrid, Spain

Dock House, The Waterfront, Cape Town 8001, South Africa

http://www.cambridge.org

First published 2003

This co-published edition first published 2004

Edited, designed and typeset by The Open University.

Printed and bound in the United Kingdom by Bath Press, Blantyre Industrial Estate, Glasgow G72 0ND, UK

A catalogue record for this book is available from the British Library

ISBN 0 521 83737 5   hardback
ISBN 0 521 54622 2   paperback

This publication forms part of an Open University course S282 *Astronomy*. Details of this and other Open University courses can be obtained from the Course Information and Advice Centre, PO Box 724, The Open University, Milton Keynes MK7 6ZS, United Kingdom: tel. +44 (0)1908 653231, e-mail general-enquiries@open.ac.uk

Alternatively, you may visit the Open University website at http://www.open.ac.uk where you can learn more about the wide range of courses and packs offered at all levels by The Open University.

To purchase a selection of Open University course materials visit the webshop at www.ouw.co.uk, or contact Open University Worldwide, Michael Young Building, Walton Hall, Milton Keynes MK7 6AA, United Kingdom for a brochure. tel. +44 (0)1908 858785; fax +44 (0)1908 858787; e-mail ouwenq@open.ac.uk

1.1

CONTENTS

INTRODUCTION

The sight of a star-filled sky on a moonless night cannot fail to inspire in us a sense of wonder at the beauty and scale of the Universe. Unfortunately, the true splendour of this view is often unattainable for many of us who live under the light-polluted skies of modern towns and cities. The first experience of a truly dark site, revealing thousands of stars and the sky-circling glow of the Milky Way (Figure 0.1), is often cited as the defining moment at the start of a lifetime's interest in astronomy. The popularity of astronomy, evident from the proliferation of amateur societies, books, television programmes and media interest, derives from this sense of wonder and desire to know, in the most fundamental sense, 'where we came from'. The daily circulation of the Sun, the monthly sweep of the Moon with its changing phases around the sky, and the annual motion of the Sun across the backdrop of the stars, played a vital role in ancient societies. Recognition of these cycles provided a clock from which daily activities could be planned and against which the seasons could be measured for crop planting, as well as being a vital aid to navigation.

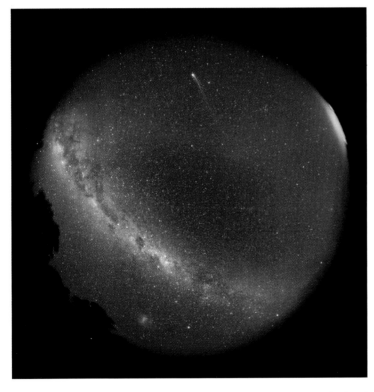

Figure 0.1 A view of the southern sky dominated by the band of the Milky Way, cut by dark interstellar clouds. Comet Hyakutake and the glow of lights from a nearby town are also visible. (G. Garrad)

In ages when religion and superstition were intimately linked to heavenly phenomena, any unexpected phenomena were often regarded with great trepidation. The desire to *understand* these phenomena led to what we now call the 'scientific method'. By observing the motions and properties of objects in the sky, it was possible to understand more about the Universe.

The scientific method, when applied to phenomena we can control on the Earth or in laboratories, includes experiments that we can design and conduct to test a hypothesis (a proposed explanation). For example, if we hypothesize that plants grow towards light we can perform an experiment with plants exposed to light from different directions (and some deprived of light) to test if this is true. Of course we must also ensure that our experiment is well designed so that we can rule out other explanations, such as that plants grow 'up' against gravity. Astronomers, generally, do not have this luxury. We can only *observe* phenomena in the sky and attempt to explain them. A hypothesis can be tested only by predicting further phenomena which may be observed. This inability to change conditions beyond the Earth to test our ideas, can lead to erroneous theories (self-consistent representations of reality) becoming established. For example, the apparent lack of movement of stars relative to each other was used to infer that the Earth was stationary at the centre of the Universe, a consensus which held sway for almost two thousand years in Europe. Only better observations of the motions of the planets, and eventually, the motions of stars, could provide proof that the Earth and Sun do not occupy a special place in the Universe. The observation of *change* in the Universe provides us with phenomena to help us understand and explain its properties. We can now measure the motions of stars and detect changes in their brightness on timescales of milliseconds to decades. However, we cannot *observe* the evolution of a star which proceeds on timescales of millions or billions of years; we must deduce this evolution by observations of many stars, which are at different stages of their evolution, and try to piece together this information to test our theories.

Astronomy has been the driving force, and the Universe the natural laboratory, for many of the greatest advances in science. Prominent examples include: Newton's laws of motion, the theories of gravity due to both Newton and Einstein, and the understanding of atomic structure and nuclear reactions.

This book is concerned with the properties and evolution of the most immediately visible objects in the Universe – stars. What are the basic physical properties of stars? How bright are they? How massive are they? What are their sizes? How are they distributed in space?

The simplest observation we can make is that the stars do not all appear to have the same brightness. The most brilliant star in the night sky is 100 million times fainter than the Sun and around 1000 times brighter than the faintest star which can be glimpsed with the unaided eye on a moonless night from a dark observing site. Several thousand stars are observable over the whole sky, but binoculars or a telescope reveal many more. The larger the telescope (i.e. the size of the mirror or lens which collects the light) the fainter and more numerous the observable stars become. If the eye (which 'stores' its image for a fraction of a second) is replaced by a sensitive detector (which records the faint starlight over long periods) attached to one of the world's largest telescopes, stars which are more than 100 million times fainter still are detectable.

The immediate question which a scientist may then ask is 'Does this range of brightness reflect the true variety of intrinsic stellar brightness (luminosity) or merely that the stars are at different distances?' We can only answer this question by determining the distances to the stars. These distances are so vast that conventional units of distance (metres or kilometres) are completely inadequate for expressing them. The Sun lies at a distance of 150 million kilometres from the Earth. The next nearest star is several hundred thousand times further away. Determining the distances of stars is one of the most fundamental problems of astronomy and is essential for deriving many of their intrinsic physical properties.

In fact, stars exhibit a wide range of properties and the Sun is a rather average star. The brightest stars are a million times more luminous than the Sun and the least luminous are tens of thousands of times fainter. Their masses, however, cover a much more restricted range, from just under a tenth of the Sun's mass to around one hundred times more massive. The smallest stars have radii of only around 10 km, but with densities so high that a thimbleful of material has a mass of more than a hundred million tonnes! The largest stars, if placed at the position of the Sun, would entirely engulf the Earth's orbit. The temperatures of the surfaces of stars (the only part we can measure directly) range from more than 50 000 K to less than 3000 K. For many stars, these properties appear to be related. The most luminous are also the hottest, the most massive and large, while the least luminous are the coolest, least massive and small. There are exceptions to this sequence. For example, there are highly luminous, large but cool stars, and hot but faint small stars (see Figure 0.2), while others undergo observable changes to these properties. Any theories of the structure and evolution of stars must be consistent with all these observations.

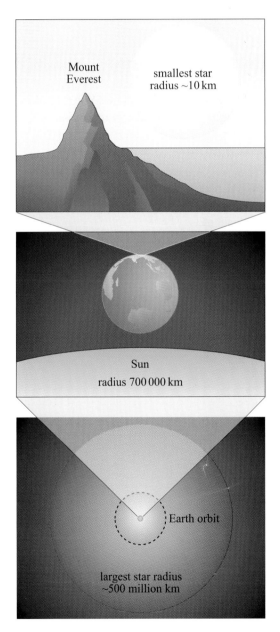

Figure 0.2 The relative sizes of stars.

(a)

(b)

Figure 0.3 Spiral galaxies, similar to our own Milky Way galaxy, seen (a) edge-on and (b) face-on. (D. Malin/IAC/RGO)

As is easily seen in Figure 0.1, stars are not distributed uniformly across the night sky. Even a small telescope will reveal that the Milky Way consists of vast numbers of stars too faint to be distinguished individually by the unaided eye. In fact there are around a hundred thousand million stars in our Milky Way galaxy. This in turn is just one of many billions of galaxies in the Universe. Although we are located inside it, careful observations have revealed its structure to be that of a flattened spiral, like the galaxies shown in Figure 0.3. The Sun is approximately one-third of the way out from the centre of the Galaxy (see Figure 0.4) which is over 6000 times further away than the nearest star. This flattened spiral (called the galactic disc) with a central bulge, contains stars, gas and dust and is surrounded by a spherical halo consisting mostly of stars or clusters of stars.

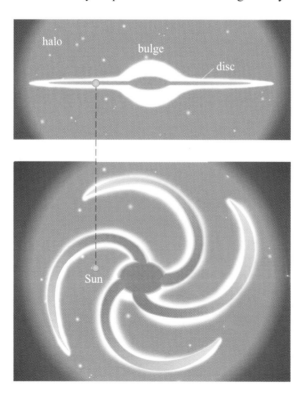

Figure 0.4 The spiral structure of our Milky Way galaxy.

We are unable to see the centre of our Galaxy due to clouds of intervening dust which partially or totally absorb the visible light from stars and can cause errors in our derivation of stellar properties if we are not aware of its presence. However, visible light is not the only type of radiation we can use to sense the Universe. Figure 0.5 illustrates views of the Milky Way obtained with telescopes from Earth or space that detect other kinds of radiation and provide us with alternative techniques for probing the properties and formation of stars.

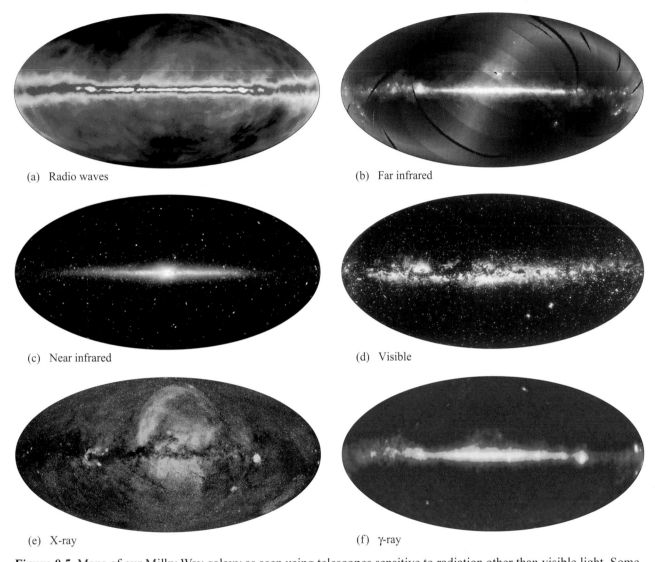

(a) Radio waves

(b) Far infrared

(c) Near infrared

(d) Visible

(e) X-ray

(f) γ-ray

Figure 0.5 Maps of our Milky Way galaxy as seen using telescopes sensitive to radiation other than visible light. Some are most sensitive to the gas (a, e, f) or dust (b) which pervades the plane of the galaxy while others are more effective at detecting radiation from stars (c). Stars, gas and dust are all apparent in the visible image (d). ((a), (d) Astronomical Society of the Pacific; (b), (c), (f) NASA; (e) Snowden *et al.*, 1997)

If we are to understand the properties of stars, most of which are so far away that they cannot be seen as anything other than points of light, the best place to start is by observing the one star which is easy to see – our own Sun. (Remember that you must ***never look directly at the Sun*** without special eye protection and never under any circumstances directly through a telescope.) The Sun is sufficiently close that we can observe structure on its surface and atmosphere and study its total output of radiation in great detail. In Chapter 1 we introduce the Sun and its observed properties. Chapter 2 is concerned with how we use these observations and the laws of physics to deduce its internal structure and energy source as well as how the Sun interacts with the Earth and the rest of the Solar System.

In Chapters 3 and 4 we examine the fundamental properties of stars, how they are distributed in space, and how we classify different types of stars. These properties have been determined from interpretation of observations of stars, but the answers form only the first step in our understanding. As scientists, we want to understand *why* stars exhibit this range of properties, how they came to exist and how they evolve. We can start to understand the internal structure and energy sources of stars if we apply the laws of physics to build mathematical models that are consistent with the observed properties. We have only a snapshot of the distribution of properties of stars at one time, but we know that any theory of the formation and evolution of stars must be consistent with observation. Once we understand the processes which give stars their observed properties we can then begin to piece together their formation, evolution and eventual demise. Chapters 5 to 8 lead us through the evolutionary process from stellar formation, through middle age to the often catastrophic death throes of stars of different masses. In Chapter 9 we look at some of the most bizarre objects in the Universe: white dwarfs, neutron stars and black holes. The physical conditions in and around such objects tend to be extreme and so they offer the opportunity to test physical theories in environments that could never be reproduced in a laboratory. Thus, the quest to understand the stars and the Universe they inhabit continues to drive scientific discovery.

CHAPTER 1
SEEING THE SUN

1.1 Introduction

As seen from the Earth, the stars seem cold and remote. Yet one star is so blindingly present in our lives that many people do not realize that it is a star at all. That star is the Sun, and it is the subject of the first two chapters of this book. This chapter concentrates on the observed properties of the Sun, and as you will see, the majority of observations depend in some way on the light or, more generally, the electromagnetic radiation that the Sun emits. Hence a secondary theme of this chapter is to introduce some concepts about the nature of light that are vital not only for solar science, but for astronomy in general.

Given that the Sun is, by far, the closest star to the Earth, it offers us unique opportunities to carry out very detailed studies of its behaviour. We will review some of the most important features of the Sun by concentrating on those parts of the Sun that can be observed directly. This sets the scene for Chapter 2 in which we explore the physical mechanisms that give rise to the Sun's observed characteristics.

1.2 Seeing the Sun's surface

1.2.1 Introducing the photosphere

Figure 1.1 is a photograph of the Sun. At first sight it looks like a pretty ordinary astronomical photograph – you might even mistake the Sun's image for that of some other body, a planet perhaps, or even a moon. But the Sun is neither a planet nor a moon; it is a star – our star – the only star that is sufficiently close for us to be able to examine its visible surface in great detail. For this reason a full discussion of Figure 1.1 is a good starting point for a book dealing with the stars.

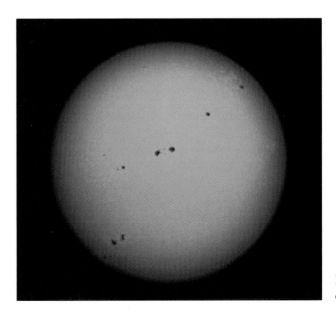

Figure 1.1 A photograph of the Sun. (NOAO)

There are a number of ways in which the Sun of Figure 1.1 differs from a planet or a moon. The most important of these differences is simply that the Sun is very much brighter than any planet or moon, from which it can be concluded that the Sun is a prodigious source of energy. The energy emitted by the Sun is crucial to our very existence, since it provides us with warmth and light. Without it, life might never have arisen on the Earth, and life as we know it today could certainly not be sustained. The rate at which energy is radiated by the Sun (i.e. the total amount of energy radiated per second) is called the **solar luminosity**. It is denoted by the symbol $L_\odot$, and is about $3.84 \times 10^{26}\,\mathrm{J\,s^{-1}}$. (The $\odot$ symbol is an ancient astrological sign for the Sun. Despite the distaste with which most astronomers view astrology, the symbol is widely used to denote 'solar' quantities, i.e. quantities pertaining to the Sun.)

QUESTION 1.1

Note that a table of often-used values is given in Appendix A1.

What is the value of the solar luminosity, $L_\odot$, in terms of the SI unit of power – the watt (W)? Given that a typical large power station produces energy at the rate of $2.5 \times 10^9\,\mathrm{W}$, work out the number of such power stations that would be required to match the energy output of the Sun.

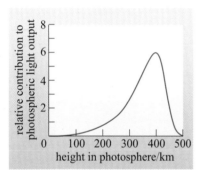

Figure 1.2 The relative amounts of photospheric light originating at various heights in the photosphere. Heights are measured from a precisely defined reference level that roughly corresponds to the greatest depth that can be 'seen'.

A second major difference between the Sun and the planets that orbit it concerns size. The visible disc portrayed in Figure 1.1 is actually about $1.4 \times 10^6\,\mathrm{km}$ in diameter. That's nearly ten times greater than the diameter of Jupiter, the largest planet, and about a hundred times greater than the diameter of the Earth. Clearly, the Sun is very much larger than any planet in the Solar System.

A third difference between a picture of the Sun and a picture of a body like the Earth or the Moon is that the visible surface of the Sun is not really a surface at all, but rather a thin, semi-transparent shell of gaseous material. Photographing the Sun is rather like photographing a cloud or a bank of fog; the light that makes up the photographic image comes from a range of depths and not from a single well-defined surface. In the case of Figure 1.1 almost all the light comes from a layer about 500 km thick called the **photosphere** (meaning 'the sphere of light'). Now, 500 km may sound pretty thick but, remember, the Sun is about 1.4 million kilometres across, so in comparison with the total solar diameter the photosphere really is a thin spherical shell. When we photograph the Sun we see into the photosphere but not through it. Figure 1.2 gives a rough idea of the relative proportions of photospheric light coming from different parts of the photosphere.

In visual terms, the photosphere is the closest thing to a surface that the Sun has to offer, but in physical terms the photosphere is much more like an atmospheric layer. In fact, the air you are breathing right now is more than a thousand times denser than the material that makes up the photosphere. If a space probe were sent into the photosphere the frictional resistance it would encounter would be almost negligible. A far greater impediment to such a mission would be the temperature. The photosphere gives off light because it is hot. Typical temperatures range from about 9000 K in the lowest parts of the photosphere to about 4500 K at the top. Most of the photospheric light comes from a region where the temperature is between 5800 K and 6000 K, so it is conventional to use values such as these to represent the 'surface temperature' of the Sun: however, the term is not really very meaningful and must be treated with some caution.

1.2.2 Large-scale features of the photosphere

Now that you are acquainted with the broad features of the photosphere (size, thickness, temperature, gaseousness) it makes sense to take a more detailed look at Figure 1.1. There are two main points to note. First, it should be fairly obvious that the edges of the photosphere are darker than the centre: there is a steady reduction in brightness as the distance from the centre of the image increases. Because the edge of the solar image is known as the **solar limb**, this gradual fall-off in brightness is called **limb darkening**. The second feature to note is that the photosphere is marked by dark blotches called **sunspots**. It turns out that both these features are important in understanding the Sun, so we shall discuss them in turn, starting with limb darkening.

Although limb darkening can be seen in Figure 1.1, it is more clearly demonstrated by using a graph to show the relative brightness of points along a line that crosses the solar disc, passing through the centre. Such a line is shown in Figure 1.3a and the corresponding graph in Figure 1.3b. As you can see from the graph, the brightness falls off very rapidly near the limb. This means that the solar limb is quite well defined, which is important because it makes it possible to use the location of the limb in various kinds of measurement.

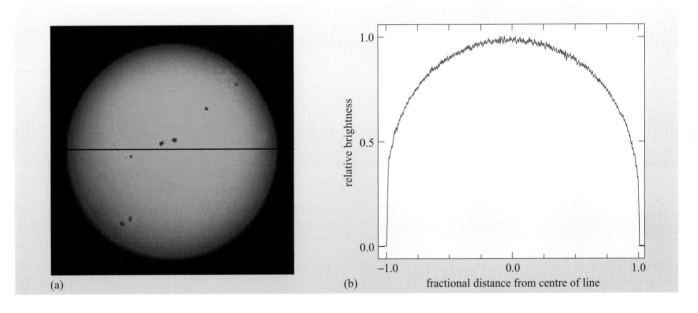

(a)   (b)   fractional distance from centre of line

Now, why should the limb of the Sun be darkened in the way indicated by Figure 1.3? In order to understand this, it is again necessary to recognize that the solar photosphere is something we look into rather than something we simply look at. As a first step towards understanding limb darkening, let us consider a scenario that is less complex than the Sun. In this simplified case we have a spherical cloud of gas in which the gas has uniform density. When we look at this cloud, our line of sight penetrates a certain distance into the cloud. Because this cloud has a uniform density, the distance along the line of sight from the edge of the cloud to the furthest point that we can see is the same for any line of sight. However, as Figure 1.4a shows, when we look along different lines of sight into the cloud, the depth below the surface of the furthest point that we can see (i.e. as measured along a line from the surface of the cloud to its centre) is *not* constant.

Figure 1.3 (a) The visible solar disc, crossed by a straight line. (b) The relative brightness of the photosphere at various points along the straight line shown in (a). ((a) NOAO; (b) Foukal, 1990)

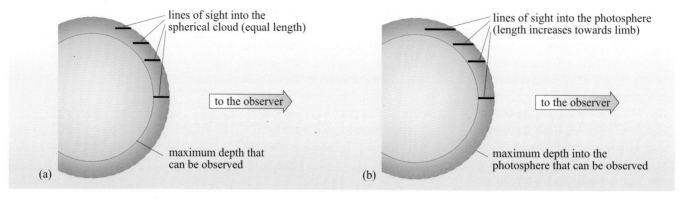

Figure 1.4 (a) For a gaseous spherical cloud of uniform density, we can see a uniform distance along the line of sight into the cloud. The depth below the surface of the furthest point that is visible becomes shallower as lines of sight approach the limb of the cloud. (b) In the Sun, the density changes with depth, and the lengths of lines of sight within the cloud increase as they approach the solar limb. However, the overall result is similar to case (a) in that the depth below the surface of the furthest point that we can see again becomes shallower as lines of sight approach the limb of the Sun. (Note that the relative distances that the lines of sight penetrate into the Sun are grossly exaggerated in this diagram.)

■   From Figure 1.4a, at which position do we see to the greatest depth below the surface of the cloud, and at which position do we see to the shallowest depth below the surface of the cloud?

❏   The line of sight that penetrates to the greatest depth is that which is directed towards the centre of the cloud. The lines of sight that reach the shallowest depths are those near the limb of the cloud

So there is an important geometrical effect here; lines of sight near the centre of the disc penetrate to a greater depth than lines of sight that are nearer the limb of the disc. Let us now consider a situation in which the temperature of the gas drops with distance from the centre of the cloud. Lines of sight toward the centre of the cloud will penetrate deepest into the cloud and hence allow us to see the hottest regions that are visible. Towards the limb, the lines of sight will only reach shallower and hence cooler regions of the cloud. An image of this cloud would then show limb darkening because the cooler material seen close to the limb simply gives off less visible light than the hotter material seen at the centre.

The situation in the Sun is somewhat more complex than this simplified scenario. While the temperature of the photosphere of the Sun does indeed increase with depth, the density of gas also increases with depth. This change in density has the effect that the distance along a line of sight from the surface to the furthest point that we could see is not constant. In fact, this distance actually increases as lines of sight move away from the centre of the solar disc. However, it turns out that despite this increase in distance of the line of sight within the photosphere, the geometrical effect that you have already seen in the simplified case is more important. This situation is illustrated schematically in Figure 1.4b. The overall result is that lines of sight towards the centre of the solar disc penetrate to a depth of about 500 km below the top of the photosphere, while lines of sight towards the solar limb penetrate to shallower depths. Because, as has already been pointed out, the temperature of the photosphere increases with depth, this variation in the depth to which we can see gives rise to limb darkening across the solar disc.

Turning now to the sunspots seen in Figure 1.1, it's worth noting straightaway that their darkness is also a consequence of temperature. Sunspots are large, relatively cool regions of the photosphere; the temperature at the centre of a sunspot is typically 4200 K, which is much less than the 6000 K or so of the surrounding photosphere. Consequently, sunspots are seen as dark patches against the bright background of the photosphere.

QUESTION 1.2

Using Figure 1.1, roughly estimate the diameter of a large sunspot.

Warning: do not attempt to look directly at the Sun.

Sunspots have been observed since ancient times, but their serious study really began in 1610, or shortly thereafter, when telescopes were just beginning to be used for astronomical purposes. Early solar observers, such as Galileo Galilei (1564–1642), David Fabricius (1564–1617) and the appropriately named Christoph Scheiner (1575–1650), soon discovered that sunspots appeared to move across the face of the Sun. This was eventually accepted as clear evidence that the Sun rotates, carrying the sunspots with it as it turns on its axis.

Individual sunspots are transient phenomena, but their lives are sufficiently long – typically a few weeks – that it is often possible to observe them crossing the entire solar disc. Sometimes, particularly long-lived spots, or groups of spots, can even be seen re-appearing over the limb of the Sun after they have crossed the far side

GALILEO GALILEI (1564–1642)

Galileo Galilei (Figure 1.5), who is usually referred to solely by his first name, was born in Pisa in 1564. He studied medicine but did not take his degree. Instead he developed an interest in mathematics and physics which he studied at home. In 1589 he was appointed professor of mathematics at Pisa, and three years later moved to a similar post in Padua. Galileo had many interests and arguably his most significant work was that which laid the foundations for the scientific study of motion.

Figure 1.5 Galileo Galilei.

Although Galileo did not invent the telescope, he was the first person to recognize its potential in astronomy. After learning of the telescope in 1609 he soon built one of his own and began making observations of the Sun, planets and stars. He reported his discoveries in a book published in 1610 called *Siderius Nuncius* (*The Starry Messenger*) which brought him widespread fame and the opportunity to advance his career under the patronage of the Grand Duke of Tuscany in Florence. However, controversy soon followed, when in 1613 he publicly supported the Sun-centred model of the Solar System that had been put forward by Nicolaus Copernicus (1473–1543). The Church deemed this view heretical, and Galileo then became embroiled in a long-running dispute over his beliefs. Events came to a head in 1633, when, under the threat of torture from the inquisition, Galileo was forced to reject the Copernican view, and was kept under house arrest for the remainder of his life. Following a campaign by prominent astronomers and historians of science, the sentence that had been passed on Galileo was formally retracted in 1992.

of the Sun. A sequence of photographs illustrating this effect is shown in Figure 1.6. In this particular case the spots were observed to take about 28 days to make a complete circuit. You might well think that this implies a 28 day rotational period for the Sun, or at least for the photosphere, but things are not quite so simple. In the first place, because the Earth orbits the Sun once a year (moving around the Sun in the same sense that the Sun rotates around its own axis), the rotational period observed from Earth is actually slightly longer than the **sidereal period**; that is, the period as measured with respect to the stars.

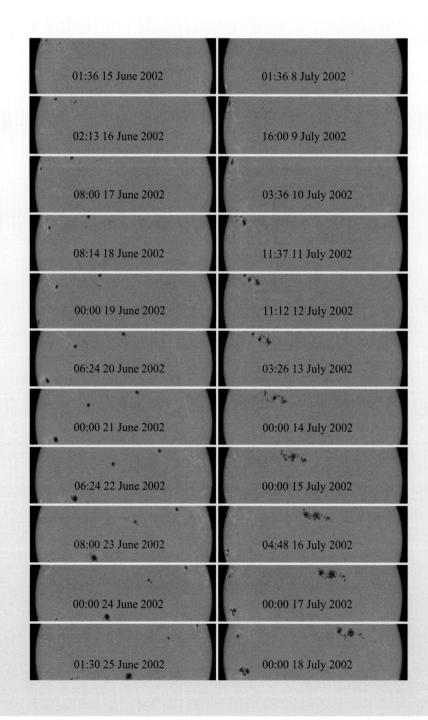

Figure 1.6 A sequence of images showing the apparent motion of sunspots across the face of the Sun. Note that each image is dated and that the whole sequence covers a period of about five weeks. (SOHO (ESA and NASA))

Second, owing to the fact that the Sun is a gaseous body, the nature of its rotation is very different from the rotation of a solid body such as the Earth. On a solid body everything 'rotates together', each part of the surface keeping in step with every other part, but on a gaseous object it is quite possible for different parts to rotate at different angular speeds, and this is just what happens on the Sun. Studies of sunspots and other indicators of **solar rotation** show that points on the solar equator have a sidereal period that is just under 26 days, whereas points further north or south have considerably longer periods: more than 26 days at a latitude of 30°, and about 30 days at a latitude of 60°. It is actually very difficult to measure the rotational period close to the poles, but it seems to be about 36 days. This rather complicated state of affairs is described by saying that the Sun exhibits **differential rotation**. More precise information about the varying rate of photospheric rotation is given in Figure 1.7.

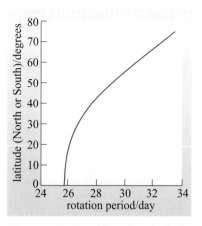

Figure 1.7 The sidereal period of the solar photosphere at various latitudes.

Apart from their role as tracers of rotation, sunspots are also good indicators of another large-scale phenomenon: **solar activity**. Data collected over many decades (see Figure 1.8) clearly show that the fraction of the solar disc covered by sunspots changes with time in a more or less regular way. A period of roughly 11 years separates each occurrence of maximum coverage, and hence of maximum solar activity, from its successor. Images of the Sun recorded at a time of maximum activity and at a time when activity is minimal differ markedly – as can be seen from Figure 1.9. Many other solar phenomena, some of which will be discussed later, also participate in the 11-year **solar activity cycle**, but none is as easy to observe as sunspots.

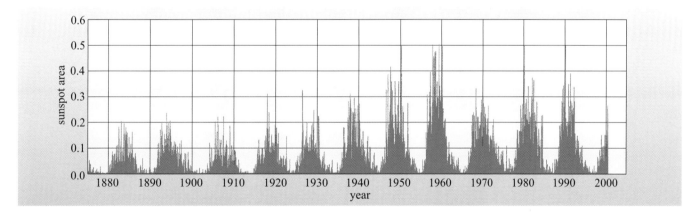

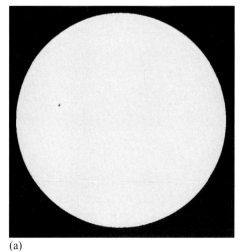

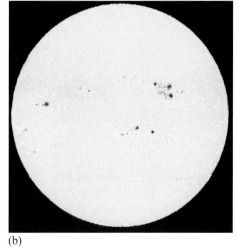

(a)   (b)

Figure 1.8 The 11-year variation of solar activity with time, as indicated by the percentage of the area of the Sun's visible disc covered by sunspots. (D. Hathaway, NASA/MSFC)

Figure 1.9 The solar photosphere, at times of low and high solar activity. (a) Low solar activity (22 June 1996). (b) High solar activity (23 December 2001). (Note that these images have been processed to remove the effects of limb darkening.) (Data provided by NSO)

1.2.3 Small-scale features of the photosphere

Although the origin of the solar activity cycle remains one of the Sun's outstanding mysteries, a good deal is known about sunspots. For example, it is well established that sunspots have stronger magnetic fields and different patterns of motion from their surroundings. Knowledge of this kind partly results from the study of magnified views of localized regions of the photosphere rather than photographs of the full solar disc. A magnified view of just this kind, showing the detailed structure of a sunspot, is shown in Figure 1.10. Such views are of great importance; the Sun is the only star sufficiently close to allow such detailed imaging of its surface.

Another small-scale phenomenon that is thought to be common in stars, but which can actually be seen only in the Sun's photosphere, is shown in Figure 1.11. The figure provides an instantaneous snapshot of the **solar granulation** – a seething pattern of bright cell-like **granules** that covers the photosphere. Each granule is typically about 1000 km across and lives for five to ten minutes. Detailed studies of the granules show that they are the tops of rising columns of hot material coming from deeper regions of the Sun where the temperatures are higher. The rising material travels upwards at a speed of $1 \, \text{km s}^{-1}$, or thereabouts, and then spreads out horizontally, radiating away its excess thermal energy. The dark 'lanes' between granules are regions where the cooled material descends back into the solar interior. The significance of these motions will be more fully explored in Section 2.2.

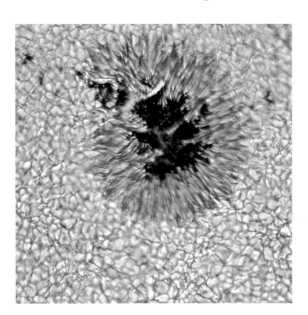

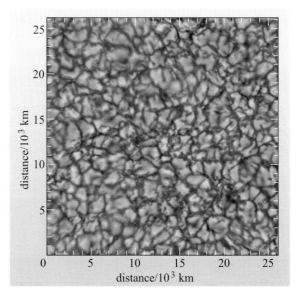

Figure 1.10 A highly magnified view of a sunspot. The sunspot itself consists of the very dark area and the pattern of fronds that seem to emanate from this region. The 'background' pattern on which the sunspot is located is the solar granulation. This image shows a region that is approximately 75 000 km × 75 000 km in extent. (G. Scharmer, Royal Swedish Academy of Sciences)

Figure 1.11 A highly magnified view of a small part of the photosphere clearly showing the solar granulation. (NOAO)

The blurring effects of the Earth's atmosphere make it difficult to carry out detailed studies of the solar granulation. Nonetheless, it is sometimes possible to obtain sequences of photographs that clearly show the formation or disappearance of individual granules. Such a sequence is shown in Figure 1.12.

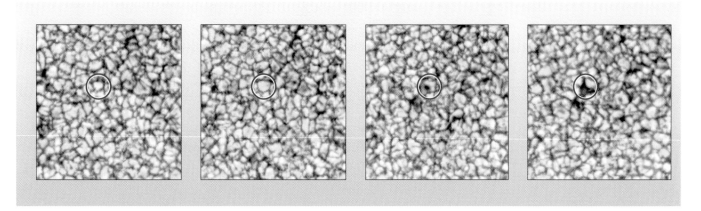

1.2.4 The nature of light

Note: throughout this chapter the term 'light' refers specifically to 'ordinary' visible light.

Although our discussion of the photosphere has concentrated on its visual appearance, it is important not to forget that the photosphere is the source of the light that illuminates the Earth and, through the effect of photosynthesis in plants, is the source of energy that keeps us alive. From our human perspective, light is the main product of the photosphere and will be a major concern throughout this book, so this is a good point at which to gather together a number of basic facts about the nature of light. Because these facts constitute 'essential scientific background' rather than a continuation of the astronomical storyline we have been developing so far, they will be separated from the rest of the section by enclosing them in a blue toned box. Such boxes will be used throughout this book to enable you to identify items of background science wherever they arise. One such example is Box 1.1.

Figure 1.12 A sequence of photographs of the solar granulation. Note that one particular granule has been highlighted by a red circle. This granule grows and decays over the 8 minute interval covered by the photographs. (BASS 2000)

BOX 1.1 THE NATURE OF LIGHT

The electromagnetic wave model of light

It is well known that a magnet is able to influence certain objects (e.g. other magnets) without touching them. This phenomenon is 'explained' by saying that the magnet produces a **magnetic field**, which occupies the space around the magnet and gives rise to the forces that act on the affected objects. To account for these forces, the magnetic field at any point must have both a strength and a direction. Consequently, the magnetic field at any point can be represented by an arrow, since an arrow has a length that can represent the strength of the field, and an orientation that can be made to correspond to the direction of the field. The use of arrows to represent a magnetic field is illustrated in Figure 1.13a. In a similar way, a suitable distribution of

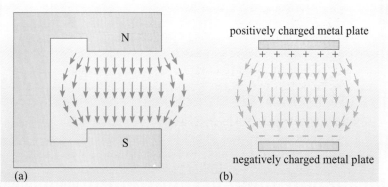

Figure 1.13 (a) A magnetic field. (b) An electric field.

positive (+) and negative (−) electric charges will give rise to an **electric field**, which can also be represented, at any point, by an arrow of appropriate length and orientation, as in Figure 1.13b.

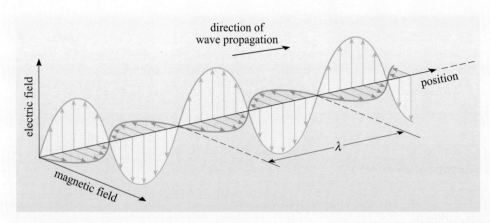

direction of wave propagation

position

electric field

magnetic field

λ

Figure 1.14
An electromagnetic wave.

In the 1860s, while carrying out a mathematical investigation of electricity and magnetism, the Scottish physicist James Clerk Maxwell showed that it was possible to create self-sustaining patterns of fluctuating electric and magnetic fields that could oscillate together through space at a certain pre-determined speed. These fluctuating field patterns are called **electromagnetic waves**, and an instantaneous 'snapshot' of a small part of the simplest such wave is shown in Figure 1.14. The electric and magnetic fields that make up the wave are always at right angles to each other, and both fields are at right angles to the direction in which the wave is travelling. Remember, Figure 1.14 is only a snapshot, so you should imagine the whole pattern moving in the direction of travel rather like an ocean wave moving across open water. Maxwell found that the speed at which electromagnetic waves had to travel was very close to the best values then available for the speed of light, so he suggested that rays of light were nothing other than electromagnetic waves.

An important characteristic of any electromagnetic wave is the distance between successive maxima of the electric or magnetic field (that is, the distance from one peak of the wave to the next). This distance is called the **wavelength** and is denoted by the Greek letter λ (pronounced 'lambda'). The wavelengths of all forms of visible light are very tiny, and different wavelengths correspond to different colours. Red light has a wavelength of about 700×10^{-9} m (= 700 nanometres, nm), and the wavelength of violet light is around 420×10^{-9} m (= 420 nm). White light, which is a mixture of all the possible colours, contains all the wavelengths between these rough limits. (You will find more on this subject in Section 1.3.2.)

Imagine yourself observing a wave like that illustrated in Figure 1.14 as it moves past some fixed point. If the speed of the wave is v and its wavelength is λ, then you should be able to convince yourself that the number of wavelengths that will pass the fixed point in a second is just v/λ. This quantity, the number of wavelengths passing a fixed point in one second, is called the **frequency** of the wave; it is measured in SI units called **hertz (Hz)** (equivalent to s^{-1}) and is denoted by the letter f. Thus, for any wave, $f = v/\lambda$. Or, more conventionally,

$$v = f\lambda \qquad (1.1)$$

For electromagnetic waves the value of v depends on the **medium** (e.g. air, glass and water) through which they move. The maximum value of v occurs when the waves travel through a vacuum (that is, empty space). Under these conditions the speed of the waves is very nearly 3.00×10^{8} m s^{-1} (the exact value is $2.997\,924\,58 \times 10^{8}$ m s^{-1}). This quantity is of such importance that it is given its own symbol, c, and is referred to as the **speed of light in a vacuum**. Thus, for electromagnetic waves travelling through a vacuum, we can write

$$c = f\lambda \qquad (1.2)$$

When an electromagnetic wave travels from one medium to another, its frequency remains the same, while, in general, its speed will change. This has the consequence that the wavelength also changes as an electromagnetic wave travels from one medium to another.

The identification of light with electromagnetic waves was a major development in the history of physics and was the source of much progress. However, scientists now recognize that the identification was not entirely correct. Electromagnetic waves can account for many of the properties of light, but not all of them. For this reason, rather than saying that light is electromagnetic waves, we prefer to say that electromagnetic waves provide a model of the phenomenon of light. **The electromagnetic wave model of light** helps us to understand the behaviour of light but it is not the whole story. Other models are also useful.

The photon model of light

At present, the most complete scientific account of light involves a branch of physics called **quantum theory**. The full quantum theory of light is much too complicated to describe here, but during its development another simple model of light emerged that was quite different from the electromagnetic wave model yet, under the right circumstances, just as valuable. This alternative model is known as the **photon model of light**. According to the photon model, a ray of light of frequency f can be thought of as consisting of a stream of separate particles called **photons**. Each of these photons carries an identical amount of energy, which we can denote by the Greek letter epsilon, ε, that relates directly to the frequency of the ray, and is given by

$$\varepsilon = hf \qquad\qquad (1.3)$$

The quantity h in this equation is one of the fundamental constants of physics; it is the **Planck constant** and it is given by $h = 6.626\,069 \times 10^{-34}\,\text{J s}$ although $6.63 \times 10^{-34}\,\text{J s}$ will do for most calculations. It is often convenient to measure the energy of photons in terms of a quantity called the **electronvolt** (eV) rather than in joules. In order to convert between joules and electronvolts, note that $1\,\text{eV} = 1.602 \times 10^{-19}\,\text{J}$. For the purposes of applying Equation 1.3 to obtain the photon energy in eV rather than joules, it is also useful to note that $h = 4.14 \times 10^{-15}\,\text{eV s}$.

The photon model of light is of particular importance when considering the interaction of light with atoms. Individual atoms can absorb or emit only entire photons. Thus, when a cloud of atoms is illuminated by a beam of light there is no possibility of a single atom acquiring half a photon's worth of energy directly from the beam. This is a subject to which we shall return in Section 1.3.2, when we further develop the 'background science' of light.

It is important to realize that neither the electromagnetic wave model nor the photon model should be regarded as 'true'. Light is neither a wave nor a particle but, under the appropriate conditions, it may exhibit wave-like or particle-like behaviour; both possibilities are encompassed by the quantum theory.

JAMES CLERK MAXWELL (1831–1879)

Maxwell (Figure 1.15) came from a well-to-do Scottish family. He was a shy child and his experience of school was not a happy one. Despite this, he maintained a curiosity about the natural world, and at the age of 15 he made the first of many discoveries: in this case the technique of drawing an ellipse using two pins and a thread. He went on to the University of Edinburgh and then to Trinity College, Cambridge. His interests were broad and he made significant advances in such diverse fields as colour vision, the stability of Saturn's rings, and the kinetic theory of gases. However, his outstanding achievement was the development of the mathematical theory that linked electricity and magnetism. His academic career took him to posts in Aberdeen and at King's College in London, before the death of his father in 1865 at which time he returned to the family home to become a gentleman-farmer engaging in scientific research. In 1874 he was invited to become the first Cavendish Professor of Experimental Physics at Cambridge, and was instrumental in establishing a world-class laboratory for physics at that university.

Figure 1.15 James Clerk Maxwell. (Science and Society Picture Library)

QUESTION 1.3

The overall shape of the graph in Figure 1.3b clearly indicates the phenomenon of limb darkening. Additionally, the graph includes a good deal of small-scale structure: it is not smooth but contains many tiny peaks and troughs. How can you account for this on the basis of the information given in this section?

QUESTION 1.4

The electromagnetic waves used to model the various colours of visible light have wavelengths in a vacuum in the approximate range 400 nm to 700 nm. What is the corresponding range of frequencies?

QUESTION 1.5

In terms of the photon model of light, what is the approximate range of photon energies corresponding to the range of wavelengths discussed in Question 1.4? Express your answers in joules and in electronvolts.

1.3 Seeing the Sun's inner atmosphere

1.3.1 Introducing the chromosphere

Although the majority of the Sun's light comes to us from the photosphere, we also receive small amounts of light from layers of hot, thin gaseous material that surround the photosphere. These outer layers of the Sun may be regarded as the Sun's 'atmosphere', though the term must be treated with the same degree of caution that we used when referring to the photosphere as the Sun's 'surface'.

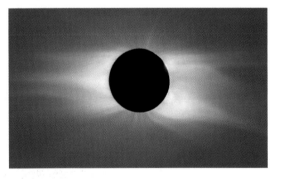

Figure 1.16 The solar corona during a total solar eclipse. (J. Durst, Schonenberg)

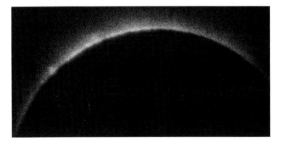

Figure 1.17 The solar chromosphere during a total solar eclipse. (G. East)

In a **white light image** of the Sun, recorded using ordinary visible light (such as Figure 1.1), the feeble light from the Sun's atmosphere is normally drowned out by the overwhelming brilliance of the photosphere. Nonetheless, under the right circumstances, it is possible to see the Sun's atmosphere, as Figure 1.16 shows. The picture was taken during a **total eclipse of the Sun** – an infrequent but predictable event that occurs somewhere on Earth on average every eighteen months or so, when the Moon passes between the Earth and the Sun, and entirely blocks the light from the photosphere for a few minutes. (It is a remarkable coincidence that, although the Sun and Moon differ greatly in size and distance from the Earth, their diameters and distances are just right to allow such a very precise blockage to occur.)

The black circle in the middle of Figure 1.16 is the silhouette of the Moon. The bright halo that surrounds it is the solar atmosphere. As you can see, the atmosphere is very extensive. For the most part it is a pearly white, but very close to the photosphere there is a narrow region with a pink or reddish tinge: this can be clearly seen in Figure 1.17. The coloured layer, which is a few thousand kilometres thick, is called the **chromosphere** (meaning the 'sphere of colour'), and constitutes the inner (or lower) solar atmosphere. The chromosphere will be our main concern in this section. The extensive outer (or upper) solar atmosphere is called the *corona*. This will be the subject of Section 1.4.

Although eclipse studies led to the initial identification of the chromosphere, and continue to play a role in its scientific investigation, they are not the only source of chromospheric information. Fortunately, much may be learnt from observations of the full solar disc, provided they are restricted to wavelengths where the chromosphere is more prominent than the photosphere. Just such a restricted wavelength view is shown in Figure 1.18. In this particular case the image was produced by red light in a narrow range of wavelengths centred on 656.3 nm. At these wavelengths, for reasons that will be explained shortly, hydrogen atoms throughout the chromosphere are highly effective absorbers and emitters of radiation. As a result, the chromosphere absorbs most of the 656.3 nm radiation coming from the photosphere, but its own emissions at that wavelength are quite prominent. It is these emissions that are mainly responsible for the reddish hue that gives the chromosphere its name. The particular kind of restricted wavelength view shown in Figure 1.18 is called an **Hα image** (pronounced 'aitch alpha'): the H indicates that the emitted light is coming from hydrogen atoms, and the α indicates that this is the first (longest) wavelength of visible light at which hydrogen atoms have this particular effectiveness as absorbers and emitters. The other, successively shorter, visible wavelengths at which hydrogen behaves in this way are 486.1 nm, 434.0 nm and 410.1 nm, which are respectively denoted by Hβ, Hγ and Hδ (β, γ, δ are the Greek letters 'beta', 'gamma' and 'delta' respectively). Restricted wavelength images at other wavelengths, associated with other kinds of atom, are also of great value, particularly calcium H (396.8 nm) and calcium K (393.3 nm) images. As their names imply, these kinds of image involve light emitted from calcium atoms. (The letters H and K historically indicated the sequence of wavelengths and have nothing to do with the usual chemical symbols for hydrogen and potassium.) A calcium K image is shown in Figure 1.19.

The chromospheric Hα (Figure 1.18) and calcium K (Figure 1.19) images are clearly very different from the photospheric white light image that we looked at earlier (Figure 1.1). In an Hα image the chromosphere is mottled with bright specks, some of which are gathered together into extensive bright regions called **plages** (the French word for 'beach' that is pronounced 'plah-je'). These are often seen in parts of the chromosphere that are directly above the active regions of the photosphere that contain sunspots and are especially prominent in calcium K images. Visible in the bottom half of the Hα image (Figure 1.18) is a long, winding dark feature called a **filament**. Filaments are quite common in Hα images; they are caused by huge clouds of relatively cool gas held high above the

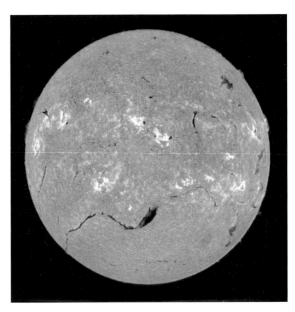

Figure 1.18 An Hα image of the Sun, produced by light from a narrow range of wavelengths centred on 656.3 nm. (Big Bear Solar Observatory)

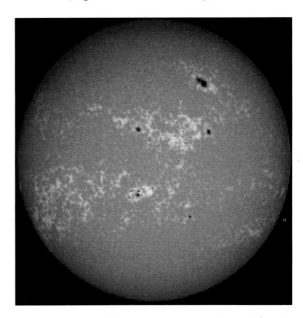

Figure 1.19 A calcium K image of the Sun. The image shows emission in a narrow range of wavelengths centred on 393.3 nm. As in the Hα image (Figure 1.18), this image reveals the structure of gas in the chromosphere. (Marshall Space Flight Center/NASA)

Figure 1.20 An Hα solar prominence on the limb of the Sun. (Big Bear Solar Observatory)

chromosphere by magnetic forces. These are the same clouds that account for the **prominences**; such as the one seen above the limb of the Sun in Figure 1.20. Overall, the chromosphere is much less uniform than the photosphere. One solar physicist has even described the chromosphere as 'a layer of froth stirred up by the photosphere'.

To get a better understanding of the chromosphere, we really need to gain some insight into the processes that give rise to Hα images and to the other, similar, images obtained at different wavelengths. Only with the aid of such insight can the radiation emitted by the chromosphere be properly interpreted and used as a source of information about the physical conditions of the Sun, such as temperature, pressure and structure. The starting point for this kind of investigation is the technique of *spectroscopy*.

1.3.2 Spectroscopy and sources of light

Most sources of light emit over a range of wavelengths. (The one well-known exception to this is the laser, which is a device that produces light at a single wavelength.) Such a range of wavelengths is commonly called a **spectrum** (plural *spectra*). **Spectroscopy** concerns the production and study of spectra.

Continuous spectra

Many sources of light emit over an unbroken range of wavelengths. Such sources are therefore said to have **continuous spectra**.

If a narrow beam of light passes through a glass prism, the beam will split up in such a way that different wavelengths travel in different directions. (This process is shown in Figure 1.21.) If the original beam contained just a few well separated wavelengths the result would be a set of quite separate and distinct images, each with its own characteristic colour (wavelength). However, if the beam came from a source which produces a continuous spectrum it would typically contain all visible wavelengths, and the result of passing it through the prism would be a multicoloured band somewhat similar to a rainbow.

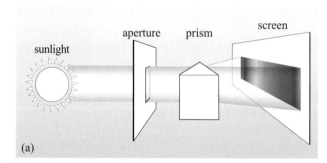

(a)

Figure 1.21 (a) The formation of a continuous spectrum by use of a glass prism. (b) A naturally occurring spectrum – a rainbow. Here the light from the Sun is dispersed by passing through raindrops. ((b) G. Garrad/Science Photo Library)

(b)

In the past, scientists engaged in spectroscopy spent a good deal of their time examining the kinds of multicoloured band described above. Nowadays things are different in that it is more common for spectral information to be given in the form of a graph. When spectra are presented in this way the horizontal axis of the graph usually shows wavelength or frequency (or sometimes photon energy). The vertical axis of the graph normally indicates the 'brightness' of the spectrum at any given wavelength. Figure 1.22 shows the relationship between this graphical representation of a spectrum and the coloured band spectrum. Unfortunately, the quantities used to measure spectral brightness have rather complicated definitions. In this chapter the vertical axes of some graphical spectra are labelled **spectral flux density**. At any given wavelength λ, the spectral flux density, F_λ, can be determined by the following procedure.

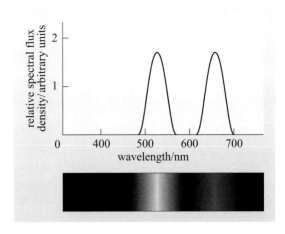

Figure 1.22 The relationship between the band spectrum and a graphical representation of a spectrum.

(a) Using an appropriate detector of area 1 m^2, pointed directly towards the source, measure the rate at which energy from the source is delivered to the detector by electromagnetic waves with wavelengths in a fixed narrow range, $\Delta\lambda$, centred on λ.

(b) Divide the measured rate of energy detection by the wavelength range $\Delta\lambda$ to obtain the detected power per square metre per unit wavelength range, typically measured in units of W m^{-2} μm^{-1} or in W m^{-2} nm^{-1}. This is the value of F_λ at wavelength λ.

In fact, most of the graphical spectra in this book will have vertical axes that show **relative spectral flux density**. In such cases, the spectral flux density at any wavelength is expressed as a *fraction* of some arbitrarily chosen reference value and there will be no SI units shown on the axis.

The black-body spectrum

There are several different types of physical process that can give rise to continuous spectra. One particularly important class of continuous spectrum is called the **black-body spectrum** (or sometimes a Planck spectrum, after the German physicist Max Planck (1858–1947)).

There are two key features of sources that produce black-body spectra. The first is that the emission of light arises as a result of the material that makes up the source being at a relatively high temperature. From everyday experience, you can probably think of examples that support the statement that 'hot things glow'. In such a source, the energy that is emitted as light, has its origin in the internal, or thermal, energy of the material that makes up the source. Not surprisingly, such sources are termed **thermal sources** of radiation. It is important to appreciate that not all sources of light are thermal in origin; those sources in which the energy emitted as light does not derive from the thermal energy of the source would be called **non-thermal sources** of radiation. An everyday example of a non-thermal source of light is a television screen, which produces a copious amount of light but is not hot.

The fact that a source of light is thermal is not a sufficient condition for it to generate a black-body spectrum; indeed, thermal sources need not even produce continuous spectra. To generate a black-body spectrum, a second condition must be met in that light within the source is much more likely to interact with the material of the source than to escape. This means that light generated within the source is in some sense 'trapped', and is only likely to escape after there has been considerable interaction with the material within the source. Therefore, a common feature of sources that approximate well to being black-body sources is that they are opaque. This is because it is unlikely that a photon could cross the material that makes up the source without undergoing an interaction.

■ Why is a source of light that is transparent unlikely to form a black-body spectrum?

❑ If the material that is emitting light is transparent then it is more likely that photons that are generated would escape from the source rather than interact with the material of the source. Hence the source is unlikely to produce a black-body spectrum.

An everyday example that satisfies the conditions for forming a black-body spectrum reasonably well is the tungsten filament of a light bulb. The filament is heated and light is produced not just at the surface, but throughout the bulk of the filament. Because solid metals are far from being transparent, most of the light that is produced by the filament does not escape to be seen, but interacts with the material of the source. Near the surface of the filament however, the assumption that light is much more likely to interact with the material of the source than to escape starts to break down, and this results in the spectrum only being an approximation to the ideal black-body conditions.

The concept of the black-body spectrum is useful because many astronomical sources produce continuous spectra that are a reasonably good approximation to the black-body form. However, it should also be noted that there are several other types of physical mechanism that can produce continuous spectra.

Figure 1.23 shows a set of black-body spectra from sources at temperatures between 3000 K and 6000 K. Note that the shape of the curve varies with temperature, but that the form of the curve is relatively simple: it has just one broad maximum or peak.

QUESTION 1.6

Figure 1.23 clearly indicates that the observed spectral flux density from the hottest source is greater than that from the coolest. Is it necessarily the case that in a fixed wavelength range more energy will arrive per second from any black-body source at 6000 K than from *any* black-body source at 3000 K?

In any graph that shows how relative spectral flux density varies with wavelength for a black-body source, the height of the graph will depend on factors such as the size and distance of the source, and the selected reference value. However, in all such graphs the overall shape of the curve is solely determined by the *temperature* of the source. This means, in particular, that the peak of each curve occurs at a wavelength, λ_{peak}, that characterizes the source's temperature, irrespective of the height of the peak. In fact, there is a simple law, called **Wien's displacement law**, that relates the value of λ_{peak} to the temperature, T, of the source:

$$(\lambda_{\mathrm{peak}}/\mathrm{m}) = \frac{2.90 \times 10^{-3}}{(T/\mathrm{K})} \tag{1.4}$$

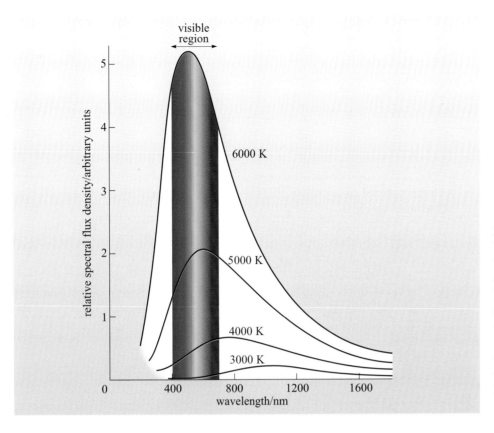

Figure 1.23 Graphical spectra for black-body sources at temperatures between 3000 K and 6000 K. The black-body sources all have the same surface area and are at the same distance from the detector. (Note that the spectra extend well beyond those wavelengths that correspond to the various colours of visible light. More will be said about these other emissions in Section 1.4.)

With the aid of Wien's displacement law it is a simple matter to determine the temperature of any source of light, *provided that it is a black-body source*. Such sources are not common, but many real sources, including the Sun and other stars, are reasonable approximations to black-body sources, so their temperatures may be estimated by this spectral technique.

QUESTION 1.7

If you were to heat a metal ball, so that its temperature steadily increased, you would find that above a certain temperature the ball would start to emit a dull red glow. As the temperature increased further the ball would become brighter and the colour would change from red to orange-white to yellowish-white to white. How would you explain these changes in appearance?

The shapes of the curves that describe the spectra of black-body sources are of great importance in science. Such curves are usually referred to as thermal radiation curves; or **Planck curves** or **black-body radiation curves**. While the black-body spectrum is important, it should also be noted that other physical processes may also produce continuous spectra, but those spectra will generally have a different shape from those of black-body sources.

Line spectra: absorption and emission

If a beam of light from a black-body source passes through a thin (low-density) gas of atoms, the spectrum of the emerging beam will generally include a number of narrow dark lines. These lines are called **absorption lines** and correspond to narrow ranges of wavelength that have been wholly or partly absorbed by atoms in the gas (a process described in Box 1.2). This situation is illustrated in Figure 1.24, which includes a graph of the so-called **absorption spectrum** that arises.

If, instead of the emerging beam, the light emitted by the gas itself is examined, it will be found that its spectrum consists of a number of narrow *bright* lines. These lines are called **emission lines**, and a spectrum composed of them is called an **emission spectrum**. For many gases, the bright lines emitted cover the same narrow wavelength ranges as the dark lines in the absorption spectrum. The graph of such an emission spectrum is also included in Figure 1.24.

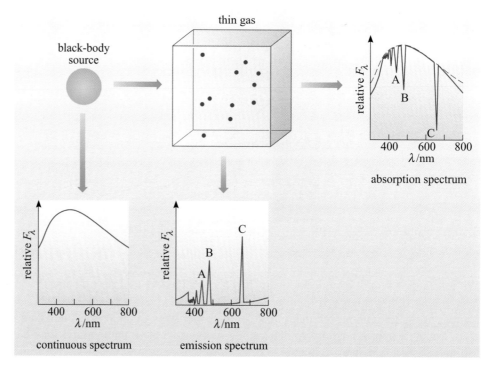

Figure 1.24 Three kinds of spectrum – continuous, absorption and emission – seen by observing a black-body source and a thin gas of atoms from various directions. The dashed line in the absorption spectrum shows the continuous spectrum that would have been observed in the absence of the gas.

Emission spectra and absorption spectra are both examples of **line spectra**. Figure 1.25 shows the line spectra produced by a number of different gases. The occurrence of such spectra is a consequence of the fact that each of the gases is composed of a characteristic type of atom that has its own internal structure. (It is also relevant that the gas is sufficiently thin to ensure that the atoms do not significantly influence one another.)

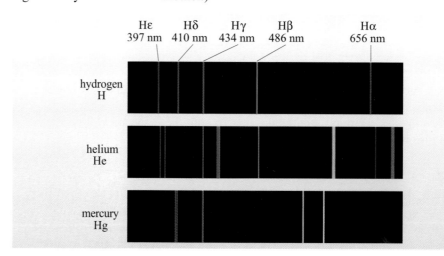

Figure 1.25 The line spectra produced by hydrogen, helium, and mercury. Note that in all cases the spectra are produced by the elements in gaseous form.

BOX 1.2 SPECTRAL LINES AND ENERGY-LEVEL DIAGRAMS

A typical atom is about 2×10^{-10} m in diameter and consists of a tiny, dense, positively charged *nucleus* surrounded by one or more negatively charged electrons. According to quantum theory, each of the electrons belonging to a particular atom may be in any of a number of allowed states, each of which is associated with some fixed amount of energy. When an electron occupies a particular state in a particular atom, the atom has the energy associated with that state. Thus, changes in the pattern of occupied states within an atom entail changes in the total amount of energy possessed by the atom. A diagram showing the energy associated with each of the allowed states in a particular kind of atom is called an **energy-level diagram**. The simplest energy-level diagram, that of a hydrogen atom (which has just one electron), is shown in Figure 1.26.

When a hydrogen atom has the lowest energy level, labelled E_1, it is said to be in its **ground state**. Above this energy, when the atom has energy E_2, E_3, etc., it is said to be in an **excited state**.

To see how the behaviour of atoms can account for the existence of line spectra, consider an atom which initially occupies a state of energy E_i (the subscript i reminds us that this is the initial energy). Such an electron may make a transition to some other state of higher energy E_f (f for the final, higher energy state), provided that the atom can acquire the necessary additional energy, $E_f - E_i$. If the atom is bathed in light from a source that produces a continuous spectrum, as in Figure 1.24, the requisite energy can be absorbed directly from the light. Now, you might think that energy absorption of this kind would involve a wide and continuous range of wavelengths, but that is not the case. The absorption of light by an atom is one of those processes that is best described by the photon model of light discussed in Box 1.1. The energy is acquired in one gulp, as it were, by the absorption of a single photon of just the right energy $\varepsilon_{fi} = E_f - E_i$. It follows from Equation 1.3 that such a photon

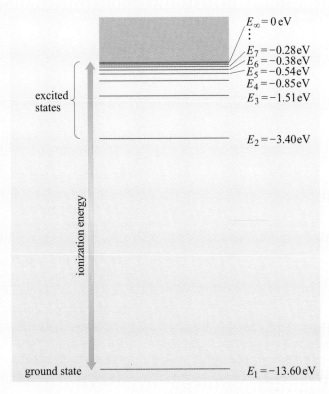

Figure 1.26 The energy-level diagram of a hydrogen atom. There are infinitely many energy levels but those of highest energy are too closely crowded together to be shown separately on a diagram of this kind. Note that the energies are given in terms of electronvolts (1 eV = 1.602×10^{-19} J).

corresponds to electromagnetic radiation of a specific frequency, namely

$$f_{fi} = \frac{1}{h}(E_f - E_i) \qquad \text{absorption} \qquad (1.5)$$

There is a limit to the amount of energy that a hydrogen atom can absorb without the electron and proton being split apart. This limit is known as the **ionization energy** and for the hydrogen atom it has a value of 13.60 eV.

QUESTION 1.8

Write down a formula for the corresponding wavelength, λ_{fi}, of the radiation that must be absorbed if a transition requiring an energy increase $E_f - E_i$ is to take place.

Applying the result of Question 1.8 to the gas of atoms shown in Figure 1.24, it should be clear that if a sufficient number of atoms in the gas increase their energy by $E_f - E_i$ every second, then a good deal of incoming radiation at the corresponding wavelength λ_{fi} will be absorbed, with the result that the spectrum of the emerging beam will have a dark absorption line at that wavelength. Since there are many possible values of E_i and E_f for any particular kind of atom, it follows that there are many possible values of $(E_f - E_i)$ and hence many, quite distinct wavelengths at which absorption lines might occur. (Whether a particular line is seen or not depends on the *rate* at which the relevant transition is occurring in the gas.) Furthermore, since each kind of atom has its own characteristic energy-level diagram, different gases will produce different absorption spectra – just as you saw in Figure 1.25.

The above account of absorption spectra requires a large number of atoms to absorb energy every second, at many different wavelengths. What happens to all this absorbed energy? The answer is that most of it is very quickly emitted again, and it is this emission that accounts for the emission spectrum that an illuminated gas produces. Once an electron occupies a state associated with an energy E_f that is higher than the energy associated with one or more other states, it may be possible for the electron to make a spontaneous transition to one of those states of lower energy. In particular, if the electron originally came from a state of energy E_i, one of many possibilities is for it to return to that same state. Under these circumstances the atom would have to shed an amount of energy $\varepsilon_{fi} = E_f - E_i$. Once again, quantum theory demands that if the energy released in a transition of this kind is given off as light then it must take the form of a single photon corresponding to electromagnetic radiation of frequency

$$f_{fi} = \frac{1}{h}(E_f - E_i) \qquad \text{emission}$$

A gas is therefore capable of producing emission lines at exactly the same wavelengths that it produces absorption lines. However, and this is a crucial point, the emission is equally likely to occur in any direction. Thus, energy absorbed from the incoming beam in Figure 1.24 will not be entirely replaced by emitted energy; rather emitted energy will be given off in all directions, allowing the emission spectrum to be seen from many different angles, whereas the absorption spectrum can be seen only by looking into the beam.

1.3.3 Solar spectroscopy and the structure of the chromosphere

A good deal of what is known about the Sun has been learnt by studying its spectrum. A photograph of the visible spectrum is given in Figure 1.27. As you can see it is essentially an absorption spectrum; the coloured bands are crossed by a large number of narrow absorption lines corresponding to various atomic transitions. In fact, something like 25 000 lines have been identified in the visible region; many originate in the photosphere, which is also responsible for nearly all the light between the lines, but some of the lines carry information about the chromosphere. Unfortunately, owing to the problems of reproducing colour images, not all of the features of Figure 1.27 are as clear as they could be, so a sharper

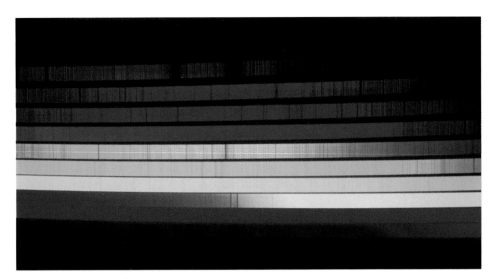

Figure 1.27 The solar absorption spectrum. Note that for convenience of display, the spectrum has been cut into sections and consecutive sections have been stacked vertically in sequence. (National Solar Observatory)

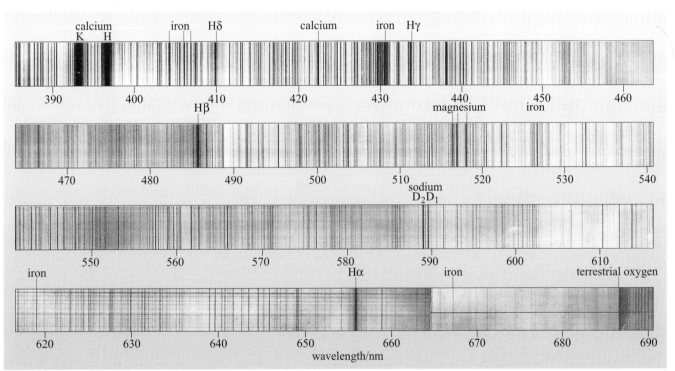

black and white image of the Sun's spectrum is given in Figure 1.28. Some of the darkest lines in this image have been labelled and a wavelength scale (in nanometres) has been included to help you locate them. It is worth noting that even the darkest lines are not completely black; they are simply darker than their surroundings. Indeed, it is often the faint emissions found within the darkest lines that are most informative about the chromosphere. A particularly strong absorption line of this sort can be seen around 656.3 nm, that is the wavelength of the Hα image we looked at earlier. Not surprisingly, this is called the **Hα absorption line**. Another prominent absorption line due to hydrogen (Hβ) at 486.1 nm, and two lines due to calcium at 396.8 nm and 393.3 nm, were all mentioned earlier as important wavelengths for studies of the chromosphere.

Figure 1.28 A black and white image of the solar spectrum. Note that for convenience of display, the spectrum has been cut into sections and consecutive sections have been stacked vertically in sequence. (The horizontal streaks on the spectra are artefacts.) (Kitt Peak National Observatory)

Figure 1.29 Sir (Joseph) Norman Lockyer (1836–1920). Lockyer developed an interest in astronomy when he was working as a civil servant, and he went on to become a pioneer of solar astrophysics. His contribution to science was much wider than his chosen specialism: he founded the journal *Nature*, which he edited for fifty years. (Royal Astronomical Society)

QUESTION 1.9

Why is it not surprising that studies of chromospheric emission lines often involve wavelengths that correspond to dark absorption lines in the solar spectrum?

Chromospheric emission lines can be observed in particular detail during total solar eclipses, when the light of the photosphere is blocked by the Moon. One of these emission lines, at 587.6 nm, has a particularly interesting history. When it was first observed, during the eclipse of 18 August 1868, its origin could not be explained in terms of any of the chemical elements then known. This led Sir Norman Lockyer (1836–1920) (Figure 1.29) to propose the existence of a new element, which he named helium after the Greek word for the Sun, *helios*. Lockyer's interpretation was correct, but confirmation did not come until 1895, when the British scientist Sir William Ramsay (1852–1916) showed that helium exists on Earth.

By comparing absorption and emission lines of solar origin with lines observed in terrestrial laboratories, it is possible to determine which kinds of atom are present in the visible parts of the Sun. In this way the presence of more than 65 of the 100 or so known elements has been established. This is a notable achievement, but even more can be accomplished by examining the details of the lines – a task best done with the aid of the graphical form of the spectrum.

Figure 1.30 is a graphical representation of the solar spectrum. The dips corresponding to some of the more prominent absorption lines can be discerned. In order to explain the detailed appearance of this kind of graph it is not enough simply to identify which atoms are represented; the relative depths of the various lines must also be explained. Line depths depend on a number of factors, including the relative abundance of each kind of atom, the inherent likelihood of each atomic transition, and the proportion of atoms of a given kind that have an electron in the appropriate initial state to give rise to a particular line. This last factor will depend sensitively on temperature because, generally speaking, the higher the temperature the more likely it is that states corresponding to higher energies will be occupied. Thus, solar absorption lines provide information about chemical abundance, temperature and related quantities such as density and pressure. Similar comments apply to the chromospheric emission lines; in fact, these lines provide some of the most direct information about the composition and structure of the chromosphere.

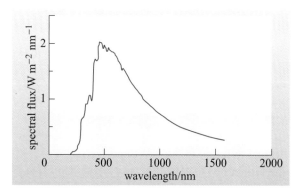

Figure 1.30 A graphical representation of the Sun's spectrum, as measured from above the Earth's atmosphere. (Phillips, 1992, from the original data of Labs and Neckel, 1968)

Information gained from spectral studies shows that the chromosphere is composed mainly of hydrogen and helium. The evidence indicates that there are between 40 and 100 helium atoms for every 1000 hydrogen atoms. (Other kinds of study, to be discussed in Section 2.2, imply that the correct value is probably about 87, corresponding to about 25% of the mass.) In view of the prominence of the calcium lines, you might also expect that calcium atoms would be abundant, but this turns out not to be the case. The relative strength of the calcium lines results from the combination of two factors: chromospheric conditions are just right to ensure that a comparatively high proportion of calcium atoms have electrons in the correct initial states to produce those lines, and the relevant transitions have a relatively high likelihood of occurring, given electrons in the correct initial states. In fact, there are probably no more than three calcium atoms for every *million* hydrogen atoms.

As far as temperature and density are concerned, it has already been emphasized that these quantities vary with height and from one region of the chromosphere to another. Nonetheless, typical values for a 'quiet' part of the chromosphere are shown in Figure 1.31. The rapid increase of temperature with height is very clear. Note that the height is measured from the usual reference level at the base of the photosphere (roughly, the greatest visible depth), so only data pertaining to heights greater than about 500 km are truly 'chromospheric'. It is thought that the chromosphere is heated mainly from below (by energy coming from the photosphere) so it is interesting, and not a little perplexing, that the temperature should rise with increasing height. Even more astonishing is the enormous rate at which the temperature rises, especially towards the top of the chromosphere. This surely indicates a very high temperature indeed for the Sun's outer atmosphere – the corona. The high temperature of the corona and upper chromosphere, its cause and its influence will be our major concern in the next section.

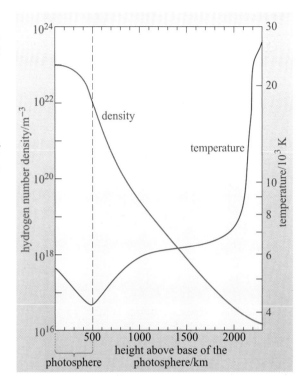

Figure 1.31 Variation of temperature and hydrogen number density (i.e. number of hydrogen atoms per cubic metre) with height throughout a 'quiet' part of the chromosphere. For comparison, at sea-level the Earth's atmosphere has on average a temperature of about 288 K and a number density (of oxygen and nitrogen molecules) of about $2 \times 10^{25} \, m^{-3}$.
Note that this figure uses *logarithmic scales* on the vertical axes.

QUESTION 1.10

Using a certain detector it has been found that a particular source of electromagnetic waves produces a black-body spectrum with a temperature of 6000 K. Use Figure 1.23 to find the ratio of the detector reading at 400 nm to that at 700 nm. What would the ratio have been if the spectrum had been that of a black body at 5000 K?

QUESTION 1.11

Treating the Sun as a good approximation to a black body, estimate its temperature from the graphical spectrum of Figure 1.30.

1.4 Seeing the Sun's outer atmosphere

1.4.1 Introducing the corona

As you saw at the end of the last section, the temperature rises steeply in the upper parts of the chromosphere and the number of hydrogen atoms per cubic metre declines. These trends continue through a narrow and highly irregular **transition region** and on into the **corona** – the Sun's outer atmosphere.

The corona is very extensive, very tenuous, and has an extremely high temperature.

The extent of the corona can be gauged from photographs taken during eclipses or from measurements made with a **coronagraph** – a special kind of telescope that uses an opaque disc to block the light from the photosphere and thus produce a sort of artificial eclipse. (Building a coronagraph that works is much harder than it sounds.) High quality images, such as that in Figure 1.32 (which was taken from a coronagraph on a space-based solar observatory), show the corona stretching out to several times the radius of the photosphere. They also reveal detailed structures, often in the form of arches or rays. These features change with time, sometimes quite rapidly, and respond to the general level of (sunspot) activity visible on the photosphere. At times of low activity the corona is usually rather quiescent and elongated at the Sun's equator (Figure 1.33a); at times of high activity it is much more lively, with streamers jutting out in all directions (Figure 1.33b).

Figure 1.32 The solar corona as observed using a coronagraph on a space-based solar telescope. The white circle represents the size and location of the solar disc. (SOHO (ESA and NASA))

Figure 1.33 The solar corona during an eclipse at a time of (a) low, and (b) high solar activity. (NCAR)

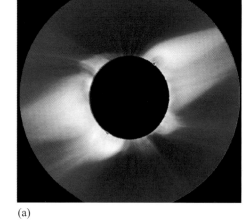

(a)

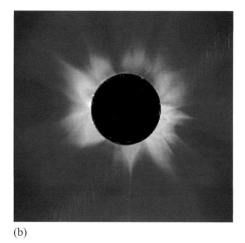

(b)

In comparison to the photospheric emission, the total amount of light produced by the corona is small. Even the brightest parts of the corona are almost a million times less luminous at visible wavelengths than an equal area of the photosphere. Most of the observed coronal light is simply white photospheric light scattered by particles (mainly electrons) in the corona – that's why the light has a characteristic pearly white colour. In the outer parts of the corona, beyond about two solar radii from the centre of the Sun, it is even possible to see the photospheric and chromospheric absorption lines in the coronal spectrum. However, in addition to features attributable to the photosphere and chromosphere, the coronal spectrum also includes some emission lines that originate in the corona itself. The strongest of these is a green line at 530.3 nm, but there is also a prominent yellow line at

569.4 nm and a red line at 637.4 nm. An image of the corona in the light of the green 530.3 nm line is shown in Figure 1.34.

When the strongest of these coronal emission lines was discovered during the eclipse of 7 August 1869, its origin was unknown. It was suggested that it might be due to a previously undiscovered element, which was tentatively named coronium. But, in contrast to the story of helium, no other evidence was found to support this hypothesis, and the cause of this and other coronal lines remained one of the most challenging problems in solar spectroscopy for over 70 years. It was finally solved in the early 1940s by the Swedish

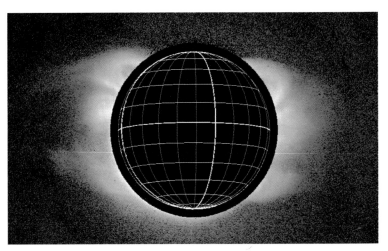

Figure 1.34 An image of the corona taken at a wavelength of 530.3 nm. (SOHO (ESA and NASA))

astrophysicist Bengt Edlén (1906–1993). In a series of experiments Edlén proved that an ionized iron atom that has lost half of its normal complement of 26 electrons has its energy levels altered in such a way that there is a transition capable of producing 530.3 nm radiation (no such transition exists in a neutral iron atom that retains all 26 electrons). Since neutral iron (chemical symbol, Fe) has 26 electrons in each of its atoms, it follows that iron atoms that have lost half their electrons will have a net positive charge that is equal in magnitude but opposite in sign to the charge of the 13 lost electrons. Such atoms, or, to be more precise, such ions are conventionally denoted by the symbol Fe^{13+}. Thus, Edlén had established that the green coronal emission was due to Fe^{13+}. The other coronal emission lines known at the time were also shown to be due to highly charged ions.

The presence of ions in the Sun is not surprising. Temperatures throughout the Sun are so high that most of the atoms are ionized to some extent. Some of the effects due to the presence of these ions are already familiar to you: the prominent H and K lines due to calcium (chemical symbol, Ca) in the chromosphere come from Ca^+ ions. In addition, much of the light from the photosphere is emitted when free electrons, liberated when atoms are ionized, combine temporarily with neutral hydrogen atoms to produce H^- ions. However, the discovery that a significant amount of Fe^{13+} exists in the corona, was a surprise. The presence of such highly ionized atoms implied that the temperature of the corona was very high indeed – at least 10^6 K. Previously it had been thought that temperatures decreased as distance from the photosphere increased; the discovery that this was not the case replaced the enigma of the emission lines by the mystery of the mechanism that could be responsible for such high temperatures.

In part, the very high temperature of the corona is due to a rather peculiar property of gases at low densities and high temperatures. Over a range of temperatures between about 10^5 K and 10^6 K, the rate at which coronal gas would lose energy by radiating electromagnetic waves is lower than at the temperatures at either end of this range (i.e. at about 10^5 K or 10^6 K). This is quite different to our everyday experience of how hot material behaves, in that we usually expect, and usually find, that the rate at which a body loses energy simply increases with temperature. Coronal gas that is at a temperature between 10^5 K and 10^6 K is in an unstable state: if the gas is heated such that its temperature exceeds 10^5 K, then the gas is unable to cool efficiently and its temperature quickly rises to a temperature in excess of 10^6 K. However, this instability only explains why the temperature is so high; it does not explain the source of the heating for the corona.

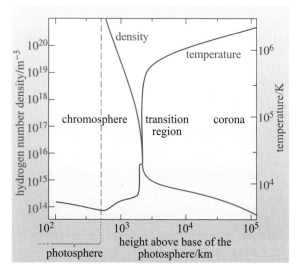

Figure 1.35 Variation of temperature (red) and hydrogen number density (blue) with height above the base of the photosphere. Note that all the scales are logarithmic. This figure incorporates the information shown in Figure 1.31, but extends to show how temperature and density vary from the photosphere to the high regions of the corona. (Adapted from Gabriel, 1976)

The corona cannot be heated simply by energy radiated by the photosphere or by heat conducted through the chromosphere; a basic law of physics (the **second law of thermodynamics**) prevents the transfer of energy from a cooler body to a hotter one by either of these methods. Nonetheless, there can be little doubt that the necessary energy does come from the lower regions of the Sun; the question is, how? It seems that the behaviour of the magnetic field in the lower parts of the corona plays an important role in heating the coronal gas. However the precise mechanism is still elusive and the absence of a universally agreed solution to the problem of coronal heating is not due to a lack of ideas but is indicative of the extreme difficulty of producing detailed theoretical descriptions of such complex regions.

Like the chromosphere, the corona is highly non-uniform and conditions vary enormously from place to place within it. Nonetheless it is possible, using spectroscopic information and basic physical principles, to obtain a good idea of 'average' conditions in a quiet part of the corona. Figure 1.35 shows the variation of temperature and hydrogen number density with height under such quiet conditions in the chromosphere and the inner part of the corona. The graphs are essentially extensions into the corona of the chromospheric data shown in Figure 1.31. Note that the location of the transition region separating the chromosphere and the corona is defined by the very rapid temperature rise just below 2500 km. The precise location depends on local conditions; it might well be as high as 8000 km on some occasions. Also note that, at the greatest heights shown in Figure 1.35, the temperature of the corona is still increasing. Temperatures of 3×10^6 to 4×10^6 K are not uncommon, and even higher temperatures are sometimes attained over limited regions.

You might think that such enormous temperatures would make the corona a powerful source of electromagnetic radiation, but this is not so. The density of the corona is very low, and such a relatively small amount of matter, even at a very high temperature, is a very poor emitter compared with the cooler but much denser photosphere. Nevertheless, the high temperature of the corona does have important implications for the electromagnetic radiation that is emitted.

▪ Why is the spectrum of the corona unlikely to be a black-body source of radiation?

❑ The corona is a very low-density region that is far from being opaque. Sources that produce spectra that are a good approximation to black-body spectra tend to be highly opaque. Hence it is unlikely that the corona would exhibit a black-body spectrum.

Despite the fact that the corona is *not* a black-body source of radiation, it is still possible to estimate the wavelength at which it will radiate. A rule of thumb that can be applied to *any* thermal source is that if the source is at a temperature T, then the typical photon energy ε that would be emitted by the source is given by

$$\varepsilon \sim kT \tag{1.6}$$

where k is the **Boltzmann constant**, which has a value of $1.38 \times 10^{-23}\,\mathrm{J\,K^{-1}}$. The symbol '~', which is often called 'twiddles', is used to express rather approximate relationships – the answer may be correct only to within a factor of ten or so. In applying Equation 1.6 to a real problem, the equation is evaluated as if the '~' were an equals sign, but it must be remembered that the answer is only an approximate solution. There are no hard and fast rules about how to quote the numerical answers from this sort of equation: while it is usual practice to give results to one significant figure, it must be borne in mind that the implied level of precision is often not justified. Equation 1.6 is very approximate because it refers to *any* thermal source, not just the special case of black-body emission. In general, a thermal source will produce a spectrum that is more complex than the black-body form; it may, for instance, appear to be a mixture of a continuous spectrum and a line spectrum. Equation 1.6 allows us to estimate, within a factor of ten or so, the typical photon energies that are likely to be observed in such a spectrum.

Let's return to the solar corona with an example of the use of Equation 1.6 that will allow us to estimate the wavelength of its thermal emission.

EXAMPLE 1.1

Assuming that the solar corona is a thermal source with a temperature of $2 \times 10^6\,\mathrm{K}$, calculate (a) the typical energy (in eV), and (b) the corresponding wavelength of photons that would be emitted.

SOLUTION

(a) The typical photon energy can be found by applying Equation 1.6, using a value of $T = 2 \times 10^6\,\mathrm{K}$,

$$\varepsilon \sim kT$$

$$\varepsilon \sim (1.38 \times 10^{-23}\,\mathrm{J\,K^{-1}}) \times (2 \times 10^6\,\mathrm{K}) = 2.76 \times 10^{-17}\,\mathrm{J}$$

which, in terms of electronvolts, is

$$\varepsilon \sim (2.76 \times 10^{-17}\,\mathrm{J})/(1.602 \times 10^{-19}\,\mathrm{J\,eV^{-1}}) = 172\,\mathrm{eV}$$

Since the symbol $\sim$ implies a very approximate relationship, the answer should be quoted to just one significant figure. Hence the typical photon energy is $2 \times 10^2\,\mathrm{eV}$.

(b) The wavelength that is typical of emission from the corona can be found from the photon energy ε. The first step is to write an equation for λ in terms of ε. Equation 1.2 can be rearranged to make λ the subject,

$$\lambda = c/f$$

and from Equation 1.3,

$$f = \varepsilon/h$$

Combining these two equations gives the required equation,

$$\lambda = hc/\varepsilon$$

Now the wavelength can be calculated using the value of ε obtained in part (a). (It does not matter whether the value used is the one in joules or electronvolts provided that the appropriate value of the Planck constant h is also used.)

$$\lambda = (6.63 \times 10^{-34}\,\mathrm{J\,s}) \times (3.00 \times 10^8\,\mathrm{m\,s^{-1}})/2.76 \times 10^{-17}\,\mathrm{J}$$

$$\lambda = 7.21 \times 10^{-9}\,\mathrm{m}$$

Since this result was obtained using the approximate relationship given in Equation 1.6, the answer needs to be expressed to one significant figure. Thus the typical wavelength of a photon that will be emitted by the corona is 7×10^{-9} m or 7 nm.

This example has shown that the thermal emission from the corona will occur at a wavelength which is about a hundred times less than visible wavelengths. Does electromagnetic radiation with such a tiny wavelength exist? If so, what is it? Once again, it's time for some more background science (Box 1.3).

BOX 1.3 THE ELECTROMAGNETIC SPECTRUM

In terms of the electromagnetic wave model, introduced in Box 1.1, visible light spans a range of wavelengths from approximately 400 nm to approximately 700 nm. Electromagnetic waves with wavelengths outside this range cannot, by definition, represent visible light of any colour. However, such waves do provide a useful model of many well-known phenomena that are more or less similar to light. For example, everyone is familiar with radio waves; we all rely on them to deliver radio and TV programmes. Radio waves are known to have wavelengths of about 3 cm or more; their well-established properties include the ability to be reflected by smooth metal surfaces and to travel through a vacuum at the same speed as light. Both of these properties, and many others that could have been

quoted, are also exhibited by electromagnetic waves of the same wavelengths. Thus we can say that radio waves can be 'well modelled' by electromagnetic waves with wavelengths greater than about 3 cm. Indeed, because the relatively long wavelength of radio waves makes their wave-like nature so obvious, it is really rather pedantic to talk about an electromagnetic wave *model* of radio waves at all. Most people would happily accept the statement that radio waves *are* electromagnetic waves.

The wide range of phenomena that can be modelled by electromagnetic waves is illustrated in Figure 1.36. As you can see, the full **electromagnetic spectrum**, as it is called, ranges from long wavelength **radio waves**,

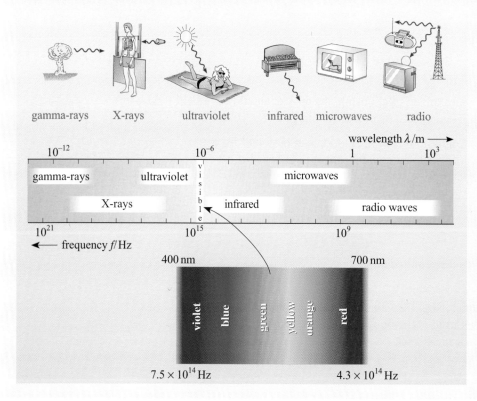

Figure 1.36 The electromagnetic spectrum. Note that the frequency and wavelength scales are logarithmic. Note also that the ultraviolet (meaning 'beyond the violet') adjoins the visible violet, and the infrared (meaning 'below the red') adjoins the visible red.

through **microwaves** and **infrared radiation**, across the various colours of **visible light** and on to such short wavelength phenomena as **ultraviolet radiation**, **X-rays** and **γ-rays**; (pronounced 'gamma rays'). These various kinds of **electromagnetic radiation** arise in a wide range of contexts (as illustrated) but fundamentally they differ from one another only in the wavelength (or frequency) of the electromagnetic waves used to model them.

The boundaries of the various regions are deliberately vague; scientists and technologists often draw the divisions somewhat loosely.

The photon model of light that was introduced in Box 1.1 can also be applied throughout the electromagnetic spectrum. In fact, when it comes to γ-rays, their very short wavelengths make it quite difficult to demonstrate their wave-like properties, and it is much more conventional to speak of them as though they were particles.

Of course, the true situation is that all forms of electromagnetic radiation are, at the present time, most accurately described by the quantum theory mentioned in Box 1.1. No part of the electromagnetic spectrum 'really' consists of waves or particles but any part may, under the appropriate conditions, exhibit wave-like or particle-like behaviour; both possibilities are encompassed by the quantum theory.

Many of the ideas in Section 1.3.2 concerning spectroscopy and sources of light are also applicable to the entire electromagnetic spectrum. Every kind of electromagnetic radiation may come from thermal or non-thermal sources, and the continuous spectrum produced by a black-body source always extends beyond the wavelength range of visible light. (This last point is already implicit in Figure 1.23, which shows the Planck curves stretching into the ultraviolet and the infrared.) Similarly, Wien's displacement law applies throughout the electromagnetic spectrum. Also, spectral lines may arise at any wavelength provided transitions of the appropriate energy exist to cause them. Such transitions are not necessarily confined to atoms; they may involve nuclei, molecules or many other systems.

QUESTION 1.12

Complete Table 1.1.

Table 1.1 A partially completed table showing the properties of six different electromagnetic waves. For use with Question 1.12.

Wavelength, λ/m	3×10^{-14}	6×10^{-10}		5×10^{-3}	
Corresponding frequency, f/Hz	1×10^{22}	5×10^{17}			3×10^{7}
Corresponding photon energy, ε/J	7×10^{-12}			4×10^{-23}	2×10^{-26}
Corresponding photon energy, ε/eV	4×10^{7}		4	2×10^{-4}	1×10^{-7}
Temperature, T/K, of a black body that has a peak in its spectrum at this value of λ		5×10^{6}	3×10^{2}		3×10^{-4}
Corresponding part of the electromagnetic spectrum	γ-ray	X-ray			radio wave

1.4.2 The Sun's electromagnetic spectrum and the corona

Figure 1.37 shows the electromagnetic spectrum of the entire Sun across a range of wavelengths from 10^{-13} m to 10 m, or at least it attempts to do so. Such a spectrum, i.e. one that covers a wide range of wavelengths is often called a **broadband spectrum**. The central region of the graph in Figure 1.37, roughly from 10^{-7} m to 10^{-4} m, which accounts for most of the emitted power (note the logarithmic scales), is not difficult to interpret. Emission in this region is dominated by the photosphere and, apart from some absorption lines that do not really show up on this scale, the solar spectrum is well approximated by the Planck curve of a black-body source at 6000 K. Provided the Sun is in a quiet state with relatively little activity taking place, the emission of microwaves and short wavelength (i.e. less than 10 m) radio waves would be as expected from the Planck curve. However, at longer radio wavelengths the emissions come predominantly from the hotter regions higher in the Sun's atmosphere. One consequence of this is that the observed diameter of the Sun increases as the wavelength of observation increases, another is that the 6000 K Planck curve provides a progressively less satisfactory approximation to the observed spectrum. Moreover, if there is a significant amount of activity on the Sun, as is often the case, the related bursts of radio emission may

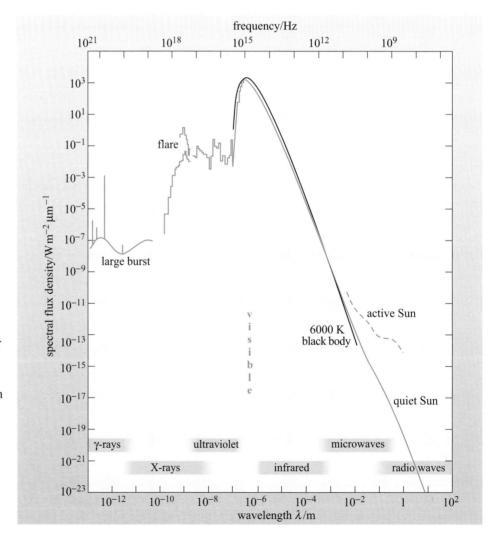

Figure 1.37 The solar electromagnetic spectrum. At any given wavelength λ, the height of the graph represents the amount of energy that would be received in 1 s by a detector, of area 1 m^2, pointed directly towards the Sun, in a range of wavelengths of width 1 μm centred on λ, *provided* the detector was above the Earth's atmosphere. The shape of the graph is discussed in the text. The black line is the Planck curve of a black body at 6000 K. (Nicholson, 1982)

well dominate the spectrum over a broad range of wavelengths. This is especially true of explosive events called **solar flares**, one of the most energetic kinds of solar activity. We will return to discuss solar flares in more detail in Section 2.3.2. The existence of these outbursts makes the radio spectrum highly variable with time and difficult to represent on a simple graph.

Similar problems exist in the X-ray and γ-ray parts of the spectrum. The black-body emission from the photosphere falls off very rapidly with decreasing wavelength and is essentially negligible below a wavelength of about 10^{-8} m. Solar radiation is seen at shorter wavelengths, but it comes mainly from the hot coronal material and especially from the active regions where the temperatures are greatest. At the very shortest wavelengths (below 10^{-10} m) the individual X-ray and γ-ray photons each carry a relatively high energy, and it is quite normal to see spectra presented as 'counts' of the numbers of photons observed in various narrow wavelength (or energy) ranges in a given time.

Prior to the space age, it was impossible to obtain observational data about the broadband spectral behaviour of the Sun. The Earth's atmosphere is a very effective absorber of certain wavelengths of electromagnetic radiation, with the consequence that Earth-based solar observations simply cannot be made for wide ranges of wavelength. Figure 1.38 provides a clear indication of the effectiveness of the Earth's atmosphere as a barrier to solar radiation. The atmosphere is largely transparent to visible light and to certain longer wavelengths, particularly to narrow bands in the infrared and to radio waves, but everything else is effectively stopped. In order to overcome this barrier, many observations have to be made from high-altitude rockets and balloons, or satellites and space probes. Such ventures are complicated and costly, but they have been responsible for many of the most important recent advances in solar science (and in many other fields of astronomy).

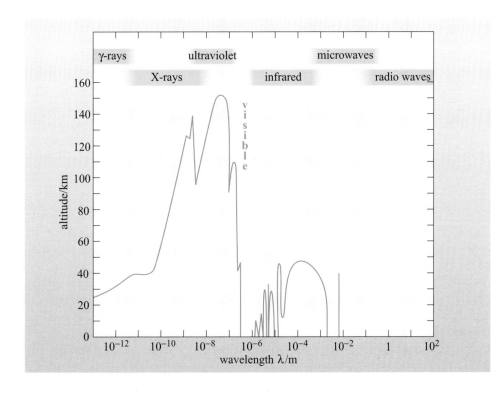

Figure 1.38 Effectiveness of the Earth's atmosphere as a barrier to incoming solar radiation. For a given wavelength λ, the graph shows the altitude at which the spectral flux density (F_λ) due to the Sun is reduced to 50% of its value at the top of the atmosphere. (Nicholson, 1982)

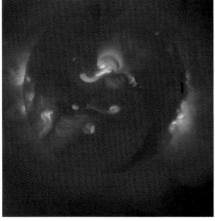

(a) X-ray (08:18 on 14 Dec 2001)

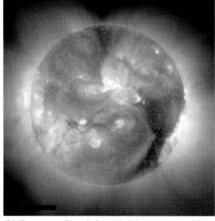

(b) Extreme ultraviolet (13:06 on 14 Dec 2001)

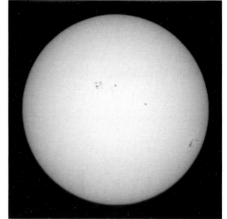

(c) White light (16:07 on 13 Dec 2001)

(d) Hα (18:27 on 13 Dec 2001)

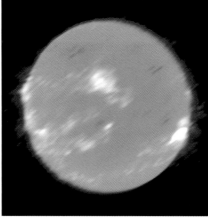

(e) Radio (23:05 on 13 Dec 2001)

Figure 1.39 The Sun at various wavelengths.
(a) X-ray ($\lambda = 0.3$ to 5 nm),
(b) extreme ultraviolet ($\lambda = 28.4$ nm),
(c) the whole visible spectrum,
(d) Hα ($\lambda = 656.3$ nm),
(e) radio ($\lambda = 1.8$ cm). These images were all taken within a 24-hour period on 13 and 14 December 2001.
((a) Yohkoh (ISAS); (b) SOHO (ESA and NASA); (c), (d) Big Bear Solar Observatory; (e) Nobeyama Radioheliograph)

Five solar images, representing different parts of the electromagnetic spectrum, are shown in Figure 1.39. It should be clear from the above discussion of the solar spectrum that the images formed at very long and very short wavelengths will have been produced largely by radiation from the corona. The X-ray image (Figure 1.39a) is essentially confined to temperatures above 1.6×10^6 K and is particularly informative about coronal structure. Bright regions associated with solar activity are easy to see, but more intensive study reveals that the corona is generally composed of 'loops' of various sizes. These loops are magnetic in origin; they are known to result from the interplay of the highly ionized coronal gases and the Sun's magnetic field. We will return to study these interactions in more detail in Chapter 2.

Another feature commonly seen in X-ray images is known as a **coronal hole**. Some idea of its nature can be obtained from Figure 1.40, which shows a sequence of X-ray images taken at intervals of 27 days from the orbiting Skylab space station in 1973. The rotation of the coronal hole – the dark boot-shaped object stretching from the pole to the equator – with the Sun is quite clear. The hole's shape and size change with time, and sometimes it seems to fragment and then merge together again. Coronal holes are devoid of the large loops seen elsewhere in the corona: they are regions where the solar magnetic field opens outwards to interplanetary space rather than looping back on to the Sun. The existence of such regions is of great importance, since they are thought to be a major source of the **solar wind** –

a gusty stream of high-speed particles that spreads out from the Sun, carrying traces of the Sun's magnetic field with it. We will return to look in more detail at the origin and effects of the solar wind later in Section 2.4.

■ Why were the Skylab images in Figure 1.40 taken 27 days apart?

❏ The Sun rotates; 27 days were required for the coronal hole to return to the centre of the field of view of the Skylab detector.

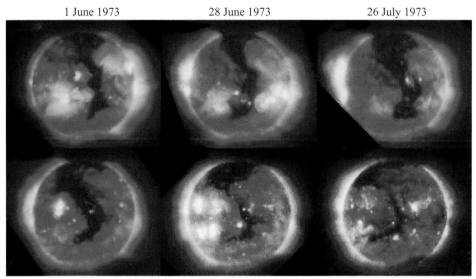

1 June 1973   28 June 1973   26 July 1973

21 August 1973   17 September 1973   14 October 1973

Figure 1.40 X-ray images of the corona taken at 27 day intervals (in 1973), and showing the evolution of a coronal hole. Such images must be obtained from space because the Earth's atmosphere absorbs electromagnetic radiation in this wavelength range. (Stanford University)

1.5 SUMMARY OF CHAPTER 1

Basic characteristics of the Sun

- The Sun is the closest star to the Earth.
- The solar luminosity (3.84×10^{26} W) is the rate at which energy is radiated by the Sun.
- The Sun emits radiation across the entire electromagnetic spectrum, from radio waves to γ-rays. The central wavelength range (10^{-7} m to 10^{-4} m), which accounts for nearly all the radiated power, is approximately described by a 6000 K black-body curve, and is dominated by photospheric emissions. The extreme wavelengths are more strongly related to solar activity and are often dominated by coronal emissions.

The nature of light

- Quantum theory provides the best available account of the nature of light, but simple models such as the electromagnetic wave model

$$c = f\lambda \tag{1.2}$$

and the photon model

$$\varepsilon = hf \tag{1.3}$$

are still of great value.

- A black-body source of radiation emits a characteristic continuous spectrum. The wavelength at which such a curve attains its maximum is determined solely by the temperature of the source, in accordance with Wien's displacement law:

$$(\lambda_{peak}/m) = \frac{2.90 \times 10^{-3}}{(T/K)} \tag{1.4}$$

- For *any* thermal source of electromagnetic radiation (not necessarily a black-body source), the typical photon energy can be estimated by

$$\varepsilon \sim kT \tag{1.6}$$

- The absorption and emission lines found in the spectra of many sources of radiation are caused by electrons making transitions between states of different energies within atoms.

- The electromagnetic spectrum encompasses such seemingly diverse phenomena as radio waves, microwaves, infrared radiation, visible light, ultraviolet radiation, X-rays and γ-rays; all these forms of electromagnetic radiation can be modelled by electromagnetic waves of decreasing wavelength or by photons of increasing energy, respectively, under the appropriate circumstances.

The photosphere

- The photosphere is the visible 'surface' of the Sun. It is a thin semi-transparent shell of gaseous material about 500 km thick, 1.4×10^6 km in diameter and characterized by a temperature of about 6000 K.

- The photosphere exhibits limb darkening and differential rotation. The rotation can be traced by sunspots, which also indicate the level of solar activity.

- The Sun is the only star sufficiently close to Earth to permit the detailed study of small-scale phenomena, such as sunspots and granulation.

The chromosphere

- The Sun's chromosphere is a patchy layer of gaseous material, a few thousand kilometres thick, that overlies the photosphere. About 75% of the mass of the chromosphere is made up of hydrogen; helium accounts for most of the remainder.

- Detailed studies of solar spectra provide information about the chemical composition of the chromosphere and its physical condition (temperature, number density, etc.). Such studies show that the temperature of the chromosphere increases rapidly with height, changing by about 20 000 K over 2000 km.

- Chromospheric spectra show a number of emission lines, notably an Hα line at 656.3 nm. Hα images of the Sun reveal details of the structure of the chromosphere and various signs of solar activity, such as plages and filaments.

The corona

- The outer atmosphere of the Sun is called the corona. It is tenuous and extensive, and separated from the chromosphere by the narrow, irregular transition region. The structure of the corona is highly variable and responds to the level of solar activity.

- Spectral signatures of highly ionized atoms reveal that temperatures of several million degrees are common in the corona. Accounting for these high temperatures is still a major challenge for solar scientists.

Questions

QUESTION 1.13

By reference to a calcium K image of the Sun, estimate the size of a plage.

QUESTION 1.14

Would it be possible to identify the material that makes up a black-body source by examining its spectrum?

QUESTION 1.15

The energy levels of the hydrogen atom (labelled E_1, E_2, E_3, etc. in Figure 1.26) have energies that are well approximated by the formula

$$(E_n/\text{eV}) = -\frac{13.6}{n^2}$$

Note that all the energies are given negative values; E_1 is the lowest.

where n is the energy subscript, 1, 2, 3, etc. (Note that the photon energy is given in terms of electronvolts, so take care in the following question to use the Planck constant in the appropriate units.)

(a) Insofar as you can, confirm that the absorption occurring at the wavelength of the Hα line can be attributed to transitions from E_2 to E_3.

(b) Which transition accounts for the Hβ absorption line? (*Hint*: attempt this by trial and error, with n no larger than 5. A useful skill to develop is that of bracketing an answer by guesswork and then homing in on it.)

(c) Which transition would account for an emission line at the same wavelength as the Hβ absorption line?

(d) Why is there no line corresponding to the transition from E_1 to E_2 in the spectrum of Figure 1.28?

QUESTION 1.16

In addition to the green emission line due to ionized iron, the corona produces two other prominent emission lines in the visible part of the spectrum: a yellow line at 569.4 nm due to calcium atoms that have lost 14 electrons, and a red line at 637.4 nm due to iron atoms that have lost 9 electrons.

(a) How can the ions responsible for the yellow and red lines be represented symbolically?

(b) Calculate the difference in energy between the relevant energy levels for each of the two ions responsible for the yellow and red lines. Express your answer in electronvolts.

QUESTION 1.17

A thermal source of radiation has a temperature of 5×10^6 K.

(a) Calculate the (approximate) wavelength at which this source would emit electromagnetic radiation.

(b) If this source is not only a thermal source, but is also a black-body source, calculate the wavelength at which the peak of emission occurs.

(c) Explain why the answers to parts (a) and (b) differ, and state whether or not the answers are consistent with one another.

QUESTION 1.18

At visible wavelengths, images of the Sun exhibit the phenomenon of limb darkening, as described earlier. At radio wavelengths the opposite effect, limb brightening, is seen. Explain the origin of limb brightening. (*Hint*: recall that temperatures in the solar atmosphere increase with height.)

CHAPTER 2
THE WORKING SUN

2.1 Introduction

You have already seen that the Sun is a prodigious source of energy. The most distinctive physical property of the Sun is that it has a steady luminosity of about 3.84×10^{26} W, which is emitted mainly in the form of visible and infrared radiation from the photosphere. The aim of this chapter is to investigate some of the physical mechanisms that are associated with the Sun's energetic processes. In addition to the steady photospheric emission, the Sun is also a source of energetic outbursts; solar flares that may emit about 10^{25} J, predominantly in X-rays, over a period of a few minutes. Less obviously, there are outflows of material from the Sun, dramatic eruption events and a steady solar wind, and these too involve large releases of energy.

The first question will be the source of the solar luminosity; this will require us to consider the extreme environment that lies at the centre of our nearest star, and to review the techniques that astronomers and solar physicists use to probe regions that are hidden from our direct view. We will then move out to the surface layers of the Sun, and investigate phenomena that are linked to solar activity. The wealth of observational detail that is available about energetic processes on the surface of the Sun has helped scientists to unravel some of the processes that drive solar activity. We will see how various seemingly diverse phenomena that are indicative of solar activity can all be attributed to the behaviour of the Sun's magnetic field. We shall also consider how the Sun interacts with its environment through the outflow of material. The interaction between the Earth's magnetic field and out-flowing solar material is a source of both beauty and danger: it gives rise not only to the magnificent aurorae (the Northern or Southern Lights), but also to magnetic effects that can disrupt electrical distribution systems on the Earth. Finally, we investigate the boundaries of the Sun's dynamic environment and find that this takes us far beyond the most distant planets of the Solar System to the realms of interstellar space.

2.2 Inside the Sun

2.2.1 Introducing the solar interior

In Chapter 1 we examined in some detail the outer parts of the Sun and the radiation they emit. In particular, we saw that most of the energy received at the Earth from the Sun is carried by radiation from the photosphere. There are three factors that account for the photosphere's effectiveness as a source of energy. First, the atmospheric layers above the photosphere are, for the most part, transparent at the visible wavelengths which dominate photospheric emission. Second, the photosphere itself is opaque in the sense that although we can see a short distance into it, we cannot see through it. Third, the photosphere is sufficiently large and at a sufficiently high temperature, around 6000 K, to be a powerful thermal source of radiation.

In our quest to understand the nature and origin of the Sun's radiation, the task that now confronts us is that of explaining why the photosphere has such a temperature and how that temperature is maintained despite the prodigious rate at which solar energy is radiated into space. You may already be familiar with the broad answer to these questions: the photosphere is heated from below by energy coming from deeper and hotter regions of the **solar interior**. This state of affairs persists because the **core** of the Sun – roughly the central 2% or 3% of the Sun's volume – is a steady and long-lived energy source powered by nuclear processes. This section is devoted to expanding these answers and giving you some feel for the nature of the solar interior.

2.2.2 The internal structure of the Sun

Discussions of the solar interior are bound to be largely theoretical because the whole region is hidden from view by the photosphere. Most of what we know about the interior is based on a number of theoretical **solar models** that have gained wide acceptance amongst solar scientists. These solar models differ from one another only in rather small matters of detail: they all agree about general principles and all give very similar results. Solar models have become increasingly refined as techniques to probe the solar interior have developed (Section 2.2.6). Indeed, by the 1990s, the reliability of solar models was such that many scientists claimed that certain anomalous results could only be explained if an aspect of our understanding of fundamental physics was incorrect – a viewpoint that we now know to be vindicated.

Each solar model is based on a few fundamental physical principles, some plausible assumptions about the interior, and some observed properties that are termed 'boundary conditions'. The *physical principles* include the requirement that the rate at which the Sun radiates energy should equal the rate at which nuclear processes produce energy in the interior, and the need for the solar material at a given depth to be able to support the weight of the matter that sits on top of it. The *plausible assumptions* relate to many issues, including the importance of internal magnetic fields and the rate at which the internal layers of the Sun rotate about the Sun's axis. The *boundary conditions* specify certain observed properties of the Sun, such as the radius ($R_\odot$), total mass ($M_\odot$), luminosity ($L_\odot$) and chemical composition.

A solar model that is constructed in this way provides numerical values for the temperature, pressure and density at any given distance from the centre of the Sun. This latter quantity is usually expressed as a fraction of the Sun's radius and is often denoted $R/R_\odot$, where $R_\odot$ is the radius of the photosphere. It is called the **fractional radius**. Representative results from a particular solar model are displayed graphically in Figure 2.1. More will be said about the route that leads to such results in the next section.

The rise of temperature, pressure and density with increasing depth indicated by Figure 2.1 is not surprising. The pressure should be expected to increase with depth, owing to the growing weight of overlying material, and the temperature to rise, because of the increasing proximity to the central energy source. Nevertheless, the details, which emerge from a lengthy computer-based calculation, are surprising. Temperature, pressure and density all change very rapidly near the photosphere, but it's not until a fractional radius of about 0.5 that the density is equal to that of water on Earth (1.0×10^3 kg m^{-3}). Even at the centre of the Sun, where the temperature is 15.6×10^6 K, the density is predicted to be only fourteen times that of lead, though the pressure is more than 10^{10} times that of the Earth's atmosphere at sea-level.

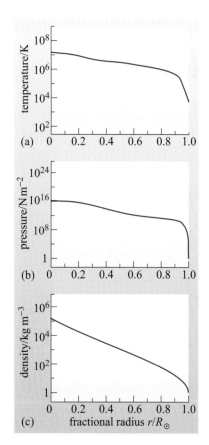

Figure 2.1 Variation with fractional solar radius of (a) temperature, (b) pressure and (c) density in the solar interior. Note that the vertical axis is logarithmic in each case.

Assuming that the core of the Sun occupies a fractional radius of 0.3, calculate the fraction of Sun's volume that is occupied by the core. Express your answer as a percentage.

2.2.3 The internal composition of the Sun

When dealing with the composition of the solar interior, it is traditional to divide the constituents into just three categories: hydrogen, helium and everything else. In this context, all those materials that fall into the 'everything else' category will be referred to as **heavy elements**, irrespective of precisely how heavy they are compared with hydrogen and helium. Thus elements such as boron, oxygen and carbon will be called heavy elements, along with iron, copper and gold. Accepting this convention, at least for the moment, the composition of a sample of material at any depth within the Sun can be defined by assigning numerical values to three simple parameters:

The **hydrogen mass fraction** (X):

$$X = \frac{\text{mass of hydrogen in sample}}{\text{mass of sample}}$$

The **helium mass fraction** (Y):

$$Y = \frac{\text{mass of helium in sample}}{\text{mass of sample}}$$

The **metallicity** (Z):

$$Z = \frac{\text{mass of heavy elements in sample}}{\text{mass of sample}}$$

The term metallicity arises from the fact that many astronomers use the term 'metals' to refer to all heavy elements, regardless of whether they exhibit the properties normally associated with metals. Since this usage of the term 'metals' is potentially confusing, in this book we refer instead to 'heavy elements'.

■ What can you say about the value of $X + Y + Z$ at any depth within the Sun?

❏ $X + Y + Z = 1$ at any depth. This follows from the fact that we defined Z to be 'everything else' apart from hydrogen (X) and helium (Y).

As you saw earlier, spectroscopic studies of the outer parts of the Sun can provide detailed information about the relative abundance of the various chemical elements found there. However, such information is unlikely to be a reliable guide to the constitution of the solar interior. Instead, determinations of X, Y and Z in the interior, like structural determinations, are usually based on calculations that involve a solar model.

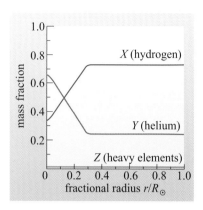

Figure 2.2 Variation of the mass fractions X, Y and Z with the fractional radius, $R/R_\odot$, for the Sun.

It is universally accepted that the nuclear processes that power the Sun convert hydrogen into helium. Thus, the values of the hydrogen and helium mass fractions (X and Y) change with time. Now, any solar model that provides a reasonably full account of energy production must take into account these changes, and the changes they produce in temperature, pressure and density. Consequently, a detailed solar model is capable of simulating the evolution of the Sun – including the evolution of its composition. By adjusting the initial values of X, Y and Z, until the simulated evolutionary process results in a structure that agrees with our knowledge of the present-day Sun, it is possible to obtain estimates for the current values of X, Y and Z at any depth.

The results of one such calculation of the Sun's present internal composition are shown in Figure 2.2. In this particular case, the value of Z was initially set at 0.017 and was assumed to be independent of depth and constant in time. The value of X, which was initially set at 0.735, is now expected to vary from 0.341 at the core, where a good deal of hydrogen has been converted into helium, to 0.735 at the surface, where there has been no conversion. Since $X + Y + Z = 1$, the present-day helium mass fraction, at any depth, is given by $Y = 1 - 0.017 - X$. It follows that, throughout the upper parts of the interior, where $X = 0.735$, Y retains its initial value, 0.248. This sort of result is common to many solar models and is generally taken to provide a precise determination of solar helium abundance.

To summarize, throughout most of its interior the Sun is approximately 73% hydrogen, 25% helium and 2% everything else (by mass). However, in the central 30% or so of its radius the percentage of hydrogen is increasingly depleted (and helium compensatingly increased), with about half the hydrogen initially present in the core of the Sun having been converted into helium.

Before leaving the subject of the Sun's composition, one point, mentioned earlier, deserves special emphasis. Most of the atoms in the Sun are *ionized*. This is particularly true in the hot, dense interior, where essentially all the hydrogen and helium atoms are completely ionized. Such a highly ionized gas is called a **plasma**. So, although it is quite common to see the Sun referred to as a gaseous body, a more specific description is that it is made of plasma. In this case, the plasma consists mainly of hydrogen and helium ions, together with the electrons that were liberated when those ions were produced.

QUESTION 2.2

Assuming that solar material pretty much stays in one place during the life of the Sun (that is, there is no significant internal mixing of solar constituents), what can you deduce so far about the location of the nuclear processes that convert hydrogen into helium?

Indeed, the location of the nuclear processes in a central core is a *direct* and continuing consequence of the increase of temperature with depth in the models: the rates of nuclear reactions rise very rapidly as the temperature increases.

2.2.4 The energy source of the Sun

This section addresses two questions:

- What are the main energy-releasing nuclear processes that take place in the core of the Sun?
- At what rate do those nuclear processes occur?

The answers to these questions provide the ultimate solution to our search for the true origin of the Sun's electromagnetic radiation. But, as you will see, they also raise other questions.

What are the energy-releasing nuclear processes?

The basic energy-releasing process taking place in the Sun is *nuclear fusion*. This is the process in which nuclei of relatively low mass are fused together to form nuclei of somewhat greater mass. The fusion is brought about by a sequence of nuclear reactions in which colliding nuclei combine and fragment, to produce new nuclei together with other particles. No energy is actually created in these reactions: it is simply that energy is liberated from the *reactants* and is redistributed amongst the *products* in such a way that some of it replaces the energy radiated by the Sun. This replacement of lost energy maintains the high temperature of the core, thus sustaining the nuclear reaction rates.

A nuclide is a nucleus with a particular atomic number, Z, and mass number, A.

A full account of the nuclear processes taking place in the Sun would be very complicated indeed. Our discussion will be limited to the one process that is thought to be responsible for the bulk of the Sun's radiant energy – the so-called **ppI chain** (note that the 'I' in ppI is the roman numeral 'one'). The details of this process were first described by Hans Bethe and Charles Critchfield in the 1940s. The name 'pp' indicates that it is the first of several different chains of reactions that start with colliding protons – p is the symbol for the proton, as is ^1_1H, the proton being the nucleus of the common nuclide of hydrogen. The nuclei involved in the ppI chain are the hydrogen nuclides ^1_1H and ^2_1H (which is called deuterium) and the helium nuclides ^3_2He and ^4_2He. The other particles that are involved are of three types:

γ-rays (denoted by γ)   These are just energetic photons of electromagnetic radiation, a concept that should already be familiar from the discussions in Boxes 1.1 and 1.3.

Positrons (denoted by e^+)   These are particles similar in many ways to electrons (denoted by e^-); they have the same mass for instance. However, some of their properties are radically different. Of particular importance is the fact that positrons have positive charge (hence their name) whereas electrons have negative charge. Positrons are sometimes called *anti-electrons*.

Neutrinos (denoted by ν, pronounced 'new')   These are electrically neutral particles (hence the name) which have a very low mass. Neutrinos travel at essentially the speed of light and interact with other particles so weakly that they are able to travel through ordinary matter with great ease. Day and night, enormous numbers of neutrinos stream through the Earth with hardly any impediment. While you are reading this sentence, more than a million million neutrinos will pass through your own head. There are three types of neutrino, and it is the type called an 'electron neutrino' ($ν_e$) that is created in the ppI reaction.

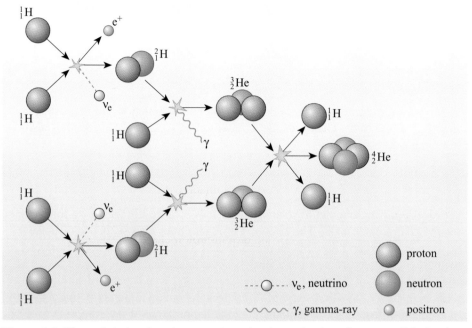

Figure 2.3 The ppI chain of nuclear reactions that is predominantly responsible for the conversion of hydrogen into helium in the Sun. Note that six hydrogen nuclei are required to initiate the chain but two are released again at the end.

The various steps in the ppI chain are shown in Figure 2.3. Pruned of its details, the overall effect of the chain is the following:

$$4\,{}^1_1\text{H} \rightarrow {}^4_2\text{He} + 2\text{e}^+ + 2\nu_e + 2\gamma$$

Thus, four protons are consumed and a helium nucleus containing two protons and two neutrons is produced along with two positrons, two neutrinos and two γ-rays. Each occurrence of the ppI chain is accompanied by a number of supplementary reactions that result, amongst other things, in the annihilation of the two positrons (along with two of the Sun's many electrons – this reaction does, of course, conserve charge), and the release of yet more γ-rays. So, apart from the production of helium nuclei and neutrinos, the final outcome of the ppI chain is the release of γ-rays. It is these γ-rays that are the ultimate source of much of the Sun's electromagnetic radiation.

At what rate do the nuclear processes occur?

We have now identified the process that is mainly responsible for the Sun's radiant energy. How common is this process? In order to get a rough idea of the answer to this question all we have to do is to divide the Sun's luminosity (the energy radiated per second) by the radiant energy liberated by each occurrence of the ppI chain. Making that estimate is our next goal. To start with, we need to know the energy released per occurrence.

■ Why does the above procedure provide only a rough idea of the rate at which the ppI chain occurs?

❑ Because we are neglecting all the other nuclear reaction chains, apart from the ppI chain, that also contribute to the generation of energy in the Sun.

HANS BETHE (1906–)

Hans Bethe (Figure 2.4) was born in Strasbourg (which was in Germany at the time, and is now in France) and educated at the Universities of Frankfurt and Munich. After the Nazi party came to power in 1933, Bethe lost his post at Tübingen on account of his mother being Jewish. He worked briefly in Britain before moving on to Cornell University in the United States in 1935.

During the 1930s it was suspected by most physicists that the key to the source of energy of the Sun must be some type of nuclear reaction. Bethe originally suggested that hydrogen might undergo fusion in a process in which carbon acts as a catalyst – a process that is now called the CNO cycle and is recognized to be important in stars of higher mass than the Sun (Chapter 6). The details of the proton-proton chain were discovered a few years later by Bethe in collaboration with Charles Critchfield. Bethe's contribution to understanding the way in which stars are powered was recognized by the award of the Nobel Prize for physics in 1967.

Figure 2.4 Hans Bethe. (Robert Barker/Cornell University)

All nuclear reactions are constrained by a number of regulating principles called *conservation laws*. These dictate which reactions are possible and determine how much energy the various particles may carry away from a reaction. Two quantities that are conserved in nuclear reactions are the electric charge (which is usually expressed as a multiple of the charge on a proton e) and the baryon number. The term **baryon** refers to a family of subatomic particles that includes the proton and the neutron, but does *not* include electrons, positrons, neutrinos or photons. The baryon number of each proton and each neutron is +1, and is zero for any particle that is not a baryon. The following example illustrates these two conservation rules in action.

EXAMPLE 2.1

Consider the initial nuclear reaction of the ppI chain as shown in Figure 2.3.

(a) Write out this reaction symbolically using the notation introduced in the discussion of the overall effect of the ppI chain.

(b) For the reaction you have written down, work out the total amount of electric charge entering and leaving the reaction, and confirm that the reaction obeys the principle of conservation of electric charge.

(c) Work out the total baryon number entering and leaving this reaction, and confirm that the principle of conservation of baryon number is obeyed.

SOLUTION

(a) The initial step of the ppI chain involves a reaction between two protons (^1_1H) and results in the formation of the hydrogen nuclide (^2_1H) as well as a positron and a neutrino. This can be written symbolically as:

$$2\,^1_1\text{H} \rightarrow\ ^2_1\text{H} + e^+ + \nu_e$$

(b) The incoming electric charge is that of the two hydrogen nuclei. Since each $_1^1H$ nucleus is simply a proton, its charge is e. Thus the total charge of the two incoming nuclei is $2e$. The outgoing hydrogen nucleus, $_1^2H$, carries charge e (as indicated by the subscript, the atomic number) and the positron carries charge e. Thus the total outgoing charge is $2e$ and hence electric charge is conserved.

(c) The baryon number entering or leaving is equal to the total number of protons and neutrons entering or leaving, and this total is given by the superscript, the mass number. The baryon number of the incoming nuclei is +2 (each has a baryon number of +1). The total baryon numbers of the products are +2 (for $_1^2H$), 0 (for the positron) and 0 (for the neutrino), giving a total baryon number of +2. Thus the baryon number is conserved in this reaction.

QUESTION 2.3

Example 2.1 considered the first nuclear reaction in the ppI chain. For the remaining two reactions:

(a) Write out each of the reactions symbolically.

(b) Show that these reactions obey the principle of conservation of electric charge.

(c) Show that these reactions obey the principle of conservation of baryon number.

Just as nuclear reactions conserve electric charge and baryon number, so they must also obey another principle, that of *conservation of energy*. The total energy emerging from a reaction must be equal to the total energy that entered the reaction. In applying this principle to nuclear reactions, it is important to include any **kinetic energy** (i.e. energy by virtue of motion) that the particles may have, but it is also important to remember Einstein's famous discovery that even particles at rest have a certain amount of energy, called the **rest energy**. The rest energy of a particle of mass m is given by

$$E = mc^2 \tag{2.1}$$

By using this formula, together with the principle of conservation of energy, it is possible to estimate the amount of radiant energy ultimately released (as γ-rays) by each occurrence of the ppI chain. In order to make the estimate we shall assume that, once all the supplementary reactions are taken into account, there is no significant change in the kinetic energy of the particles present before and after each occurrence of the ppI chain. We shall also assume that the energy associated with the two neutrinos is negligible (even though it's actually about 2% of the total). With these assumptions, we have to consider only the rest energies involved in the ppI chain – though we must remember to add the contribution arising from the annihilation of the two positrons with two of the Sun's electrons to produce yet more radiant energy. Thus,

radiant energy eventually resulting from each occurrence of the ppI chain $= [(4 \times \text{rest energy of } _1^1H) - (\text{rest energy of } _2^4He + 2e^+) + (\text{rest energy of } 2e^+ + 2e^-)]$

Having got this far, there is no reason why you shouldn't do the rest of the work for yourself, so here's your chance.

QUESTION 2.4

The mass of the helium nucleus, $_2^4\text{He}$, is 6.645×10^{-27} kg. That of the hydrogen nucleus, $_1^1\text{H}$, is 1.673×10^{-27} kg, and that of each electron or positron is 9.110×10^{-31} kg.

(a) Use this information to find the radiant energy eventually resulting from each occurrence of the ppI chain, under the assumptions given above.

(b) Use your answer to part (a), together with the value of the solar luminosity of 3.84×10^{26} J s^{-1}, to estimate the rate at which the ppI chain occurs.

(c) Just for fun, use your answer to part (b) to estimate the mass of hydrogen consumed per year by the ppI chain.

The Sun is such a massive body – 1.99×10^{30} kg – that, despite the consumption of an enormous amount of hydrogen every year, it has been able to shine fairly steadily for about 4.5×10^9 years and will probably continue to do so for another 4 or 5×10^9 years.

The above discussion should have given you a clear general picture of the main processes leading to the release of energetic γ-rays in the hot, dense conditions of the Sun's core, and hence of the origin of solar radiation. However, you should also be aware that there are some major questions that still need to be answered. A fairly obvious one that arises whenever conditions in an inaccessible region are described is 'how do you know it's really like that?' This will be addressed shortly in the section entitled 'Testing theories of the solar interior'. But an even more pressing question is this: 'How does the energy liberated in the core of the Sun account for the visible light that emerges from the Sun's surface?' This issue, which relates directly to our concern about how the Sun shines, is the subject of the next section.

2.2.5 How energy reaches the surface

The core of the Sun is a hot dense plasma – more than ten times denser than metals in our everyday world. Photons are not likely to travel very far through this material before they encounter an electron or an ion. Therefore the γ-rays produced by the ppI chain and other solar nuclear reactions are quite unable to travel directly to the Sun's surface and onward into space. Instead, there must be some mechanism, or set of mechanisms, whereby energy is *transported* from the core of the Sun to its surface. There are three fundamentally different methods of transferring energy from place to place that might, in principle, be involved – *radiation, conduction* and *convection*. Before explaining which are important in the Sun, we shall examine the basic principles of all three as a piece of background science in Box 2.1.

BOX 2.1 METHODS OF ENERGY TRANSFER: RADIATION, CONDUCTION AND CONVECTION

Radiation

In the case of **radiation**, the energy is carried from place to place by waves, rays or streams of particles that are emitted and absorbed. We usually think of *radiative energy transport* in terms of electromagnetic radiation, as in the case of the infrared radiation from an electric fire, or the light from the Sun. However, radiative energy transport is not really quite so limited. In some circumstances, for instance, streams of neutrinos can transfer substantial amounts of energy. A characteristic of radiation is that it may operate across a vacuum, though it does not necessarily have to do so.

Conduction

In contrast to radiation, **conduction** can transfer energy only between places that are physically linked by a material medium. In conductive energy transport the basic mechanism is that moving particles collide and redistribute their energy, and a common characteristic is the existence of a temperature gradient between the source of the energy and its destination. A common example of conduction is the heating of a metal spoon used to stir hot soup. The soup heats the end of the spoon with which it is in contact, causing the atoms in the metal to vibrate rapidly. The energy of these vibrations is gradually passed (mainly by electrons) to more slowly moving neighbouring atoms in cooler parts of the spoon. In this way the cooler parts are heated and some of the energy initially contained in the soup is eventually transferred to the handle of the spoon.

Convection

Convection, like conduction, requires a medium. Moreover, in the case of convection, the medium must be a fluid (a gas, a liquid or a plasma) and it must be in a gravitational field. *Convective energy transport* can also be exemplified by a common culinary experience, that of heating a saucepan of water. When a saucepan is placed on a gas ring or an electric hob the water at the bottom of the pan is heated and expands. This reduces the density of the water at the bottom of the pan, so, under the influence of the Earth's gravitational field, it starts to rise, displacing the cooler denser water above. When the heated water reaches the surface it radiates away some of its energy and cools down while the water at the bottom of the pan is warmed by more energy coming (by conduction) from below. Thus, the cycle is able to repeat itself and a pattern of **convection currents** is established within the saucepan – as indicated in Figure 2.5. The process is easy to see if there are objects such as peas in the water.

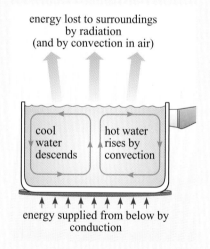

Figure 2.5 Radiation, convection and conduction in and around a heated saucepan of water.

Given the high density of the deep solar interior, you might expect that conduction would be important there, but this turns out not to be the case. The main mechanism of energy transport throughout the central 70% or so of the Sun's radius is actually radiation involving photons. Of course, the radiative energy transfer that takes place is not the sort with which we are familiar. Because of the conditions there, the photons have to make their way through the solar material via a lengthy sequence of encounters with other particles, in which they are either scattered, or absorbed and re-emitted. The photons emerging from each of these encounters have an almost equal chance of travelling in any direction, so their progress up to the photosphere takes the form of a **random walk**, as indicated schematically in Figure 2.6. A typical step in the random walk is only a few centimetres long (even less in the core), so the outward spread of energy from the core is a gradual 'diffusive' process.

■ In terms of the electromagnetic spectrum, what is the difference between photons that are produced in the core of the Sun and the majority of photons that are produced at the photosphere?

❏ The photons that are generated in the core of the Sun are typically in the γ-ray part of the electromagnetic spectrum, whereas the majority of the photons emitted from the surface correspond to the visible part of the spectrum.

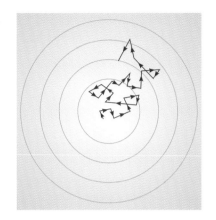

Figure 2.6 Radiant energy emerging from the solar core initially makes its way outwards via a series of radiative processes along a random walk. The graduated tone represents increasing distance from the Sun's centre. For clarity, this schematic diagram shows the path of one photon after every interaction. In reality, more than one photon may be produced after each interaction.

Since the rate of energy generation in the core is the same as the rate at which energy is lost from the photosphere, the degradation of the average photon energy between the core and the surface implies that the number of photons must increase.

There is a two stage process by which the average energy of photons is degraded from the very high energies that result from the ppI chain to the relatively low energies associated with visible light. The first stage is that the γ-rays that are generated by nuclear reactions undergo multiple scatterings with the electrons and ions in the core. In general, each scattering will redistribute energy between a photon and the electron or ion that it scatters off. This has the effect of diminishing the initially very high photon energy, and transferring energy to the particles of the plasma.

While it would be impractical to predict what would happen to an individual photon after every single scattering event, it is possible to determine what the distribution of photon energies would be for many photons that all undergo multiple scattering events. In fact, you have already encountered this distribution of photon energies.

■ Photons which are generated in the core are much more likely to interact with electrons or ions within the core than to escape. What type of spectrum is likely to result from such conditions?

❏ A black-body spectrum. In Box 1.2 it was stated that a source in which photons are much more likely to interact with the material within the source than to escape is a condition for the formation of a black-body spectrum.

So the process of multiple scattering changes the distribution of photon energies from the γ-rays that are produced by the ppI chain to a black-body spectrum that is characteristic of the core temperature. This process is called **thermalization**. However there is another key aspect to this process: the electrons and ions within the plasma interact with each other to *produce* photons. Again, because of the high degree of interaction between matter and radiation, these photons also have a spectrum that corresponds to a black-body source at the core temperature. The overall effect is that the energy that was generated as a small number of high-energy photons is now in the form of a large number of lower energy photons whose energies are distributed according to a black-body curve. This, of course, is a black-body spectrum that corresponds to the core temperature of the Sun, and has a peak at much shorter wavelengths than the approximately black-body spectrum that is emitted from the photosphere.

QUESTION 2.5

The ppI reaction produces positrons, which undergo annihilation reactions with electrons. Each annihilation event yields two γ-rays, each of which has an energy of 0.51 MeV (1 MeV = 10^6 eV).

(a) Calculate the wavelength of the γ-ray photons produced by positron–electron annihilation.

(b) Calculate the peak wavelength of the Planck curve that corresponds to the core temperature of the Sun. In what part of the electromagnetic spectrum does this peak lie?

It should also be noted that not all of the energy released by the pp chains is in the form of γ-rays. A substantial fraction of the energy that is liberated in each reaction is in the form of the kinetic energy of the nuclei or particles that are formed. Any nuclei that are formed soon scatter off protons and electrons in the dense plasma in the core. The kinetic energy of one particle is soon redistributed to many particles moving at random in the plasma, and the energy is converted into the thermal energy of the plasma. This is also a process of thermalization, but the term now applies to nuclei or particles rather than photons. The one product of the pp chains that does not share energy with the plasma in the core is, of course, the neutrino. Neutrinos simply stream out of the Sun without interaction while carrying with them a small fraction of the energy that is liberated by the pp chains.

The second stage of the degrading of the average energy of photons arises from the gradual nature of the outward transport of energy from the core to the surface of the Sun. As we have seen, basic physical principles, the principles that provide the foundation for a solar model, lead to the conclusion that the interior temperature of the Sun decreases as the distance from the centre increases. This decrease in temperature is reflected in the distribution of speeds found amongst electrons or ions of a given type at various distances from the Sun's centre. On the whole, the lower the temperature in any given region, the lower the average speed of particles (of a given type) in that region. As photons diffuse outwards they are scattered and absorbed by interactions with electrons and ions, with the result that the energy of the average photon is gradually reduced. No energy is lost in this process; it is simply redistributed and shared amongst increasing numbers of photons. The sharing and redistribution of energy is highly effective and the local electrons and ions may be regarded as the immediate source of the radiation. This state of affairs is described by saying that the radiation is in **local thermodynamic equilibrium** with the material through which it passes. This is in fact a more formal way of stating the condition about the interaction between radiation and matter that is necessary for the formation of a black-body spectrum. Thus in a region that has a certain temperature, the electromagnetic radiation has a black-body spectrum that is characteristic of that temperature. This applies right up to the photosphere.

Detailed calculations show that, under the conditions of temperature and density found in the solar interior, convection is the main mechanism responsible for energy transport throughout the outer 30% or so of the Sun's radius. In the solar context, it is very strongly suspected that the convection currents are divided into a number of **convection cells**. The uppermost cells are quite small, typically 1000 km across, but quite deep. They account for the ever-changing solar granulation seen in the photosphere. (This was described in Section 1.2.3.) A deeper layer of larger cells, typically 30 000 km across, is thought to account for a similar but less obvious phenomenon called **supergranulation**. Supergranules are not seen as light and dark patches on the photosphere, but their presence can be deduced from detailed observations of the movement of photospheric material or from various magnetic field measurements. It has also been suggested that there might be an underlying layer of giant convection cells, but there is no strong evidence in favour of this proposal.

Figure 2.7 shows a cross-section of a sector of the Sun and is designed to emphasize energy transport. As you can see, beyond the core (where the nuclear reactions take place), it is conventional to call the region in which energy is still mainly transported by radiation the **radiative zone** and the outer region in which convection dominates the **convective zone**. The top of the convective zone roughly corresponds to the photosphere, where radiation once again becomes the dominant mechanism for energy transport. Figure 2.8 shows the location and relative sizes of these regions in a cut-away view of the Sun.

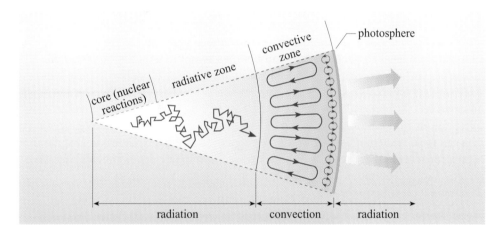

Figure 2.7 Energy transport in the Sun.

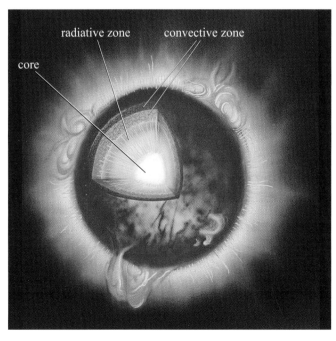

Figure 2.8 A cut-way view of the solar interior showing the location and relative sizes of the core, the radiative zone and the convective zone.

Light takes about 8.3 minutes to travel from the photosphere to the Earth. If the Sun were entirely transparent it would only take radiation about 2.3 seconds to make the journey to the photosphere from the core. However, the Sun is not transparent and the journey actually takes a lot longer. It has been estimated that energy being released in the core right now will take about two hundred thousand years to reach the surface. Fortunately, as you will learn in the next section, there is evidence that energy is still being released in the Sun's core.

2.2.6 Testing theories of the solar interior

Are our ideas about the Sun's interior right? Of course, theories are constantly under review; theorists are engaged in an unending search for inconsistencies or unseen implications within existing ideas, while at the same time looking for possible applications of new ideas. There is also a constant interplay of theory and experiment as new laboratory measurements of quantities such as nuclear reaction rates permit improved calculational precision. But, ultimately, theories must be tested against observation. It is just such observational tests that are the subject of this section.

In the case of the solar interior, current ideas can be subjected to a number of observational tests, though some are less conclusive than others. For example, some success has been achieved by those who attempt to explain the observed properties of Sun-like stars on the basis of modified solar models, but observations of other stars inevitably lack the detail and precision that are easily obtained in solar observations. Important information about the solar interior has been obtained from observations of the Sun's overall shape and surface composition, but even these data fail to provide much insight into the properties of the deep interior. Despite this catalogue of woes, recent years have witnessed two major developments that have already had an enormous influence on the study of the Sun. These two developments will now be described.

Observations of solar oscillations

It has been known since the early 1960s that the Sun oscillates – its surface moves up and down. The maximum speed of the surface during these oscillations is about $500 \, \text{m s}^{-1}$. The observed motions are not, at first sight, particularly orderly – localized regions of the photosphere rise and fall, somewhat irregularly, through distances of many kilometres in characteristic time periods of five minutes or so. Despite the lack of apparent coherence it was established in the mid-1970s that the observed movements partly result from the combined effect of many simple **global oscillations** that individually are very orderly indeed. A few of these global oscillations are illustrated in Figure 2.9. Each involves a coherent movement of the entire solar surface, and each has its own characteristic time period. The contribution of an individual oscillation to the overall motion of the surface is tiny – typically less than $0.2 \, \text{m s}^{-1}$. Some of the low-frequency global oscillations actually penetrate deep into the solar interior, as indicated in Figure 2.10. In view of this, it is not surprising that conditions in the Sun's interior influence the relative significance of the various global oscillations and thus the detailed surface movements they jointly produce.

Thanks to the existence of these deep-rooted global oscillations it is possible to learn about the Sun's interior by observing its surface. Because this is similar to the way that terrestrial seismologists learn about the Earth's interior by studying the

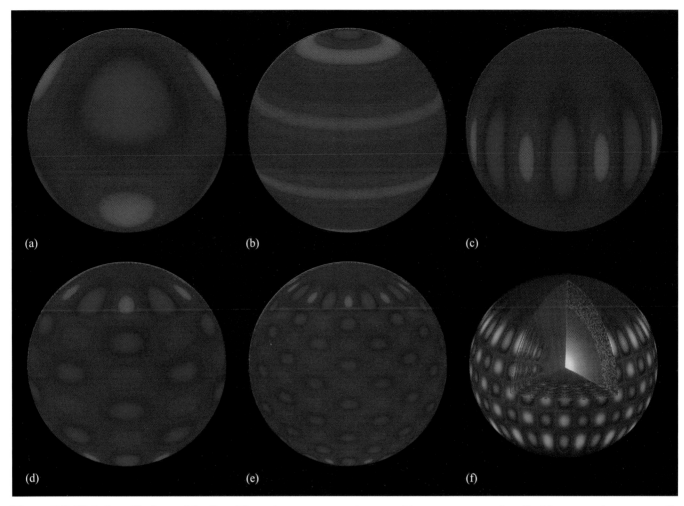

Figure 2.9 Global oscillations of the Sun. The red areas represent zones of temporary expansion, the blue areas show zones of temporary contraction. Figures (a) to (e) show the motion of the surface of the Sun for different modes of oscillation. (f) shows the surface and a schematic cut-away illustrating how one particular mode propagates within the Sun. ((f) National Solar Observatory)

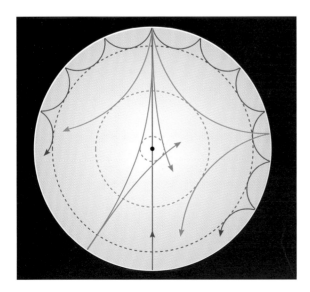

Figure 2.10 Different global oscillations will travel different distances into the Sun. This diagram shows how four modes of oscillation propagate within the body of the Sun. Some modes, such as that shown in red here, can only reach shallow depths within the Sun before being refracted back towards the surface. Other modes are capable of reaching deep within the Sun, and some, such as the modes shown here in blue and brown reach the core. (Adapted from Lang, 2001, from an original figure by J. Christensen-Dalsgaard and P. H. Scherrer.)

vibrations initiated by earthquakes, the subject has come to be called **helioseismology**. It is also possible to study, in a rather limited way, global oscillations in stars apart from the Sun, and this is referred to as **asteroseismology**. In the case of the Sun and stars, there are a variety of sources that act as the initial driver of oscillations, the most prominent being the constant churning motion of the convective zone.

Significant advances in helioseismology were made during the 1990s as a result of specialized programmes in which the Sun is observed continuously for weeks or even months at a time. Such observations are necessary if the properties of solar oscillations are to be measured well enough to be used for helioseismology.

■ Why is a single ground-based telescope likely to be unsuitable for making such long continuous observations?

▢ Observations from a single ground-based telescope will be interrupted on a daily basis as the Sun sets for the night. (Of course, this condition would not apply if such measurements were made from polar regions, where the Sun might not set for weeks on end. Observing the Sun from the South Pole has been attempted but is hampered by practical difficulties.)

In order to make continuous long-term measurements of the Sun, two different approaches have been used. A simple, yet expensive, strategy is to place the telescope in space at a location that allows an uninterrupted view of the Sun. This was the approach taken for the multinational SOHO (SOlar and Heliospheric Observatory) mission that was launched in 1995. A different approach is to construct a network of telescopes at different longitudes around the Earth, such that by the time the Sun sets on one telescope, it has risen, and is being monitored from, another telescope in the network. Several such networks exist, and provide very long-term solar monitoring programmes.

Helioseismology experiments have been used to provide confirmation that solar models do provide a good description of the Sun. One such illustration of this is shown in Figure 2.11, which shows the speed of sound within the Sun as compared to that predicted by a solar model. The speed of sound through a gas depends on its temperature, hence differences in sound speed can be used to trace differences in temperature. In Figure 2.11 the areas shown in red are hotter than predicted by the solar model, and blue areas are cooler than expected. The first point to note is that the deviation from the solar model that has been adopted here is very small; throughout most of the Sun the difference between predicted and measured sound speed is less than 0.7%. So the solar model appears to be a good description of reality. Furthermore, these results confirm that the core temperature of the Sun is 1.56×10^7 K. The deviations from the solar model are also of interest, as they highlight unexpected features within the Sun. The boundary between the radiative and convective zones at $0.68R_\odot$ is prominent; the temperature in the radiative zone is higher than expected as this boundary is approached, while the temperature in the convective zone is the same as the solar model predicts. This slight temperature enhancement in the radiative zone is an unexpected result, and one which has led to refinement and improvement of the solar model.

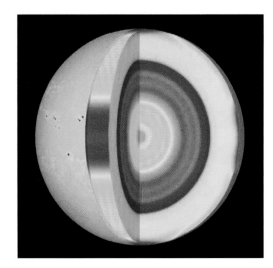

Figure 2.11 Variations of sound speed within the Sun. This figure shows the difference between a solar model and results obtained from the analysis of data from the SOHO mission. Red and blue/green areas show regions in which the sound speed is respectively higher and lower than predicted. The speed of sound is related to temperature and the red and blue/green zones correspond to temperatures that are respectively higher and lower than predicted.
(A. G. Kosovichev, Stanford University)

However, the overall conclusion from helioseismology studies has been a confirmation that solar models provide a very good description of the structure of the Sun.

Helioseismology has also been used to investigate the rotational motion of material within the Sun. We have already seen from the study of sunspots that the Sun rotates differentially at its surface and helioseismology provides a method of measuring how this pattern of rotation changes with distance into the Sun. The result of one such study using data from SOHO is shown in Figure 2.12 which illustrates that the pattern of differential rotation that is seen at the surface extends throughout the convective zone, but below this, there is an abrupt change to rotation that follows the same behaviour as that of a solid body.

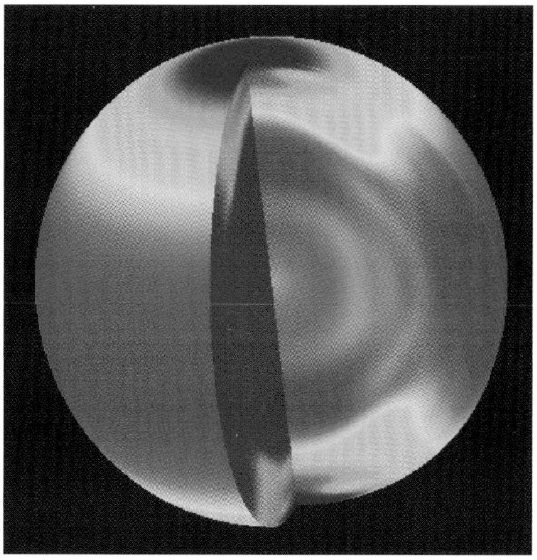

Observations of solar neutrinos

Although electromagnetic radiation has a hard time escaping from the Sun's core, the neutrinos produced there have no such difficulty. Observations of **solar neutrinos** potentially provide a very direct test of our ideas about solar nuclear reactions but, unfortunately, the low interaction rate that allows the neutrinos to escape from the Sun also makes them very hard to detect when they reach (and pass through) the Earth. Nonetheless, experiments to monitor solar neutrinos have been running since 1970. The first of these experiments was set up in the Homestake gold mine in South Dakota, USA. (The underground location has no influence on the neutrinos, but it does cut out various other particles that might otherwise lead to spurious results.) This neutrino detector (Figure 2.13) is rather unusual; it consists of a large tank containing 610 tonnes of perchloroethylene (tetrachloroethene, C_2Cl_4) – a liquid used in the dry-cleaning business. As neutrinos flood through the tank, one occasionally interacts with a chlorine nucleus to produce a $^{37}_{18}Ar$ nucleus. At the end of a typical 80 day run the tank is emptied, its contents are analysed, and the $^{37}_{18}Ar$ nuclei are counted. This is no mean feat, because usually only about 50 argon nuclei are found amongst the roughly 10^{31} nuclei in the tank.

Figure 2.12 The rotation of the solar interior as determined from helioseismological data. The figure shows the rate of rotation: regions that are the same colour have the same period of rotation. The colour coding is such that the red regions are rotating faster than average, the yellow regions are rotating slower than average, and the blue regions have a yet slower rotation rate. (Fleck *et al.*, 2000)

Figure 2.13 The tank of C_2Cl_4 used to trap solar neutrinos at the Homestake gold mine in South Dakota. (Brookhaven National Laboratory)

The results of the Homestake mine experiment have been a source of controversy for many years. The small number of observed neutrinos implies a rate of neutrino production that is only about one-third of that expected. This mismatch between observation and theory is referred to as the **solar neutrino problem**. Since the first results were published from the Homestake mine experiment, the deficit of solar neutrinos has been confirmed by other experiments that detect solar neutrinos in a variety of ways. One of the problems faced by researchers was that the early neutrino detection techniques, such as that based on C_2Cl_4, were sensitive to only the most energetic solar neutrinos, and those come from a relatively rare process involving the decay of boron (^8_5B) nuclei. It was not until the mid-1990s that experiments confirmed the Homestake mine result by measurement of the neutrinos produced from the ppI chain itself.

Up until the late 1990s opinion was divided as to what could be the most likely mechanism to produce the observed deficit. One view was that solar models were flawed in some important respect; for instance that it may have been incorrectly assumed that there is no mixing of material in the core. This view began to lose support as results from helioseismology experiments started to show that solar models were valid. The most popular alternative explanation was that the nature of the neutrino itself causes the problem. Physicists know of three kinds of neutrino, respectively referred to as 'electron neutrino' ν_e, 'muon neutrino' ν_μ and 'tauon neutrino' ν_τ. The neutrinos created in the Sun's core should all be electron neutrinos and this is the only sort of neutrino that can be detected in the tank of C_2Cl_4. As a solution to the solar neutrino problem, it was suggested that a particular kind of interaction, between the neutrinos leaving the core and the solar material through which they must pass, causes the neutrinos to change type – with the consequence that only about a third of the neutrinos emerging from the Sun are still electron neutrinos.

The resolution of the solar neutrino problem was one of the most important issues facing solar astronomy at the end of the 20th century, and in order to settle the matter, several large solar neutrino experiments were commissioned. One such experiment is the Sudbury Neutrino Observatory (SNO, Figure 2.14a) in Canada, which like the Homestake mine experiment, is located deep underground. The detector in this case is a spherical acrylic tank that holds 1000 tonnes of *heavy water* (Figure 2.14b). The tank is surrounded by 9500 photosensitive cells that detect flashes of visible light that result from neutrino interactions within the tank. There are several types of interaction that neutrinos may have with heavy water. A molecule of heavy water is identical to normal water except that the hydrogen (^1_1H) atoms are replaced by atoms of deuterium (^2_1H). A deuterium nucleus may interact with neutrinos in several ways. In one reaction, an electron neutrino can interact with the neutron in the deuterium nucleus, resulting in the production of a proton and a high-energy electron. As this electron moves at very high speed away from the site of reaction, it produces a flash of light that can be detected by the photocells. However, another reaction, in which the deuterium nucleus is essentially split into a proton and a neutron, can be triggered by any type of neutrino – muon neutrinos and tauon neutrinos, as well as electron neutrinos. This reaction causes the emission of a γ-ray as the neutron is absorbed by another nucleus within the tank. The γ-ray interacts with an electron causing it to recoil at high speed, and again a flash of visible light is produced. The flashes of light from different types of interaction differ from one another and it is possible to determine which reaction has given rise to an observed flash. The advantage of SNO over earlier experiments is that it provides a way to measure the flux of all three types of neutrino rather than just the electron neutrinos. This capability allows detailed testing of the hypothesis that neutrinos change their type. In 2001 the team running SNO announced their first results, which indicated that neutrinos do indeed change type. This result supports the view from helioseismology that solar models are a good description of the Sun, and provides an exciting new challenge to theoretical particle physicists to explain why the neutrino should behave in this way.

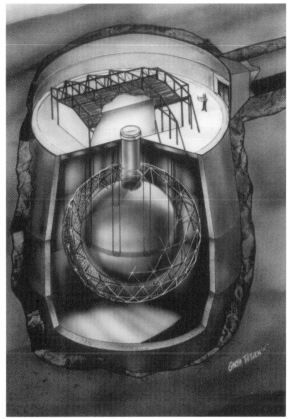

(a)

Figure 2.14 The Sudbury Neutrino Observatory (SNO). (a) An artist's impression of the deep underground chamber that houses the observatory, and (b) a photograph of the spherical acrylic tank that is filled with heavy water to act as a neutrino detector. (SNO)

(b)

Estimate the number of solar neutrinos per second passing through a $0.01\ m^2$ detector located on the Earth and pointed directly towards the Sun. You may assume that the distance from the Sun to the Earth is $1.50 \times 10^{11}\ m$, and you will find it useful to know that the surface area of a sphere of radius r is $4\pi r^2$. Take care to write down any other assumptions you make in obtaining your estimate.

2.3 Solar activity

In Chapter 1 you saw that the Sun has an 11-year cycle of activity, and that this cycle is clearly shown by the numbers of sunspots. In this section we return to investigate indicators of solar activity in more detail. We start by looking again at sunspots, and in particular, at the relationship between sunspots and magnetic fields. We shall see that other phenomena which are related to solar activity are also magnetic in origin. After considering the question of how the most energetic events linked to solar activity are powered, we finally consider the global changes to the Sun's magnetic field that occur over the activity cycle.

2.3.1 Sunspots and active regions

You saw in Section 1.2 that sunspots are a prominent feature of the photosphere, and that the number of sunspots follows an 11-year cycle. An individual sunspot usually shows the features that are clearly visible in the largest sunspot in Figure 2.15. There is a dark central region called the umbra that is surrounded by a somewhat lighter region called the penumbra. The pattern arising from granulation that is seen elsewhere in the photosphere is disrupted in the sunspot. The penumbra characteristically shows fine filaments, which tend to be directed radially away from the centre of the sunspot. (These are quite distinct from the much larger scale chromospheric filaments that were discussed in Section 1.3.1.) The umbra is notable for the complete absence of granulation and the low incidence of other features, although occasionally bright 'dots' or filamentary structures are seen within this region.

Measurements of magnetic fields reveal that all sunspots are characterized by magnetic field strengths that are much higher than elsewhere in the photosphere. While it seems clear that the properties of sunspots arise from these strong magnetic fields, the detailed physical processes are not well understood. One currently favoured idea is that the intense magnetic field suppresses convection. This reduces the rate of energy transport to the photosphere and this results in the relatively cool regions which characterize the sunspot.

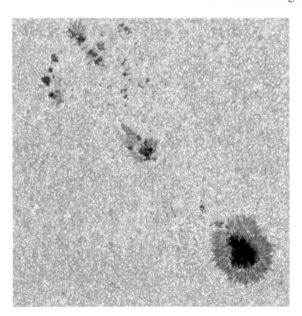

Figure 2.15 A group of sunspots observed on 12 May 1998. The extent of this image is approximately 270 000 km × 270 000 km. (Royal Swedish Academy of Sciences)

In order to describe the way in which a magnetic field varies within a sunspot, or more generally in any region, it is useful to use the concept of magnetic field lines as described in Box 2.2.

BOX 2.2 MAGNETIC FIELD LINES

We saw in Box 1.1 that the effect that a magnet has throughout a region of space can be described by using the concept of the magnetic field. The magnetic field describes both the direction and magnitude of the magnetic force at all locations in a region of space. We also saw from Box 1.1 how a magnetic field can be represented by arrows that show the direction and strength of the field. This idea can be extended by imagining the path that would be drawn in space if we were to follow one of these arrows for a short distance, and then to consider the arrow at the new position. This path is called a field line, and provides a good way of visualizing how a magnetic field varies throughout a region of space. If you look at Figure 2.16a and join the arrows that seem to be connected, this will give a good approximation to the field lines as shown in Figure 2.16b.

The interpretation of **magnetic field lines** is that their direction at any point in space is simply the direction in which a compass needle would orient itself. Furthermore the direction of the field is indicated by small arrows on the lines. By convention, a magnetic field such as that shown in Figure 2.16b is considered

to be directed *from* a north magnetic pole *to* a south magnetic pole. The strength of the magnetic field is interpreted not by the length of any arrows, but by how closely together the magnetic field lines are packed.

Note also that magnetic field lines form continuous closed loops: they do not start or end at any point in space. Although it may appear from a diagram such as Figure 2.16b that the field lines start and end at the boundary of the magnet, in fact, the field lines pass through the magnet such that every line is actually part of a continuous loop.

The magnetic field that is produced by a bar magnet is special because many magnetic fields in nature show a similar pattern. This field is called a **dipole field** because it has two poles, one north and one south, and has a pattern as shown in Figure 2.17. The field lines emerge from the north pole of the magnet and loop around to the south pole. The magnetic field of the Earth follows this pattern well in the region that is close to the surface of the planet. Rather confusingly, the magnetic pole in the Earth's *Northern* Hemisphere is actually a *south* magnetic pole.

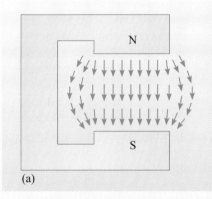

(a)

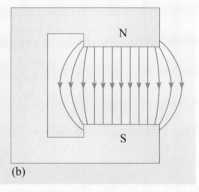

(b)

Figure 2.16 (a) The magnetic field represented by the length and direction of arrows, and (b) the magnetic field pattern of the same magnet.

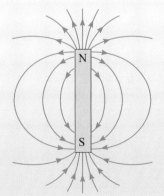

Figure 2.17 The dipole field of a bar magnet.

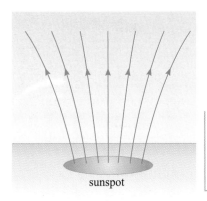

Figure 2.18 Magnetic field lines in a sunspot. (Note that in an individual sunspot, the field lines may run either into, or out of, the Sun.)

The magnetic field lines from an individual sunspot tend to run almost vertically in or out of the umbra. In the penumbra, the magnetic field curls over as shown in Figure 2.18. Since the field can be directed either upwards or downwards, an individual sunspot will resemble either a north or south pole, and these two possibilities are referred to as north (or positive) and south (or negative) **polarities** respectively.

■ A sunspot is observed with field lines in the umbra that are running into the Sun. What is the polarity of the sunspot?

❑ Since magnetic field lines run into the sunspot, it is similar to a south pole and so has south (or negative) polarity.

There are no differences in structure or behaviour between sunspots of different polarities, but the observed polarities of sunspots do provide an important clue as to the nature of the solar cycle.

It was noted earlier that sunspots often exist in groups, and in particular, there is a tendency for them to occur in pairs. When the magnetic fields of these sunspot pairs are measured it is always found that the two members of the pair have opposite polarities. A schematic illustration of the field lines associated with a pair of sunspots is shown in Figure 2.19. The field lines emanate from one sunspot and form a loop which arches above the photosphere and returns to enter the Sun at the other member of the pair. This suggests that the formation of a pair of sunspots is due to a 'bundle' of magnetic field lines breaking out from the Sun's interior. We will investigate how the magnetic field of the Sun may give rise to such a pattern when we consider the solar cycle in Section 2.3.5.

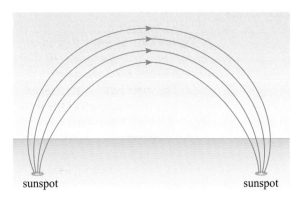

Figure 2.19 The loop of magnetic field lines that connect two members of a pair of sunspots.

Sunspots are clearly associated with very intense magnetic fields, and the sunspots themselves are an indicator of solar activity. However, when the magnetic field over the photosphere is measured, it is found that extended areas around sunspots show the same polarity patterns as the sunspots, but with lower field strengths. The strength of the magnetic field in the photosphere can be mapped because of a phenomenon (called the *Zeeman effect*) that arises when atoms or ions are subject to a strong magnetic field. The electronic energy levels in atoms and ions are altered by magnetic fields, and this has the effect that individual spectral lines split into two or three lines of slightly different wavelengths. The size of this effect is small; for example, in the strong field of a sunspot the separation between split spectral lines may only be about 10^{-3} nm. Such splitting can be measured using a specialized instrument called a **magnetograph** that can map the magnetic field strength over the visible surface of the Sun. A map produced in this way is called a **magnetogram**.

A magnetogram that was obtained at a time of high solar activity is shown in Figure 2.20. The strong fields associated with sunspots are clearly visible, as are the surrounding regions that show the same polarity as the sunspots, but have moderate field strength. In addition there are extended regions which show moderate field strengths but are not associated with any sunspots. All of these regions of enhanced magnetic field are locations in which the various phenomena that are associated with solar activity are likely to occur, and hence are called **active regions**.

■ List the phenomena associated with solar activity that you have already encountered.

❑ Sunspots (Section 1.2.2); plages (Section 1.3.1); filaments and prominences (Section 1.3.1); solar flares (Section 1.4.3).

All of these phenomena can be understood to some degree in terms of the behaviour of the underlying magnetic field. It has already been noted that the relatively low temperature of sunspots probably arises from a suppression of photospheric convection by the enhanced magnetic field. Plages are bright regions of the chromosphere, which occur in approximately the same location as bright photospheric regions called **faculae**. Both plages and faculae seem to be related to sunspots, since they appear in active regions before the emergence of a sunspot (or sunspot group) and remain visible until after any sunspots have disappeared. Hence plages and faculae are also clearly magnetic in origin, but the exact mechanisms that are at work are not well understood.

You saw in Section 1.3.1 that the terms *filament* and *prominence* refer to the same physical phenomenon – that of a strand of material that is supported above the photosphere. In fact, filaments and prominences can occur both at times of solar activity and quiescence. The type of filaments that are seen in active regions tend to be much more variable than those in quiescent parts of the Sun. Active prominences and filaments are fairly short-lived, maybe only existing for days or even just a few hours, compared to quiescent filaments that may be stable for weeks or months.

Another feature that is very commonly seen in active regions is called a **coronal loop** (Figure 2.21). These are typically seen in ultraviolet and X-ray images of the Sun and take the form of closed loops that contain plasma at temperatures that may exceed 10^6 K. As their name suggests, these features extend up into the solar corona. The shape of coronal loops suggests an association with the magnetic field within the active region.

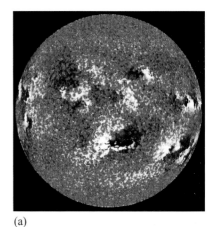

(a)

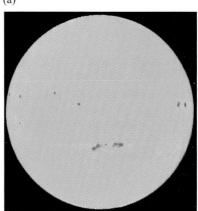

(b)

Figure 2.20 Simultaneous solar measurements of magnetic field and photospheric emission.
(a) The upper panel of this image shows a magnetogram of the solar photosphere. The colour coding of the magnetogram is: black – negative field; blue – weak field of either polarity; white – positive field.
(b) The lower panel shows the intensity of the photosphere over a wide range of visible wavelengths, and hence shows the location of sunspots. (Data provided by National Solar Observatory)

Figure 2.21 A coronal loop. This image shows the emission at 17.1 nm – this is in a part of the electromagnetic spectrum that lies at the short wavelength end of the ultraviolet band. This image was obtained from a telescope on the TRACE (Transition Region And Coronal Explorer) satellite. (TRACE (NASA))

■ Compare the image of a coronal loop (Figure 2.21) with the field lines associated with a pair of sunspots (Figure 2.19). What does this suggest about material within the prominence?

❑ The shape of the active prominence seems to follow loops of the magnetic field lines. This suggests that material in the prominence can only move along magnetic field lines.

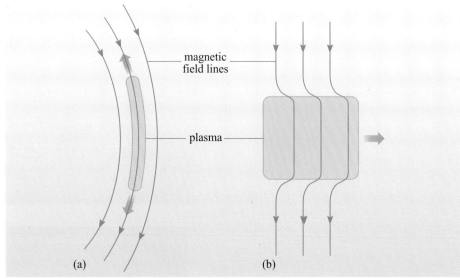

magnetic
field lines

plasma

(a)

(b)

This is indeed what happens. The material within a coronal loop is a plasma, and an important property of plasmas is that they interact very strongly with magnetic fields. In the presence of a sufficiently strong magnetic field, plasma is constrained to move along the magnetic field lines. Hence an ionized gas in a strong magnetic field may move along field lines, but cannot escape from the closed loops that field lines form. This then explains the appearance of the coronal loop: it is a loop of field lines that has trapped the plasma.

Figure 2.22 The interaction between a plasma and a magnetic field: (a) in cases where the magnetic field is strong the plasma is constrained to move along magnetic field lines; (b) when the magnetic field is relatively weak, magnetic field lines are carried with the flow of plasma.

In passing, it is also worth noting that the interaction between magnetic fields and plasmas may take on a somewhat different character when the field is relatively weak, but the flow of plasma is relatively strong. In this case the plasma will carry the magnetic field lines with it. The magnetic field in such a situation is often referred to as being 'frozen-in' to the plasma, a description that embodies the idea that the field lines have to follow the motion of the plasma. The two extremes of behaviour – a plasma that is confined to a strong magnetic field, and a magnetic field that is carried with a strong flow of plasma – are illustrated schematically in Figure 2.22.

The features of solar activity that we have seen so far, sunspots, plages, active prominences and loops, all depend on the existence of strong magnetic fields in active regions. In the following two sections we will look at phenomena that are thought to arise when those magnetic fields undergo very rapid changes.

2.3.2 Solar flares

Solar flares were mentioned briefly in Section 1.4 as being rapid and energetic outbursts of electromagnetic radiation. The duration of a flare is typically between about 100 s and a few 1000 s. During that time the flare may release up to about 10^{25} J in radiation, and material in the region of the flare may be heated to temperatures of over 10^7 K. The regions in which flares occur are small: typically less than 0.1% of the solar disc. These regions often lie between sunspot pairs or within sunspot groups, although flares may also occur in active regions in which sunspots are not present. Given their association with sunspots and active regions, it is not surprising to find that the incidence of solar flares follows the solar activity cycle. At the time of solar activity maximum, the rate of occurrence of flares is about ten times that at solar minimum.

During a flare, electromagnetic radiation is emitted over a wide range of wavelengths, but the dominant emission is in the X-ray and extreme ultraviolet (a region between the ultraviolet and the X-ray parts of the electromagnetic spectrum). This is illustrated in Figure 2.23 which shows an image taken with an X-ray telescope; the flaring region is far brighter than other emitting regions. Flares are also conspicuous in the γ-ray and the radio parts of the spectrum. At visible wavelengths the emission is usually swamped by the brightness of the photosphere, and only the most energetic bursts can be detected as increases to the broadband optical flux density. However, solar flares cause a dramatic increase in emission from spectral lines such as Hα, and so can be detected in images that are tuned to the appropriate line wavelength. The emission at different wavelengths varies over the duration of the flare. The first sign of a flare is a rapid, but short-lived outburst that is seen in very energetic X-rays (termed 'hard' X-rays) and in microwave emission. As the outburst dies away in these wavebands, emission in lower energy, or 'soft', X-rays, extreme ultraviolet light and in the Hα line increases and reaches a maximum after a few minutes and then slowly decays away.

The association between solar flares and sunspot groups suggests that flares are essentially a magnetic phenomenon. This is confirmed by observations such as the extreme-ultraviolet image of a solar flare shown in Figure 2.24. The emission comes from hot plasma (at a temperature of about 10^6 K) that is contained within the loops of the magnetic field. Detailed observations of the onset of flares show that the initial burst of hard X-rays often occurs at both foot-points of a magnetic loop. The simplest interpretation of this is that some process high in the loop gives rise to an energetic burst of particles (electrons or protons) which follow the field lines down towards the solar surface. It is the interaction of these particles with material in the chromosphere that gives rise to hard X-ray emission. This interaction also heats the chromospheric material to temperatures exceeding 10^6 K. This hot plasma then rises back up along the loop, probably filling it completely (as seen in the example shown in Figure 2.24). It is this material that gives rise to the lower energy X-ray and extreme ultraviolet emission. This scenario also explains why the soft X-rays and extreme ultraviolet light are emitted at a later time than the initial hard X-ray burst. This sequence of events is summarized in Figure 2.25. It should be stressed however, that this pattern of events is not seen in all solar flares, and that the study of these energetic outbursts is an on-going area of research.

This is only a partial model for a solar flare: it does not address the fundamental problem of how such a rapid outburst of energy can be generated. We will return to this problem later in Section 2.3.4, after considering a different phenomenon that also involves the rapid release of a large amount of energy.

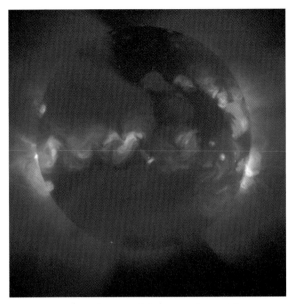

Figure 2.23 An image of the Sun taken with an X-ray telescope in which a solar flare is clearly visible on the western limb (right-hand side) of the Sun. (Yohkoh (ISAS))

Figure 2.24 An image of a solar flare taken at a wavelength of 17.1 nm (from the TRACE mission). (TRACE (NASA))

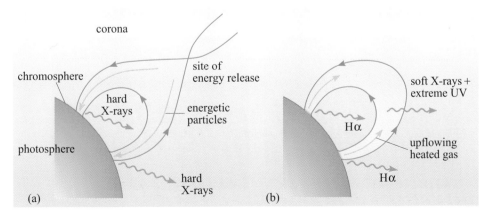

Figure 2.25 The sequence of events in a solar flare. In the first stage (a) an energetic event gives rise to streams of particles that travel down the magnetic field lines and result in hard X-ray emission. This is followed by (b) the upflow of heated material from the chromosphere giving rise to soft X-ray, ultraviolet and Hα emission from the magnetic loop. (Adapted from Lang, 2001)

2.3.3 Coronal mass ejections

We have already seen that the solar corona is a dynamic environment in which large-scale changes can occur very rapidly. While solar flares are rapid outbursts of energy that are characterized by intense bursts of electromagnetic radiation, there is a different type of event, called a coronal mass ejection (CME), that manifests itself as a violent outflow of coronal material. Some of the best observations of CMEs have been obtained using the coronagraph on board the SOHO spacecraft that is able to view the corona over a range of distances from $1.1R_\odot$ to about $30R_\odot$ from the solar centre. Figure 2.26 shows a sequence of SOHO observations that illustrate the evolution of a typical CME over an interval of about eight hours. Frame (a) shows the corona prior to the start of the coronal mass ejection. The CME itself appears in frame (b) as a 'bubble' projecting towards the top left-hand corner of the frame. Through subsequent frames this bubble expands outwards from the surface of the Sun, reaching a size of tens of $R_\odot$ in a matter of hours. After several hours the CME passes out of the field of view of the coronagraph, but continues to travel out into the interplanetary medium.

QUESTION 2.7

From the sequence of images shown in Figure 2.26, estimate (a) the speed of the front of this CME, and (b) the time that it would take for this CME to travel a distance equal to the Earth's distance from the Sun. What assumptions do you need to make in carrying out your calculations?

Figure 2.26 The evolution of a coronal mass ejection observed over an 8-hour interval on 5–6 August 1999. The dark disc blocks the Sun so that the coronagraph can observe structures in the corona in visible light. The white circle represents the size and position of the Sun. (Fleck *et al.*, 2000)

The mass of ejected material in a large CME event is typically between $5 \times 10^{12}\,\text{kg}$ and $5 \times 10^{13}\,\text{kg}$, and CME events occur with an average frequency of about one per day. The rate of occurrence depends on the level of solar activity; at solar maximum there are about three CMEs per day, while at solar minimum the rate is about one-tenth of this.

(a) 1999/08/05 18:18
(b) 1999/08/05 19:42
(c) 1999/08/05 21:18
(d) 1999/08/05 23:18
(e) 1999/08/06 00:42
(f) 1999/08/06 02:42

The loop-like shape of CMEs, such as that shown in Figure 2.26, suggest that the plasma is confined within a magnetic field. The essential feature here is that the magnetic field is moving and plasma is being forced to move with the field. The origin of these ejection events probably lies in a rapid reconfiguration of the magnetic field in the lower parts of the solar corona. This allows energy that is stored in the magnetic field to be released suddenly, causing a violent outflow of coronal material.

QUESTION 2.8

Calculate the total kinetic energy of a large coronal mass ejection. Use the estimated speed from your answer to Question 2.7 and assume that the ejection has a total mass of 5×10^{13} kg. Express your answer as an order of magnitude estimate (i.e. to the closest power of ten).

2.3.4 Magnetic reconnection

Solar flares and coronal mass ejections both require a mechanism for the sudden release of energy, and it is also clear that both phenomena are magnetic in origin. Rather than concentrating on the features that distinguish these two types of event, solar physicists take the approach that there is likely to be a single underlying mechanism that drives both solar flares and CMEs. The starting point for such theories is the idea that magnetic fields act as a store of energy. It is not too difficult to convince yourself of this; if you take two bar magnets and arrange them end on such that like poles are next to one another, you will find that you have to push the magnets together to get the poles to touch. If you release one magnet, some of the energy that you supplied by pushing the magnets together, will be returned to that magnet in the form of kinetic energy as it is repulsed. It is the interaction between the magnets that allows energy to be stored. This interaction is described by the magnetic field: as the magnets are pushed together the field pattern changes and stores energy that can be recovered when one of the magnets is released.

The energy stored by magnetic fields that are found in the solar corona can easily meet the energy requirements of solar flares and CMEs. The problem, which is not fully solved, is how the magnetic field energy can be released rapidly enough to explain the violent outbursts seen in solar flares. Despite this problem it seems likely that a process called **magnetic reconnection** is at play. The fundamental idea is that in some regions, field lines that are running in opposite directions are squeezed together as shown in Figure 2.27. The region that separates field lines that run in opposite directions has a particularly important role. Remember that all solar magnetic fields exist in a medium which is ionized – a plasma. Because they contain electrons and ions that are not bound together as neutral atoms, plasmas are able to conduct electricity. In fact, the existence of the magnetic field causes an electric current to run along the boundary that separates regions in which the field runs in opposite directions. In Figure 2.27 the electric current would, in fact, run out of the page, perpendicular to the field lines. The flow of an electrical current through a conductor usually gives rise to heating; this is the principle behind such everyday appliances as kettles and electric heaters. Similarly, the current flowing through the plasma also causes heating. Heat is generated at the boundary between the different field directions.

- What is the source of energy for this heating?

- The energy comes from the magnetic field – this causes the electrical current that results in heating.

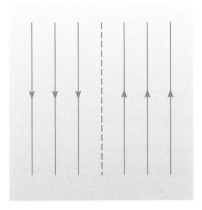

Figure 2.27 The configuration of magnetic field lines in a region where magnetic reconnection can occur. The dashed line shows the boundary between regions of opposite polarity.

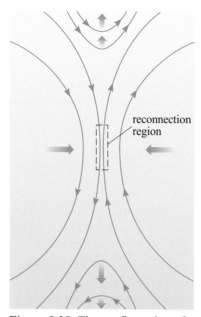

Figure 2.28 The configuration of magnetic field lines if two regions of opposite polarity are pushed together. Note that reconnection only occurs over a region of rather limited extent.

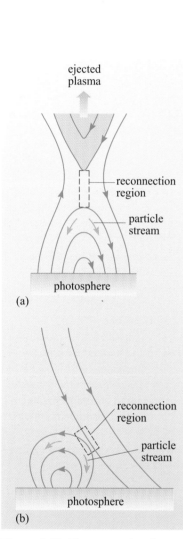

Figure 2.29 Two scenarios that have been proposed for the magnetic field configuration that may cause solar flares and coronal mass ejections. While both scenarios account for the observed features of solar flares, the ejection of a large mass of plasma is more likely in case (a) than in case (b). ((a) Sturrock, 1980; (b) Heyvaerts *et al.*, 1977)

Since energy is removed from the magnetic field, the magnetic field must drop in strength as heat is produced. In terms of Figure 2.27, field lines of opposite direction essentially cancel each other out. However, this figure does not account for the fact that all magnetic field lines are actually in the form of continuous loops: in reality there can only be certain parts of field lines where they come together as shown. If we consider a somewhat more realistic situation in which two regions come together in the way shown in Figure 2.28, then there will be heating in the region where the field lines are closest. Two opposing field lines may disappear from this region, but at the edge of the region, the lines connect up to one another to form a single field line. This field line behaves in a way that is similar to an elastic band under tension. Initially, the field line has a very sharp bend in it at the point where the reconnection took place, but immediately tries to straighten itself out, and in doing so it moves rapidly away from the reconnection region. Because plasmas are forced to move with the field lines, this rapid motion of the field lines can accelerate particles in the plasma to high velocities. Hence the process of reconnection can result in the conversion of energy stored in the magnetic field into the kinetic energy of particles.

The way in which magnetic reconnection events may give rise to solar flares or CMEs is illustrated in Figure 2.29, which shows configurations of magnetic fields that might explain these events. In both scenarios that are illustrated, a common feature is the fact that the reconnection event occurs high in a coronal loop. This results in particles being accelerated down to the foot-points of the loops to give the characteristic outbursts seen as solar flares. In the case shown in Figure 2.29a, the field reconfiguration after reconnection is such that it can drive the ejection of any plasma that lies above the reconnection point. In the case shown in Figure 2.29b, there is a somewhat different magnetic field configuration; the reconnection event causes a solar flare, but results in a smaller ejection of plasma. It should be noted that these are speculative ideas about the way in which reconnection takes place. Unfortunately, at present, it is not possible to map the magnetic fields in the corona in a way that would allow these ideas to be tested.

Even so, magnetic reconnection looks like a promising mechanism to provide the energy for solar flares and CMEs. It should be noted however, that there are many problems to be solved before a full understanding of the reconnection process is achieved. In particular, the sort of scheme that has been outlined above does not seem to release energy quickly enough to explain the observed behaviour of solar flares. This area thus remains a topic of great interest to solar physicists.

2.3.5 The solar cycle

In Chapter 1 we saw that the number of sunspots, and all phenomena associated with solar activity, undergo cyclic variability with a period of about 11 years. This recurrence of solar activity is termed the **solar cycle**. So far in this chapter we have considered how *local* variations in the solar magnetic field give rise to phenomena related to solar activity. To gain some understanding of the solar cycle, we need to consider the *global* behaviour of the solar magnetic field, i.e. the way in which the field behaves over the entire Sun.

Figure 2.30 shows a series of magnetograms taken at different times during the solar cycle. There are several important observations that can be drawn from this sequence. The first is that, as might be expected, the occurrence of local enhancements of the magnetic field depends very strongly on the solar cycle: at solar maximum there are many regions of increased magnetic field, at solar minimum these regions are scarce. These areas correspond to active regions on the Sun. A second observation relates to the polarity of sunspot pairs. You have already seen that sunspots tend to be formed in pairs of opposite polarity. These pairs tend to be aligned roughly along lines of constant latitude, but the sunspot that is further ahead in the sense of the rotation of the Sun (called the leading sunspot) is closer to the equator than the trailing sunspot. This behaviour also applies to active regions as a whole; they also show this bipolar behaviour and orientation with respect to the Sun's equator. At the start of the solar cycle, when active regions first start to appear after solar minimum, the location of active regions tends to be at high latitudes, both north and south of the equator. As the solar cycle progresses, the band in which active regions occur migrates towards the equator.

Figure 2.30 The upper panel shows a series of magnetograms taken over a 22-year period, while the lower panel shows a quantity (called the sunspot number) that measures the number and extent of sunspots. The colour coding of magnetograms is: black – negative field; blue – weak field of either polarity; white – positive field. (Data provided by National Solar Observatory)

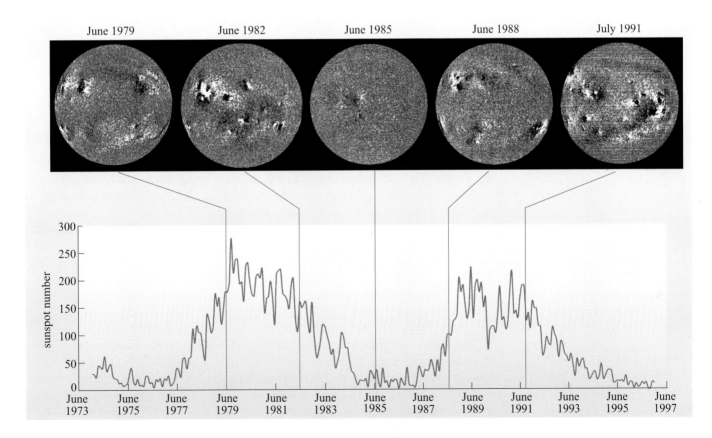

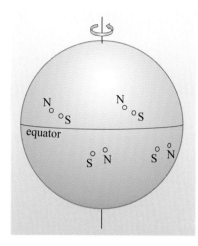

Figure 2.31 Between two solar minima, the polarity of sunspot pairs is such that one polarity always leads in one hemisphere and the opposite polarity leads in the other hemisphere.

A remarkable feature of the polarities of sunspots within a sunspot pair is that at a given time, in one hemisphere, the polarity of the leading sunspot is always the same. In the other hemisphere at the same time, it is the opposite polarity that always leads. This is illustrated schematically in Figure 2.31 and careful inspection of Figure 2.30 should allow you to conclude that this behaviour applies to active regions as well as sunspot pairs. After solar minimum, when sunspot pairs are first seen in a new cycle, the pattern of polarity switches over in the sense that what was the polarity of leading sunspots in the northern hemisphere now becomes the polarity of leading sunspots in the southern hemisphere and vice versa.

■ What is the period of the solar cycle, if it is defined as the time between two solar maxima which have the same behaviour of magnetic polarity?

❏ 22 years. The time between solar maxima is 11 years, but two maxima that are 11 years apart will show opposite polarities for the leading sunspots. Hence it will be another 11 years before the pattern of polarities is repeated.

So when the behaviour of the magnetic field is studied in detail, the solar cycle has a period of 22 years rather than 11 years.

More detailed measurement of the solar magnetic field shows that at the time of solar minimum, the Sun has a global magnetic field that is a reasonable approximation to the dipole field (see Box 2.2). Between minima the field pattern is complicated and characterized by the bipolar active regions. By the time that the dipole field reappears at the following solar minimum, the polarity of the entire field has swapped over.

Solar physicists are yet to reach a full understanding of the mechanism that drives this magnetic cycle. However a scheme which explains some of the broad features of the solar cycle was proposed in 1961 by Horace Babcock, and continues to form the basis on which more sophisticated models are based. The key elements of the scenario are the differential rotation of the Sun and the idea of magnetic reconnection (Section 2.3.4). Babcock's scheme is illustrated schematically in Figure 2.32. At the start of the cycle, at solar minimum, the external field is similar to that of a dipole. Within the Sun, the magnetic field lines run close to the surface, either in or just below the convective layer (Figure 2.32b). These field lines are frozen-in to the plasma within the Sun and so are dragged around by any motion of the plasma. The differential rotation of the surface layers of the Sun leads to a 'winding-up' of the field lines. This has the effect of changing the direction of the field lines from the original pole-to-pole direction to being roughly parallel to the equator (Figure 2.32d). The stretching of the field lines also results in a transfer of some of the energy that the surface layers have due to their motion into energy of the magnetic field. As the field gets wound up even further, loops of field lines start to emerge, and form active regions whose polarities agree with observed behaviour (Figure 2.32e). Towards the end of the cycle it is thought that the active regions that are now close to the equator will start to reconnect across the equator (Figure 2.32f and g). This removes the leading regions and forms loops between trailing regions. In Babcock's scenario, the foot-points of these loops migrate towards respective poles, such that a dipole field is regenerated, but now with an opposite polarity to the field at the start of the cycle. This is then the half-way point in the 22-year cycle, the process repeats with all polarities reversed before returning to the original configuration at the end of the cycle.

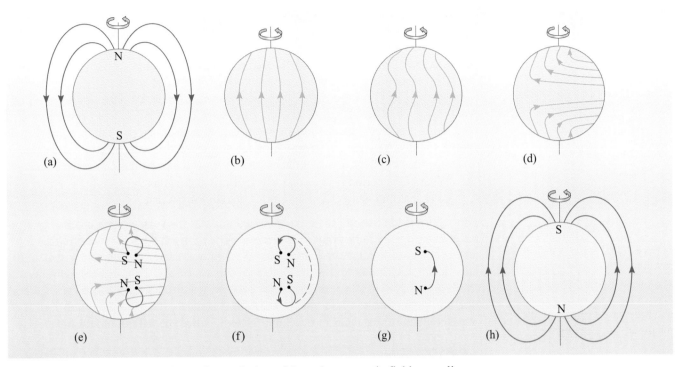

Figure 2.32 A sequence that shows the evolution of the solar magnetic field according to Babcock's scheme. Note that field lines that are interior to the Sun are shown by light green field lines and those exterior to the Sun are shown in dark green. (a) The Sun at solar minimum: the exterior field pattern is close to a dipole field. (b) Inside the Sun these field lines run from one polar region to the other but are close to the surface of the Sun. (c) The interior field lines as the process of differential rotation starts to 'wind-up' the field. (d) As differential rotation continues, the interior field lines become tightly wrapped in bands that run almost parallel to the equator. (e) Loops of field lines emerge to form active regions. (f) As the end of the active part of the cycle approaches, loops reconnect across the equator; leading regions cancel one another out leaving a loop between trailing regions. (g) The new field line now runs approximately in the direction from one pole to another. The foot-points of the new loop now migrate polewards. (h) The result of many such reconnections is a dipole field, but which now has the opposite polarity to that shown in part (a). The cycle then repeats itself but with all polarities swapped.

While this scheme has some very attractive features, in particular that it explains the observed pattern of polarities and orientation of active regions, it is far from being a complete explanation of the Sun's magnetic field. In particular, there is nothing within the model that explains why the Sun should have a magnetic field at all. It is known that magnetic fields can be formed when electrically conducting fluids flow according to certain patterns, and it seems very likely that this sort of effect is responsible for the generation of the magnetic fields. It is however a considerable challenge to theorists to model these sorts of flows within the Sun to reproduce the magnetic field as seen, and this continues to be an active area of research.

2.4 The Sun in space

In the final section of this chapter we take a step back from the Sun to examine some of the effects that it has on its surrounding environment. You have already seen that material is lost episodically from the Sun in coronal mass ejections, and here we will look in more detail at the way that the Sun loses mass continuously by the solar wind. These types of outflow interact with the planets that orbit the Sun and we will briefly look at some of the effects that solar activity can have on the Earth. Finally, we shall investigate how far into space this solar wind extends, and examine the boundary between the environment of the Sun and interstellar space.

Before discussing this in detail, it is useful to sketch out the typical distances that are involved in describing the environment of the Sun. We have already seen that the radius of the Sun ($R_\odot$) is 6.96×10^8 m, and that the solar corona extends to at least several solar radii away from the Sun. The most obvious features in the environment of the Sun are the planets. We will not discuss the planetary system in any detail here, but note that the average distances of planets from the Sun range from 5.79×10^{10} m (Mercury) to 5.90×10^{12} m (Pluto). The average distance from the Sun to the Earth is 1.50×10^{11} m, and this is used as a convenient unit for the measurement of distances in the Solar System, called the **astronomical unit** (AU).

QUESTION 2.9

What are the average distances in AU from the Sun to (a) Mercury, and (b) Pluto? What are the average distances in units of $R_\odot$ from the Sun to (c) Mercury, and (d) the Earth?

2.4.1 The solar wind

In addition to episodic coronal mass ejections, there is also a continuous outflow of solar material that is termed the solar wind. The existence of a low-density, yet high-speed solar wind was proposed in 1951 by Ludwig Biermann as a result of a study of the observed shapes of the tails of comets. This suggestion was followed by an analysis of the structure of the solar corona by Eugene Parker. In an attempt to describe mathematically the structure of the corona, Parker suggested that the atmosphere around the Sun could not simply be static, and that material in the corona must flow out into interplanetary space.

The existence of the solar wind was confirmed directly by measuring its flow using experiments on Russian and American space probes in the mid-1960s. The temperature of the gas in the solar wind is typically very high, being over 10^5 K, which leads to the gas being highly ionized and hence forming a plasma. The composition of the solar wind is dominated by electrons and protons, along with a small fraction of helium nuclei and an even smaller percentage of heavier nuclei. The density of material in the solar wind is very low; in the vicinity of the Earth the number density of protons in the solar wind is about 7×10^6 m^{-3}. The speed of the wind is high, being typically several hundred km s^{-1}. One of the surprising results that emerged from the experiments that measured the solar wind directly, was that the speed of the wind seems to vary between a rather steady fast flow of about 750 km s^{-1}, and a much more erratic slower flow that has a speed of between 300 and 400 km s^{-1}.

As has already been mentioned, the origin of the solar wind in the solar corona was something that was anticipated before direct measurements of the wind itself were available. While the exact mechanism that drives the solar wind is still an area of active research, it seems clear that an important process involved in forming the solar wind arises from the relatively high gas pressure in the corona. Gas in the corona accelerates to reach outflow speeds of a few $100 \, \text{km s}^{-1}$ within a distance of about 20 to $30 R_\odot$ from the Sun, but then only accelerates slowly at distances beyond this (see Figure 2.33). This then is a partial explanation for the origin of the solar wind; the outflow from the corona simply carries on into interplanetary space. However, to understand why the speed of the solar wind (as measured near the Earth) seems to be either fast or slow, we need to consider the effect of the magnetic field of the Sun on gas in the corona.

You saw in Section 2.3.5 that the magnetic field at the surface of the Sun varies considerably over the solar cycle. At times close to solar minimum the field over the surface of the Sun is weak but its pattern is similar to the dipole field. Close to solar maximum the field is much more complex and shows an irregular pattern that is difficult to characterize. We will concentrate here on how the solar magnetic field affects the solar wind around the minimum of the solar cycle.

> What are the two types of interaction between a magnetic field and a plasma that you have already come across?

> In cases where the magnetic field is strong and the flow of plasma is weak, any movement of plasma is along field lines. In cases where the field is weak and the flow is strong, the field lines move with the plasma.

The structure of the magnetic field of the corona at solar minimum shows both these types of behaviour as illustrated in Figure 2.34. As mentioned above, at the surface of the Sun the field follows the dipole pattern closely. Those field lines that emanate from the *polar* regions pass high into the corona and into the region in which the bulk motion of the solar wind starts. These magnetic field lines are then carried by the solar wind and distort the dipole pattern, such that these field lines become highly extended as they are carried away from the Sun. The field lines that emerge from the Sun close to its magnetic *equator* form loops that do not reach high into the corona. These magnetic field lines may be distorted but they remain as closed loops, and ionized gas is trapped within these regions.

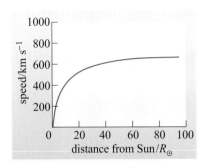

Figure 2.33 The speed of the solar wind away from the Sun increases as the distance from the Sun increases. The wind accelerates rapidly in the first few tens of $R_\odot$, and accelerates only slowly after this.

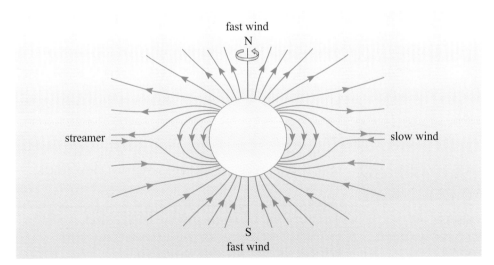

Figure 2.34 The magnetic field structure (green lines) of the solar corona and solar wind at a time close to the minimum of solar activity, showing the expected location of streamers and the location of the origin of fast and slow components to the solar wind. (Adapted from Pneuman and Kopp, 1971)

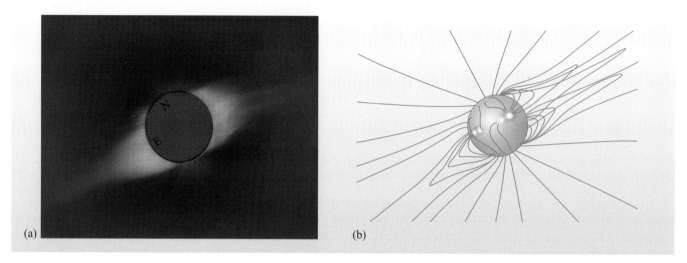

Figure 2.35 The solar corona during a total solar eclipse, at a time when the Sun is close to the minimum of solar activity: (a) in visible light; and (b) the magnetic field pattern that is deduced from the observed streamers. (The arrows showing the direction of the field lines have been omitted for clarity.) ((a) S. Koutchmy, Institut d'Astrophysique); (b) SAIC)

These structures provide an observational test for the model of how the solar magnetic field interacts with the corona and solar wind. Images of the corona taken during total solar eclipses near the minimum of solar activity often show 'helmet' streamers (so-called because their shape resembles old-fashioned pointed helmets), such as those shown in Figure 2.35a. The field pattern that is deduced from the shape of these streamers, as shown in Figure 2.35b, has similar features to the field pattern that is expected when a dipolar field interacts with an out-flowing plasma, in that some loops of field lines are distorted, but remain close to the Sun, whilst other loops become vastly extended as they are carried by the solar wind.

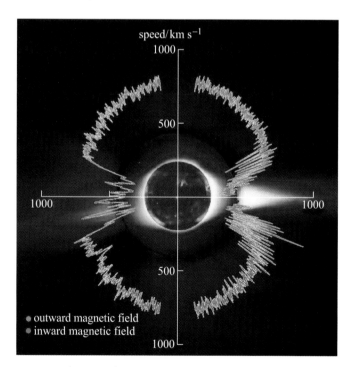

Figure 2.36 The solar wind speed as a function of heliocentric latitude as measured by the *Ulysses* spacecraft from 1992 to 1997. (Ulysses (ESA and NASA))

The fast component of the solar wind appears to originate in those parts of the corona where field lines do not loop back on themselves. These regions of open field lines correspond to the coronal holes that were mentioned in Section 1.4.3. Those regions of the Sun that are covered in closed magnetic loops seem to give rise to the more erratic slow component of the solar wind. This relationship between the two speeds of the solar wind and the magnetic field pattern was demonstrated using measurements made with the *Ulysses* space probe in the early 1990s. Most of the space probes that have explored the Solar System have stayed close to the equatorial plane of the Sun. *Ulysses* is remarkable in that its orbit was planned so that it would pass over the poles of the Sun, and therefore be used to measure the properties of the solar wind at different solar latitudes. The results of these measurements are displayed in Figure 2.36, which shows that at high latitudes, only the fast component of the solar wind is seen. However, near the solar magnetic equator, the solar wind shows the erratic behaviour that is characteristic of the slow component. Finally it should be noted that the magnetic axis and the rotation axis are not exactly aligned: there is an offset of several degrees between the two. This means that the solar wind which reaches the Earth (which orbits the Sun in a plane close to the solar equator) alternates between the fast and slow components.

Based on the density and the speed of the solar wind measured near the Earth, estimate the mass loss from the Sun due to the solar wind. Express your estimate in terms of kg s^{-1} and $M_\odot \, \text{yr}^{-1}$. Assume that the density of the solar wind is 7×10^6 protons m^{-3} (ignore the contribution to the mass from other particles) and that the speed of the wind is that which corresponds to the 'fast' component. (*Hint*: calculate the mass of material that flows out over a sphere with radius equal to the radius of the Earth's orbit.)

2.4.2 Solar – terrestrial interactions

We have seen that the Sun is a source of out-flowing material, either in the form of the solar wind or coronal mass ejections, and it is of particular interest to us to understand the consequences of the interaction between this material and the Earth. The importance of the role of magnetic fields in determining the properties of the solar wind was illustrated in Section 2.4.1, and so it should not be surprising to find that the magnetic field of the Earth plays a key role in determining how the solar wind interacts with the Earth. Consider the situation that is likely to arise when a steady solar wind impinges on the magnetic field of the Earth. The fact that the solar wind is an ionized gas means that it cannot easily cross magnetic field lines. Therefore, the Earth's magnetic field presents an obstacle to the flow of the solar wind and, as in the solar corona, this results in either material being channelled along field lines, or the field lines being swept along with the wind. Upstream from the Earth there is a region where the pressure of the solar wind is similar to the resistance to the wind provided by the Earth's magnetic field. At this point the wind 'feels' the influence of the magnetic field and forms a feature called the **bow shock** (so-called because of its similarity to the bow wave that builds up in front of a moving ship). The flow of the wind is diverted around the Earth, and in doing so, the magnetic field pattern is swept back into a long tail as shown in Figure 2.37. The tear-drop shaped region around the Earth in which the Earth's magnetic field dictates how charged particles will move is called the **magnetosphere** (the name is somewhat misleading as it is not a spherical region). For most of the time, the

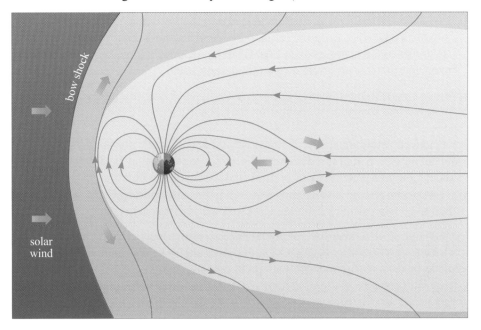

Figure 2.37 The magnetosphere of the Earth. The impact of the solar wind on the Earth's magnetic field distorts the field into the shape shown. The solar wind cannot easily cross magnetic field lines and so is 'swept' around the outer boundary of the magnetosphere (blue). However, some particles from the solar wind can enter through the 'tail' region of the magnetosphere – off to the right of the region shown in this diagram.

magnetosphere is a barrier that excludes most of the solar wind, although a small number of particles (electrons, protons and ionized atoms) continuously enter through the tail region. A further source of particles in the magnetosphere is the upper atmosphere of the Earth.

The extent of the magnetosphere depends on the balance between the force imparted by the solar wind and the resistance offered by the Earth's fixed magnetic field.

■ If the force imparted by the solar wind on the magnetosphere were to increase, what would happen to the position of the bow shock?

❑ The bow shock would move inwards towards the Earth, to a point where the force imparted by the wind is balanced by the stronger magnetic field closer to the Earth.

So the magnetosphere is a dynamic entity: its structure depends on the strength of the solar wind. Changes in the shape of the magnetosphere have an effect on the motion of charged particles that are trapped within it, and this in turn can lead to observable effects. One such effect is that a change in the shape of the magnetosphere can give rise to a variation in the local magnetic field at the surface of the Earth. Such variations are termed **geomagnetic disturbances** or, in extreme cases **geomagnetic storms**. Intense variations arise when large coronal mass ejections strike the magnetosphere. At times close to the maximum of the solar activity cycle, about ten such coronal mass ejections strike the Earth's magnetosphere every year and result in strong geomagnetic storms. The motion of charged particles in the magnetosphere can also give rise to very rapidly changing magnetic fields at the surface of the Earth, and this can induce currents in electrical power lines, leading to overloading and damage.

An important factor that determines the severity of a geomagnetic storm is the orientation of the magnetic field in the coronal mass ejection with respect to that of the Earth. We have already seen that CMEs are outflows of plasma, hence they transport magnetic field lines with them as they move away from the Sun.

■ Consider a coronal mass ejection that strikes the Earth's magnetosphere (i.e. moving to the right from the left-hand edge of Figure 2.37). If the magnetic field lines in the leading edge of the CME have opposite polarity to the Earth's magnetic field, what effect do you think may occur?

❑ Since magnetic field lines of opposite polarity are being forced together, it is possible that magnetic reconnection may occur (Section 2.3.4).

This is indeed what happens in the most severe geomagnetic storms – magnetic reconnection results in a rapid reconfiguration of the magnetic field, which causes the acceleration of electrons and protons that are located within the magnetosphere. The reconnection event also creates a breach in the magnetic barrier around the Earth, and allows particles from the CME to enter the magnetosphere.

Geomagnetic storms have several potentially adverse effects on human activity. Apart from the possibility of disruption of power supplies, the rapidly changing magnetic fields can also interfere with radio communication and navigation systems. Furthermore, the presence of energetic particles within the magnetosphere poses a health risk to astronauts and can also damage sensitive electronic equipment on satellites. An increase in the number of energetic particles in the magnetosphere may arise from geomagnetic storms, but also from solar flares. Given the practical importance of knowing, or being able to predict the effect of solar activity on the Earth's environment, the state of solar–terrestrial interactions is continually monitored and reported under the generic name of **space weather**.

The most dramatic effect that can arise when the magnetosphere is disturbed is the production of an **aurora** (plural aurorae) which are also called the Northern or Southern Lights (aurora borealis and aurora australis, respectively). Aurorae are typically observed at locations on the Earth that have latitudes of 60 to 70 degrees (North or South), although they are occasionally seen from much lower latitudes. The form of an aurora usually resembles a glowing curtain of light (Figure 2.38) which seems to move and change in a matter of minutes. Aurorae themselves are observed to last for several hours, and in extreme cases for days.

Figure 2.38 An aurora. (Lionel Stevenson/Science Photo Library)

Aurorae are formed when the varying magnetic field of the magnetosphere causes electrons and protons from the tail region to move along the magnetic field lines down into the atmosphere of the Earth. These charged particles collide with atoms in the upper atmosphere, at heights of between about 100 km and 400 km above the surface of the Earth.

■ What may happen to an atmospheric atom as a result of these collisions?

❏ The atom may be excited to a higher electronic state, or if the collision is energetic enough the atom may become ionized.

Photons are emitted as these excited atoms or ions revert to lower electronic states, and it is these photons that form the visible auroral glow. Thus the spectrum of an aurora shows prominent emission lines from those elements that are common in the Earth's atmosphere, such as oxygen (green and red emission) and nitrogen (purple and red emission).

The particles that give rise to aurorae are channelled down magnetic field lines, and tend to impinge on the Earth's atmosphere in oval shapes that surround each magnetic pole. The auroral ovals (Figure 2.39) have been imaged in their entirety from satellites in orbits around the Earth, and show the huge extent of the aurora – a display seen from the ground is just a tiny part of the full oval.

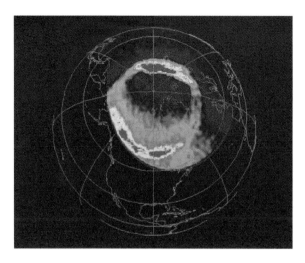

Figure 2.39 The northern auroral oval, as imaged by the POLAR satellite in ultraviolet emission. (G. E. Parks, University of Washington and the POLAR UVI team/NASA)

QUESTION 2.11

Discuss whether you might expect the Earth's aurorae to have prominent spectral lines due to the following: (a) atomic oxygen, (b) atomic iron, (c) ionized nitrogen.

2.4.3 The heliosphere

The solar wind streams outwards from the Sun into interplanetary, and eventually, interstellar space. As the solar wind gets further and further away from the Sun, its density drops while its speed remains roughly constant. The overall effect is that the pressure that the solar wind has due to its motion will drop with increasing distance from the Sun. So far, we have not discussed whether there is any gas in the space between the stars, but we can speculate that if there were, this would presumably have a pressure *of its own* due to the thermal motion of the particles in the gas. We will see later in this book that the space between the stars does indeed contain material: a mixture of gas and dust that is called the interstellar medium. Because the pressure of the solar wind continues to drop with distance from the Sun, at some point this pressure will balance the pressure of interstellar gas. The volume of space in which the pressure of the solar wind exceeds the pressure of gas in interstellar space defines the extent of the Sun's influence on its environment, a region that is called the **heliosphere**.

The boundary of the heliosphere is expected to have components as illustrated schematically in Figure 2.40. In moving outwards from the Sun, the solar wind flows freely until it reaches a region called the **termination shock** where it responds to the pressure of the interstellar medium by slowing down considerably. Further out from this shock is the boundary of the region in which the solar wind dominates the motion of matter. This boundary is termed the **heliopause**; beyond this boundary lies the interstellar medium. If the Sun is moving relative to the local interstellar gas, then it is also likely that a bow shock will form ahead of the heliopause in the direction of motion, and that there would be a build up of interstellar gas in this region.

Observational data that could confirm this view and allow space scientists to determine the extent of the heliosphere are currently (2002) rather limited. It is believed that the Sun is moving relative to the local interstellar medium with a speed of $26\,\mathrm{km\,s^{-1}}$, and hence it is anticipated that there is a bow shock and a density enhancement in the direction of the Sun's motion. The analysis of radio signals that are believed to originate when intense bursts of solar wind collide with interstellar gas, suggests that the heliopause may be between 110 and 160 AU from the Sun. The location of the termination shock is believed to be at about 85 AU from the Sun. It is hoped that the *Voyager 1* space probe, which had reached a distance of 83 AU from the Sun at the start of 2002, will pass through the termination shock within the next few years, and that within a few decades it will become the first artificial object to enter interstellar space.

In November 2003 contradictory findings about whether *Voyager 1* had passed the termination shock were published by two separate research collaborations. On the basis of one type of measurement, one group claimed that *Voyager 1* had passed the termination shock in mid-2002. The other group, who analysed different data, concluded that the spacecraft had yet to reach this feature of the heliosphere.

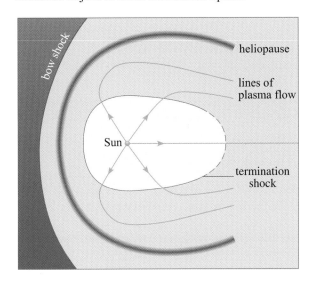

Figure 2.40 A schematic illustration of the expected structure of the heliosphere. (Adapted from Gosling, 1999)

QUESTION 2.12

Estimate the time taken for the fast component of the solar wind to travel from the Sun to the termination shock. Express your answer in days.

2.5 Summary of Chapter 2

The solar interior

- The temperature and density of the Sun increase as the distance from the centre (fractional radius) decreases. The core temperature is thought to be about $15.6 \times 10^6\,\mathrm{K}$.

- Theoretical models of the solar interior provide detailed structural information and favour a composition that is (by mass) roughly 73% hydrogen, 25% helium and 2% heavy elements, except in the central core, where the high temperatures promote chains of nuclear reactions that convert hydrogen nuclei into helium nuclei.

- The Sun's electromagnetic radiation originates in the central core, where it is released by the nuclear reactions that convert hydrogen nuclei into helium nuclei. In addition to high-energy γ-rays, these reactions also produce neutrinos.

- Energy released in the core makes its way to the surface by radiation and convection. Radiative energy transport in the solar interior is a gradual diffusive process in which the radiation is in local thermodynamic equilibrium with the material through which it passes.

- The solar neutrino problem is a deficit of the observed number of electron neutrinos in comparison to that expected from the solar model. The most likely explanation for this is that neutrinos change type as they travel from the core of the Sun to the Earth. This phenomenon is not well understood.

- The broad features of theories of the solar interior are not currently in dispute. However, observational tests involving solar oscillations and other sources of data may require some refinement in the details of those theories.

Solar activity

- The phenomena that indicate solar activity are primarily magnetic in origin. At times of high solar activity the Sun exhibits many active regions in which there are strong bipolar fields.

- Solar magnetic fields store energy that originates in the motion of the plasma in the surface layers in the Sun. This energy can be released rapidly giving rise to solar flares and coronal mass ejections. It is likely that magnetic reconnection plays a role in the process of energy release required for these type of events.

- The solar activity cycle is a cycle in which the global magnetic field of the Sun changes dramatically. The period taken for the magnetic field to return to its initial state is 22 years or twice the period inferred from sunspot number variations.

The solar wind and the heliosphere

- The corona of the Sun is not a static atmosphere, but undergoes a continuous outflow which forms the solar wind. The out-flowing material is a plasma, the composition of which is similar to the composition of the Sun, and the speed of flow, as measured near the Earth, is between about 300 and 750 km s^{-1}.

- Out-flowing coronal plasma interacts with the magnetic field of the Sun. Depending on the strength of the magnetic field, this interaction may result in the flow being restricted to move along field lines or in the transport of field lines with the plasma.

- The magnetic field of the Earth acts as a barrier to exclude most of the solar wind from a region around the Earth. This region, called the magnetosphere, may shrink or expand as a result of changes in the strength of the solar wind, and this alters the pattern of flow of charged particles within the magnetosphere. This may give rise to geomagnetic storms and aurorae.

- The pressure exerted by the solar wind drops with distance from the Sun, and eventually balances the pressure of interstellar gas. This boundary marks the limit of the heliosphere, and is thought to lie between 110 and 160 AU from the Sun.

Questions

QUESTION 2.13

Estimate the numbers of hydrogen and helium nuclei contained in the Sun.

QUESTION 2.14

Estimate the total amount of time for which the solar luminosity can be sustained at its current value by the conversion of hydrogen into helium.

QUESTION 2.15

Why are the following nuclear reactions impossible?

(a) $^1_1H + ^4_2He \rightarrow ^5_3He$

(b) $^7_4Be + ^1_1H \rightarrow 2\,^4_2He + 2e^+ + \gamma + 2\nu_e$

(c) $^3_2He + ^4_2He \rightarrow ^7_3Li + \gamma$

QUESTION 2.16

If the Sun's luminosity is entirely due to the loss of rest energy from the solar core, how much mass does the Sun lose each year?

QUESTION 2.17

Calculate the number of photospheric photons that are required to carry away the energy generated in a single ppI reaction.

QUESTION 2.18

The structure of filaments is an area of on-going investigation. One model that has been proposed to explain how a filament is supported is based on the idea that a cross-section of the filament (perpendicular to the direction in which it is elongated) may have a field pattern as shown in Figure 2.41. Explain how this field pattern supports the filament.

QUESTION 2.19

Briefly describe the energy conversions that may occur in magnetic reconnection.

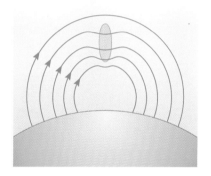

Figure 2.41 A proposed model for the way in which a filament may be supported by a magnetic field. The field pattern is shown for a plane that is perpendicular to the direction of elongation of the filament. For use with Question 2.18. (Phillips, 1992)

CHAPTER 3
MEASURING STARS

3.1 Introduction

The Sun, we have asserted, is a star. But all of the other stars are so far away that they appear as mere points of light in the sky, seemingly unchanging. And yet there have always been clues that the starry heavens are not as changeless as they appear to be. On rare occasions a new star will flare up in the sky, perhaps becoming visible in daylight, before fading in a few months back into invisibility. Such bright new stars are rare, and before the advent of the telescope in 1609, only ten or so had been recorded. Of particular importance was the one observed by the Danish astronomer Tycho Brahe ('Tie-co Bra-hay') (1546–1601) in 1572. Using the method of trigonometric parallax, which you will meet shortly, he was able to show that the new star lay beyond the Moon, thus overthrowing the then prevailing view that transient phenomena were confined to the volume of space closer to us than our satellite. So here was the first observational evidence that all regions of the cosmos are subject to change.

We now know that Tycho's new star was a *supernova*, which, as you will see in Chapter 8, is a massive star ending its life in a gigantic explosion. This has been established through painstaking studies of the radiation from those points of light in the sky. Massive stars are rare, and evolve quickly, particularly at the supernova stage. Earlier stages of their evolution, and the evolution of less massive stars, are usually too slow to be observed directly. We thus have to piece the evolution story together from observations of many stars of each particular type, at as many stages in each type's evolution as we can observe.

The observable properties are, however, almost entirely of *external* appearances and *external* events, and can take us only so far in revealing the evolution of the stars. To go further we must understand their interiors, and in Chapters 6 and 7 we shall be particularly concerned with how this understanding has been developed, and where it leads us in our understanding of stellar evolution. In this chapter, however, we shall be concerned with the essential preliminary step, namely measuring the observable properties of the stars.

The key properties that we need to characterize stars sometimes require units which are unfamiliar to us in everyday life or other branches of science. For example, the distances between stars are so vast that familiar length units such as kilometres are totally inadequate. Astronomical measurement systems and nomenclature have evolved over time as scientists' understanding and the technology have become more refined. In some cases this has resulted in an apparently arbitrary nomenclature, which is often confusing at first. However, such systems have been retained even when the basis of the system has been modified as a result of a greater understanding of the physical processes involved.

But before we look at the stars as bodies, what about the stars in space: how much further away than the Sun are the stars? Are the constellations really fixed, and how are stars distributed across the sky?

3.2 Stars in space

3.2.1 Are the stars fixed in space?

For thousands of years, and perhaps ever since the human species emerged on this planet, we have identified patterns in the stars – the constellations – and given them names. Today, we still use constellation names that originated in antiquity. Figure 3.1 shows the relative positions of the brighter stars in one of the better known constellations, that representing the mythological hunter Orion. The boundaries of the 88 constellations used today have been defined by the International Astronomical Union, the governing body of astronomical names. The brightest stars have proper names, of Greek, Roman or Arabic origin, but generally stars are referred to by letters or numbers. The brightest stars in a constellation are referred to by Greek letters (usually, but not always) in order of brightness followed by a constellation designation, often shortened to three letters. For example, Betelgeuse ('betel-jers') is α (alpha) Orionis, or α Ori, and Rigel ('rye-jel') is β (beta) Ori. The Greek alphabet and constellation abbreviations are listed in Appendix A2. When a photograph of the constellation is taken, many more stars, invisible to the unaided eye, are seen. Many of these are referred to by their designations in star catalogues but most fainter ones have no names at all.

How fixed are these patterns of stars? Let's take perhaps the best known of all, the Plough (known in North America as the Big Dipper), as an example. Figure 3.2 shows the Plough as it appears now and as it appeared to our ancestors 100 000 years ago, during the Old Stone Age. It certainly looks different. However, over a human lifespan the change is negligible. Figure 3.2c shows the Plough about 70 lifespans, or 5000 years into the past, at the beginning of written history.

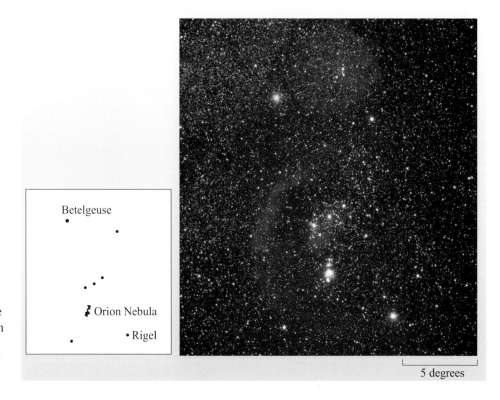

Figure 3.1 (left) The patterns of the brighter stars in Orion. (right) A photograph of the same region taken in visible light showing the range of colours of stars (see Sections 3.3.2 and 4.2). The Orion Nebula, visible in this image as a reddish glow, is a dense cloud of gas and dust within which stars are forming, and is visible to the unaided eye. (Photo: D. Malin/ AAO)

Betelgeuse

Orion Nebula

• Rigel

5 degrees

■ Compare this with the Plough today, in Figure 3.2a. Do you think that, to the unaided eye, it would have been noticeably different 5000 years ago?

❑ No, though it would have been easier to find 5000 years ago, when light pollution was a lot less!

Not surprisingly, it was a long time before such stellar motions were discovered, in 1718 by the British astronomer Edmund Halley (1656–1742). The motions continue, and Figure 3.2d shows the Plough as it will appear 100 000 years in the future.

The motion of a star across the sky is called its **proper motion** – so-called because it is intrinsic to the star and not a result of the motion of the observer or a moving reference point. It is usually expressed in seconds of arc per year, arcsec yr^{-1} (3600 arcsec = 1 degree).

■ What further information would need to be added to Figure 3.2 to enable you to calculate the proper motions?

❑ A fixed reference point is needed, with zero proper motion. (In fact, *all* the stars shown happen to have non-zero proper motions, and all the values are around 0.1 arcsec yr^{-1}, but in different directions across the sky.)

Currently, the proper motion record is held by Barnard's Star, a faint star in the constellation Ophiuchus with a proper motion of 10.4 arcsec yr^{-1}. By contrast, Rigel has a minuscule proper motion of less than 0.002 arcsec yr^{-1}. The proper motion of a star arises from its motion relative to us, in particular the component of its motion in a direction transverse to our line of sight to the star, as shown in Figure 3.3. This is called the **transverse velocity**. The magnitude of the transverse velocity, the transverse speed v_t, is given by

$$v_t = d \times \sin \mu$$

where μ is the proper motion and d is the distance to the star. Since μ is a very small angle, we can approximate $\sin \mu$ by μ measured in radians (per unit time), so

$$v_t = d \times (\mu/\text{radians}) \tag{3.1}$$

Thus, a small proper motion at a large distance corresponds to a larger transverse speed than the same proper motion at a small distance. You will learn about the measurement of stellar distances later, but v_t varies enormously, from under 1 km s^{-1}

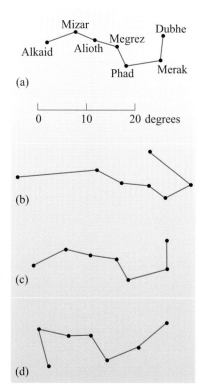

Figure 3.2 (a) The Plough; part of a larger constellation called the Great Bear (Ursa Major) as it appears today, and (b) 100 000 years ago, (c) 5000 years ago, and (d) 100 000 years in the future.

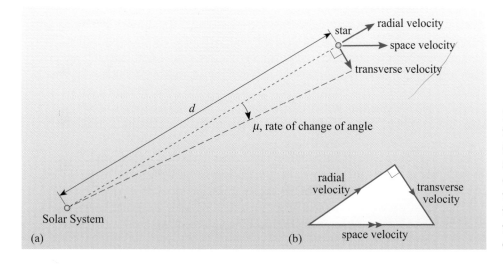

Figure 3.3 (a) A star's motion through space, relative to the Sun. (b) The overall velocity through space, the space velocity, has two components: the radial velocity in the observer's line of sight and the transverse velocity in the plane of the sky. (Space velocity will be described below.)

to well over $100\,\mathrm{km\,s^{-1}}$. (For Barnard's star and Rigel, the values are $89\,\mathrm{km\,s^{-1}}$ and $2.2\,\mathrm{km\,s^{-1}}$, respectively.)

Earlier, we gave μ in $\mathrm{arcsec\,yr^{-1}}$, and now, via Equation 3.1, we have v_{t} in $\mathrm{km\,s^{-1}}$. Let's follow this through. With μ in $\mathrm{arcsec\,yr^{-1}}$ and d in km, Equation 3.1 gives v_{t} in

$$\mathrm{km}\left(\frac{\mathrm{arcsec\,yr^{-1}}}{\mathrm{radians}}\right) \quad \text{or} \quad \left(\frac{\mathrm{km}}{\mathrm{yr}}\right)\left(\frac{\mathrm{arcsec}}{\mathrm{radians}}\right)$$

This is the same as $\quad (\mathrm{km\,s^{-1}})\left(\dfrac{\mathrm{s}}{\mathrm{yr}}\right)\left(\dfrac{\mathrm{arcsec}}{\mathrm{radians}}\right)$

Note that (s/yr) is the number of years in a second $(1/(3.16 \times 10^7))$ and (arcsec/radians) is the number of radians in a second of arc $(1/206\,265)$. Therefore (s/yr) and (arcsec/radians) are both pure numbers, and so v_{t} is in units of $\mathrm{km\,s^{-1}}$.

The component of the star's motion relative to us is called its **radial velocity**. In Figure 3.3 it is directed away from the Earth, but it could equally well have been directed towards us. Unlike the transverse velocity, the radial velocity can be obtained without knowing the distance to the star. The method relies on the Doppler effect (named after Christian Doppler, Figure 3.4), which has many applications in astronomy. The Doppler effect is described in Box 3.1.

If a star behaves in any way like a car horn, then we can use the Doppler effect to obtain the radial velocity. A star does indeed have the equivalent of a horn – in the radiation that it emits. In Chapter 1 you learned that the solar spectrum exhibits many spectral lines. This is also the case for any stellar spectrum. A spectral line is like a car horn in that it corresponds to an emission or an absorption at a specific frequency. Moreover, we can identify the atomic transition giving rise to a spectral line, and thus we know the emitted frequency ('emitted' here covers absorption lines as well as emission lines). Therefore, if the frequency we observe differs from the emitted frequency, then from the size of the difference we can use Equation 3.3 (with the speed of light, c, in place of c_{s}) to obtain the radial speed, and from the

CHRISTIAN ANDREAS DOPPLER (1803–1853)

Figure 3.4 Christian Doppler. (Science Photo Library)

Christian Doppler (Figure 3.4) was born into a family of successful stonemasons in Salzburg, but was unable to take over the business due to poor health. His scientific abilities led him to the study of higher mathematics, mechanics and astronomy at the University of Vienna where he subsequently gained a post as assistant to the professor of higher mathematics. The difficulties of obtaining a permanent academic post in Austria at the time (which involved a centralized competition, with written and oral examinations and centred on teaching rather than research abilities) led him to consider emigration to America before he finally obtained a post in Prague. His research activities were hampered by onerous teaching and examination duties (for example, in July 1847 he orally examined over 800 students) and complaints from students that his examining was too harsh. His work on the Doppler effect was clearly the high point of his career. Despite regarding light as longitudinal rather than transverse waves and erroneously using his theory to explain the colours of double stars, he predicted the future importance of the effect in determining the motions of stars. Although it was not possible to observe the effect on light at the time, experiments involving musicians playing instruments on approaching and receding trains confirmed it.

BOX 3.1   THE DOPPLER EFFECT

The **Doppler effect**, named after the Austrian physicist Christian Andreas Doppler is the name given to the observed change in frequency of the waves emitted by a source when it is moving with respect to the observer. It is familiar in the change in pitch of the sound received from a car horn as the car sweeps past. As the car approaches, the pitch is higher – the frequency is higher; as the car recedes, the pitch is lower – the frequency is lower.

Figure 3.5 shows how the Doppler effect arises. In Figure 3.5a the car is stationary with respect to the observers at A, B and C. Its horn emits sound at a frequency f, and the sound waves spread out at the speed of sound, c_s, in the air. The circles are separated by one wavelength, λ, of the sound, given by

$$\lambda = c_s/f \qquad (3.2)$$

The observers at A, B and C all hear the same frequency, f, as that emitted by the horn. There is no Doppler effect.

In Figure 3.5b the car is moving with respect to the observers, and again it emits sound at a frequency f. Once the sound is emitted it still travels away from the car through the air at the speed c_s, which is unchanged by the car's motion. Thus the motion of the car causes the waves to pile up ahead of it – since each successive wave is centred on a new position of the car – giving rise to a decrease in wavelength at A, and hence (Equation 3.2) to an increase in frequency. Behind the car the waves are spread apart, giving rise to an increase in wavelength at B, and hence to a decrease in frequency. When the motion is perpendicular to the line from the car to the observer then there is no change in frequency: the observer at C hears the emitted frequency. The change in frequency thus requires that the velocity of the car has a component along the direction from the car to the observer, that is it requires a radial velocity; a transverse velocity produces no Doppler effect. It can be shown that the radial velocity is given by

$$v_r = c_s \times (f - f')/f' \qquad (3.3)$$

where f' is the observed frequency. Thus, the Doppler effect provides us with a way to measure radial velocities. (A negative value of v_r results from a motion towards the observer, as at A, i.e. in a direction opposite to the radial direction of the source from that observer).

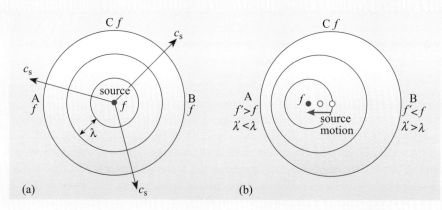

(a)   (b)

Figure 3.5 The Doppler effect with a car horn. (a) The car is stationary. (b) The car is in motion: an observer at A hears the horn at a higher pitch than the car driver; an observer at B hears it at a lower pitch than the driver; and an observer at C hears it at the same pitch as the driver.

sign of the difference we can tell whether the star is moving towards or away from us; we can thus obtain the radial velocity. (For light, or any other form of electromagnetic radiation, Equation 3.3 is an approximation requiring $v \ll c$, a condition met by stellar radial speeds.)

In the case of stars, it is more usual to work in wavelengths rather than frequencies. Using $\lambda = c/f$, we can express Equation 3.3 as

$$v_r = c \times (\lambda' - \lambda)/\lambda \qquad (3.4)$$

Where λ' is the observed wavelength.

> ■ If the observed wavelength is longer than the emitted wavelength, in which direction is the star moving?
>
> ❏ It is moving away from us.

Increases in wavelength are called **red-shifts** from the days when observations at visible wavelengths dominated astronomy; at visible wavelengths an increase in wavelength takes us towards the red end of the spectrum. Likewise, decreases in wavelength are called **blue-shifts**. These shifts are collectively called **Doppler shifts**.

Radial velocities are roughly of the same order as transverse velocities. The two velocities together specify the overall motion of the star through space with respect to us. This **space velocity** is given (using Pythagoras's theorem, see Figure 3.3b) by

$$v = \sqrt{(v_{\mathrm{t}}^2 + v_{\mathrm{r}}^2)} \tag{3.5}$$

These overall motions are not entirely random, but are partly related to the large-scale motions in our Galaxy and by the grouping of many stars in clusters, which you will meet later in Section 3.2.4. Indeed, whether a star belongs to a cluster can often be decided by comparing its motion through space with that of the cluster members.

3.2.2 How far away are the stars?

Throughout history, people have been fascinated by the question of how far away are the stars. What gulf of space separates the stars from the Sun? Are all the stars at the same distance? The stars are so remote that the measurements that provided definitive answers to these questions began to emerge only in the first half of the 19th century. In addition to helping us to establish cosmic architecture, we also need to know stellar distances in order to investigate the stars as individual bodies. For example, until their distances were known, no progress could be made on determining either the size of the stars, or how much energy they radiated. Thus, one star could appear to us to be much brighter than another simply because it is far closer.

So, how do we measure the distances to the stars? There is a range of methods, each appropriate for different distances, which build up the astronomical distance scale, from the closest stars to the most distant galaxies. You will come across a number of these later. Here we shall describe the technique which was first to yield a stellar distance, is still the most accurate for the nearer stars, and provides us with an important unit of distance widely used in astronomy.

Figure 3.6 illustrates the principle of the method. When the Earth is at point A in its orbit, the relatively nearby star appears as shown against the background of far more distant stars. Six months later, from point C, the position of the star against the distant background appears to have shifted. You can easily demonstrate this kind of shift to yourself by holding up a finger at arm's length in front of you, and viewing it against a distant background alternately with one eye and then the other.

Returning to the stars, from the apparent shift in position of the nearby star, the angle 2θ in Figure 3.6 is measured by an Earth-based observer, and the distance d to the star is then given by

$$d = b/(\theta/\mathrm{radian})$$

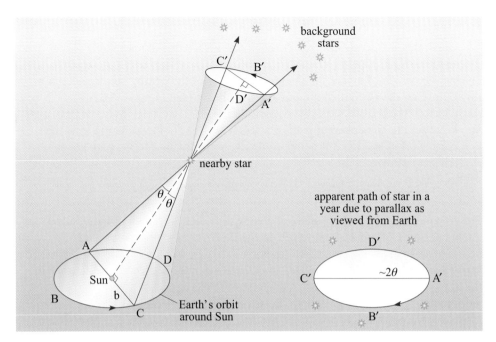

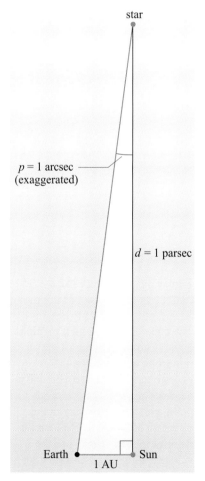

Figure 3.6 Trigonometric parallax. A nearby star's position changes relative to more distant stars as the position of the observer (on the Earth orbiting the Sun) changes. The star tracks out a path on the sky called the parallactic ellipse.

where b is the distance from the Earth to the Sun, and where we have again used the small-angle approximation (as in Equation 3.1), θ always being very small. Thus, if we know b then we can obtain d. The distance b can be obtained in a variety of ways, nowadays by measuring the times it takes radar pulses, which travel at the well-known speed of light, to return to the transmitter after being reflected off various bodies in the Solar System. The details will not concern us: the important point is that b is known, and therefore the distance d to the nearby star can be obtained. This method of obtaining stellar distances is called **trigonometric parallax**, this being the change in direction to an object as a result of a change in the position of the observer. In everyday parlance it is just called parallax.

The distance b, from the observer to the Sun is always very close to 1 astronomical unit (i.e. 1.50×10^{11} m), which is the *average* distance from the Earth to the Sun. If $b = 1$ AU we define the angle θ as equal to p, the **stellar parallax** (often abbreviated to **parallax**) and so

$$(d/\text{AU}) = 1/(p/\text{radian}) \tag{3.6}$$

In the most precise measurements a small correction must be made to account for the fact that the Earth's orbit is not precisely circular.

The important unit of distance referred to earlier is the **parsec** (pc), defined as the distance d corresponding to a stellar parallax of 1 arcsec, as shown in Figure 3.7. Thus, by definition,

$$d/\text{pc} = \frac{1}{p/\text{arcsec}} \tag{3.7}$$

With 206 265 arcsec in a radian, it is not too difficult to show, from Equations 3.6 and 3.7, that 1 pc = 206 265 AU (and hence 1 pc = 3.09×10^{13} km).

Figure 3.7 The definition of the parsec.

■ Is 'parsec' a reasonable name for this new unit?

❏ Given that it is made up from *par*allax and arc*sec*, it *is* a reasonable name.

During the course of a year the apparent position of the nearby star tracks out a tiny ellipse on the sky called the **parallactic ellipse** (see Figure 3.6).

■ What will be the apparent path of a star if it is: (a) in the same plane as the Earth's orbit, and (b) if it is in a direction perpendicular to the plane of the Earth's orbit?

❏ (a) A line of angular length 2θ, (b) a circle of angular radius θ.

Stellar parallaxes are all very small and were only discovered after many failed attempts. Some of these attempts led to the discovery of other important phenomena such as proper motion (Section 3.2.1) by Edmund Halley, and binary stars (Section 3.2.3) by William Herschel. James Bradley observed oscillations in the apparent position of the star γ Draconis over the period of a year with an amplitude of 20 arcsec. He had not, however, observed parallax since the maximum displacement occurred three months too late. He had discovered the **aberration** of starlight. The apparent direction of arrival of light from the star was the result of the combination of the speed of the Earth in its orbit and the finite speed of light from the star (analogous to running in the rain with an umbrella, the telescope had to point slightly in the direction of motion of the Earth to 'catch' the starlight – see Figure 3.8). All these effects have to be accounted for when attempting to measure parallax.

The first published measurement of parallax was by the German astronomer Friedrich Wilhelm Bessel in 1838. He found the parallax of the star 61 Cygni ('sig-nee') to be 0.314 ± 0.020 arcsec. (It is now known to be 0.287 arcsec.) This is a very small angle, 6000 times smaller than the angular diameter of the Moon. It is therefore not surprising that science had to wait so long for this first parallax measurement. It gave us the first measurement of a distance to a star (other than the Sun).

Figure 3.8 (a) A pedestrian running in the rain needs to point an umbrella forwards to keep dry. The apparent direction of the rain is the result of the sum of the raindrops' velocity and the pedestrian's velocity. (b) Similarly, an observer on the Earth needs to point a telescope slightly in the direction of the Earth's velocity in its orbit to see a star. Since the velocity of light, c, is much larger than the Earth's orbital velocity the shift in position is very small. This effect is called aberration.

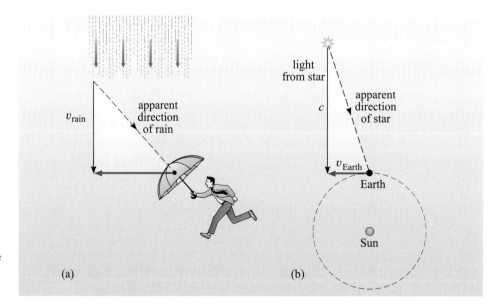

JAMES BRADLEY (1693–1762)

James Bradley (Figure 3.9) studied theology at Oxford and became a clergyman before resigning to become Savilian Professor of Astronomy at Oxford and later he succeeded Edmund Halley as Astronomer Royal. Although he did not measure the motion of stars due to parallax, his discovery of the aberration of starlight confirmed the Copernican concept of a moving Earth. He was reluctant to publish his work until he had fully confirmed his results – his careful observations, which resulted in the discovery of the wobble in the Earth's rotation axis due to gravitational interactions with the Moon, lasted almost 20 years!

Figure 3.9 James Bradley. (Science Photo Library)

QUESTION 3.1

Calculate the distance to 61 Cygni, using the current value for its parallax, expressing your answer in parsecs, AU, and metres. How much further away from us is 61 Cygni than the Sun?

The largest parallax is for the star Proxima Centauri, which is therefore the nearest star, though with a parallax of only 0.772 arcsec it is 1.3 pc away, which is nearly 270 000 times further from us than the Sun. Clearly, a large gulf of space separates us from the stars. It's not that we are particularly isolated: the distances between the stars, at least in our part of the Milky Way, are generally of the same order as the distance to Proxima Centauri.

FRIEDRICH WILHELM BESSEL (1784–1846)

Friedrich Bessel (Figure 3.10) began work in Germany as an apprentice in an exporting company but dreamed of escape though travel. His study of languages, geography and navigation led to mathematics and eventually astronomy. Heinrich Olbers (1758–1840, famous at that time for his work on the discovery of asteroids) obtained a post for him at Lilienthal Observatory after receiving a paper written in 1804 on the motion of Halley's comet. This was followed by a commission from the Prussian government to build the first large German observatory at Königsberg (now Kaliningrad, Russia). After its completion in 1813 he remained its director until his death and made major advances in the accuracy of positional astronomy. In addition to the discovery of parallax, resulting from the determination of precise positions of 50 000 stars, he developed Bessel functions (a mathematical tool which he used for studying planetary motions) and suggested the presence of Neptune from perturbations in Uranus' orbit. Olbers claimed his own greatest contribution to astronomy was to lead Bessel to become a professional astronomer.

Figure 3.10 Friedrich Bessel. (Science Photo Library)

Table 3.1 lists the ten nearest stars after the Sun. Parallaxes smaller than 0.01 arcsec are very difficult to measure from Earth-based observatories, so we can obtain distances with useful accuracy (within a range of a few tens of parsecs) for only a few hundred stars. We can do far better from space, above the troublesome effects of the Earth's atmosphere.

The Hipparcos satellite was named to commemorate the Greek Astronomer Hipparchus (c.170–120 BC) who produced the first accurate star catalogue.

The Hipparcos satellite (*Hi*gh *P*recision *Par*allax *Co*llecting *S*atellite) launched by the European Space Agency (ESA) in 1989, measured the parallaxes of 118 000 stars to an average precision of better than 0.001 arcsec (1 *milli*-arc second). In addition, proper motions were measured, with a precision of 0.001 arcsec yr^{-1}, permitting derivation of accurate three-dimensional positions and motions of stars out to beyond 100 pc. Hipparcos also measured the positions, motions, brightnesses and colours of a further 2.5 million stars with lower precision.

Appendix A3 contains information on the 100 closest stars to the Sun.

Table 3.1 The ten nearest stars after the Sun.

Name	Parallax /arcsec	Distance /pc	Distance /ly	Proper motion /arcsec yr^{-1}	Comment[a]
Proxima Centauri	0.772	1.30	4.22	3.9	
α Centauri A	0.747	1.34	4.36	3.7	triple system
α Centauri B	0.747	1.34	4.36	3.7	
Barnard's Star	0.547	1.83	5.95	10.4	
Wolf 359	0.419	2.39	7.77	4.7	
Lalande 21185	0.393	2.54	8.28	4.8	
Sirius A	0.380	2.63	8.57	1.3	binary system
Sirius B	0.380	2.63	8.57	1.3	
L-726–8A	0.373	2.68	8.73	3.4	binary system
L-726–8B	0.373	2.68	8.73	3.4	

[a] In a binary system, two stars are in orbit around each other. In a triple system there are three such stars (see Section 3.2.3).

The Gaia satellite, planned for launch by ESA in about 2012, will revolutionize our view of our Galaxy, measuring the positions and motions of the billion brightest objects in the sky. Its design goal is to determine parallaxes to an accuracy of 1 *micro*-arc second (0.000 001 arcsec) (giving distances of stars out to tens of thousands of parsecs), and proper motions to an accuracy of 1 micro-arcsec yr^{-1}. While most of the objects detected by Gaia will be stars, allowing us to map the three-dimensional structure and motion of the whole of our Galaxy, it is also expected to find, from their motions, over 150 000 new asteroids in our Solar System and 50 000 stars in our Galaxy possessing planets.

Finally, note that the parsec is not used solely to express the distances to the stars: it is commonly used to express *any* distance greater than the size of the Solar System. Another unit commonly used for such distances is the **light-year** (ly), which is the distance that electromagnetic radiation would travel in a vacuum in a year, and is equal to 0.307 pc. Neither the light-year nor the parsec is an SI unit but they result in more reasonable numbers for stellar distances; the parsec is the preferred unit for professional astronomers.

3.2.3 Binary and multiple star systems

William Herschel observed stars which were apparently very close to each other in the sky in an attempt to identify tiny relative motions that might be due to stellar parallax. This method would only work if the stars were chance 'line of sight' alignments with one nearby and one distant. He failed to measure parallax, but catalogued around 700 pairs of stars within 2 arcsec of each other, far more than would be expected by chance, and showed that several of them were orbiting each other. In fact, John Michell (1724–1793) had earlier calculated that the probability of chance alignments of close double stars was extremely small.

THE HERSCHEL FAMILY

Sir (Frederick) William Herschel (1738–1822) (Figure 3.11a) was born in Germany but emigrated to England after visiting with the regimental band of the Hanoverian guards. His musical activities funded his interest in astronomy as he constructed ever larger telescopes. He made a systematic survey of the sky, discovering the planet Uranus in 1781, which he originally named Georgium Sidum in honour of King George III, which no doubt helped him to be appointed court astronomer. In addition to his studies of double stars, he also discovered two satellites of Uranus and Saturn, accurately determined the rotation period of Mars and deduced the Sun's motion through space, relegating the Solar System from its favoured place at the centre of the Universe. He used prisms and thermometers to detect radiation in the solar spectrum but beyond the red end of the visible spectrum and hence discovered infrared radiation.

(a)

William's sister, Caroline Herschel (1750–1848) (Figure 3.11b), as a girl in Germany, was denied an education and was destined to become a housekeeper. In 1772 she moved to England to become housekeeper for William and also became his observing assistant, compiling catalogues of 2500 nebulae and 1000 double stars. By quizzing William over the breakfast table she learnt spherical trigonometry and logarithms and when he was away she searched the sky for comets, discovering eight. Like her brother, she received several prestigious awards – the first when she was 78 and the last on her 96th birthday!

(b)

William's only son, Sir John Frederick Herschel (1792–1871) (Figure 3.11c), continued the astronomical dynasty. From 1834 to 1838 he systematically mapped the Southern skies from Cape Town in South Africa. He invented an instrument which allowed him to adjust the brightness of an image of the full moon to match a star under observation – a great advance in stellar photometry. His General Catalogue of Nebulae and Clusters is still, in updated form, the standard reference (and known as the NGC). John Herschel's fame extended beyond the scientific community and he was recognized as a great public figure of his time.

(c)

Figure 3.11 (a) William Herschel, (b) Caroline Herschel, (c) John Herschel. (Royal Astronomical Society)

You may have noticed from Table 3.1 that seven of the nearest ten stars are in binary or triple star systems. In fact over half the known 'stars' are binary systems, and so such systems account for over two-thirds of individual stars. (Note that the word 'star' applies to an apparently single point of light in the sky, as well as to truly individual stars. A binary system is often called a binary star, and so a **binary star** actually contains two stars!)

■ Why do most stars appear to be single when observed with the unaided eye or a telescope?

❏ They are so far away that even if the two stars in a binary are separated by large distances their *angular* separations are so small they cannot be distinguished (we say that the binary star cannot be resolved).

When both stars in a binary system can be seen as distinct points of light, they are called **visual binary systems**, though nowadays the observations are made by electronic or photographic imaging, and not by the eye at the telescope eyepiece. One of the most famous visual binaries is Sirius, the brightest star in the night sky as seen from the Earth. (When we refer to a binary star by its name, e.g. Sirius, we have not distinguished between the two components, Sirius A and Sirius B. Often, one star (A) is much brighter than the other so the luminosity of the combined binary star is essentially the same as its brightest component.) With a large telescope we can see two stars, the brightest (primary) star Sirius A, and the much fainter Sirius B, first seen by the US telescope maker Alvan G. Clark (1832–1897) in 1862, while he was testing a new telescope lens. The orbital period of each with respect to the other is 50 years, and their angular separation on the plane of the sky has a maximum of 11.5 arcsec. Examples of binary stars visible in a small telescope are: γ And, Albireo (β Cyg) and Mizar (ζ UMa). (Mizar (ζ UMa) is also a pair with Alcor which can be distinguished with the unaided eye.) Figure 3.12 shows the orbit of one star in a binary with respect to the other (the small irregularities arise from uncertainties in the observations – the orbit is actually very smooth). In reality, both stars are moving about their common centre of mass that

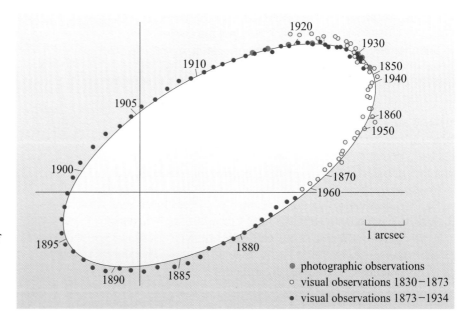

Figure 3.12 The orbit in the sky of one star relative to another in the visual binary system 70 Oph. The orbital period is 88 years. (Strand, 1973)

1 arcsec

• photographic observations
○ visual observations 1830−1873
• visual observations 1873−1934

is moving with its own space velocity. Visual binaries are of immense value in determining the masses of individual stars (Section 3.3.7) but can only be identified if they are close to the Sun and if the stars are well separated.

Other binaries reveal their nature indirectly.

Spectroscopic binaries reveal themselves from the changing Doppler shifts (Section 3.2.1) in the spectral lines of one or both stars as they orbit each other (Figure 3.13). The majority of known binary stars have been identified by this method since there is no requirement for them to be close to the Sun, only that they are bright enough for good quality spectra to be obtained.

▨　Is there any other requirement of the binary star system for it to be identified by relative Doppler shifts of the spectral lines?

❏　The orbital plane of the two stars must not be at right angles to the line of sight. If this is the case then there will be no component of motion of the stars along the line of sight and therefore no radial velocity component and hence no relative Doppler shift.

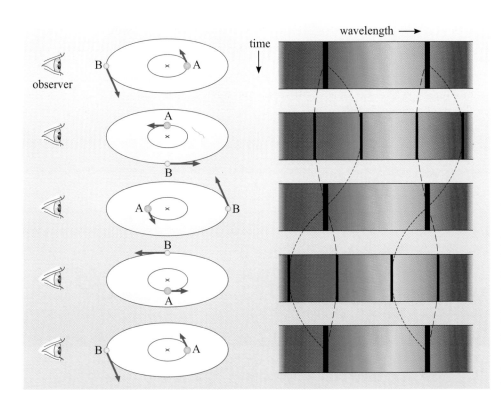

Figure 3.13 Identification of a spectroscopic binary star from Doppler shifts of spectral lines as the stars orbit their common centre of mass (marked with a cross). The diagrams on the left show the geometry of the stars in their orbits. On the right is a schematic representation of part of the spectrum of the stars (the spectra cannot be separated as the stars are too close together to be resolved).

There are also **eclipsing binaries**, in which we view the orbit so close to edge-on that one star is seen to pass in front of the other, and the observed brightness of the 'star' will dip (as you will see in Figure 3.41). These are rare but a famous example that can be seen with the unaided eye is Algol (β Per), which exhibits a rough halving of its luminosity for a few hours every 69 hours. In some mythologies, Algol is the winking eye of a demon.

3.2.4 Star clusters

A brief glance at the night sky will confirm that the stars are not distributed uniformly across the sky. If you are able to observe from a dark site (unfortunately not so simple these days due to the light pollution from street lighting) you will see that the greatest concentration of stars coincides with a faintly glowing band which circles the sky – the Milky Way. Binoculars or a telescope reveal that this is the light of myriads of individual stars. It is immediately apparent that the Galaxy is flattened, but less clear that our Sun is located well away from the centre (see Figure 0.1 in the Introduction). This is due to the presence of obscuring dust towards the centre of the Galaxy. Counting stars provides a tool for probing the size and thickness of absorbing regions and hence for mapping of the distribution of dust in the Galaxy. (The effect of interstellar gas and dust on starlight is discussed in Sections 4.3.2 and 4.3.3.)

Intermediate between the large-scale distribution of stars on the sky and the closely bound binary or multiple star systems, are groups of hundred or thousands of stars. Any group with more than a few members is called a **star cluster**. The Pleiades ('ply-a-dees'), in the constellation of Taurus visible on winter nights in the Northern Hemisphere, appears as a group of five or more stars to the unaided eye. Binoculars reveal many more; it contains around 3000 stars (see Figure 3.14).

Figure 3.14 The open cluster, the Pleiades, is easily visible to the unaided eye and is about 80 million years old. The blue light is starlight scattered by dust close to the stars (see Section 4.3.3). (ROE/AAO/ D. Malin)

The Pleiades is an example of an **open cluster**. They typically contain a few hundred stars, are sometimes irregular in shape, often contain very hot luminous stars, generally lie close to the plane of the Galaxy and are often associated with dust and gas clouds. These clusters contain barely enough mass to be bound by gravity and will eventually dissipate. As you will see later (Section 4.2.5), open clusters play a vital role in understanding the process of stellar evolution. Since all the stars were formed at more or less the same time in the same region of space they provide a 'snapshot' of a collection of stars which are all the same age, formed from the same material. They also provide a way of comparing the relative properties of stars since all the stars are effectively at the same distance from us.

Figure 3.15 The globular cluster 47 Tucanae. This cluster is visible to the unaided eye from the Southern Hemisphere as a faint smudge but requires a telescope to see individual stars. (NASA/ESA)

A different type of star cluster, with very different properties, is observable in our Galaxy. **Globular clusters** like 47 Tucanae (Figure 3.15) contain many thousands of stars, tightly bound by gravity into a spherical shape. They are distributed spherically about the Galaxy, have little or no gas and dust, and usually contain no very hot stars. The properties of globular clusters imply great age and they provide a tool for probing the structure and history of material in our Galaxy. They are also sufficiently bright to be identifiable in other galaxies, providing a method of determining the distances to those galaxies (assuming their properties are the same in external galaxies).

3.3 The stars as bodies

3.3.1 How big are the stars?

The Sun is a spherical body with a radius about 110 times that of the Earth. The volume ratio is a more impressive million or so. What of the other stars? Is there a great range of sizes? Is the Sun typical, or exceptionally small, or particularly large? Of one thing we can be certain: with a few readily identified exceptions, the stars are spherical, just as we expect for massive bodies dominated by their own gravity. Thus we can characterize size by radius.

For a star of known distance d, the most direct way to calculate the radius is to measure the angular diameter α. Then, as Figure 3.16 shows, the star's radius is given by

$$R = [(\alpha/2)/\text{radians}] \times d \qquad (3.8)$$

provided that α is small, which it certainly always is!

If we have a telescope that can show a star as a disc then we can measure α. Alas, this is possible for only those stars with particularly large values of α. This is because a telescope, even when looking at an object of *negligible* angular size, will produce an image that is not a point but a blur of finite size. Thus, a star has to

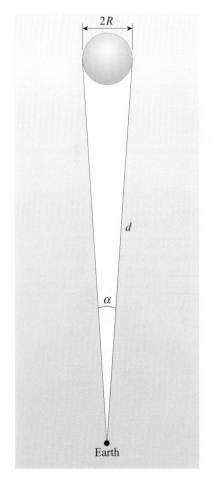

Figure 3.16 A star's angular diameter and radius.

have a sufficiently large angular diameter to produce an image with a size that considerably exceeds that of the point blur. This blur arises partly from turbulence in the Earth's atmosphere (called 'seeing') and partly from fundamental optical limits that are more severe the smaller the main mirror or lens in the telescope. Modern techniques such as adaptive optics can be used to overcome some of the effects of atmospheric turbulence by constantly adjusting the shape of the telescope's mirror. Alternatively, telescopes can be combined to make, in effect, a single large telescope using a technique called long baseline inteferometry. Using such techniques, the angular diameters of a few hundred stars have been measured, from the star with the largest angular diameter, namely Betelgeuse, prominent in Orion (Figure 3.1), with $\alpha = 0.050$ arcsec, down to as little as 0.0004 arcsec for ζ Puppis – equivalent to the width of a human hair at a distance of a few kilometres! (The diameter of Betelgeuse is so large that surface details have been imaged – see Figure 3.17.)

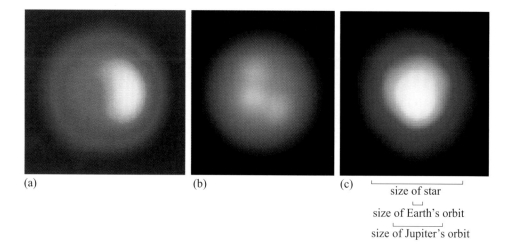

(a) (b) (c)

size of star

size of Earth's orbit

size of Jupiter's orbit

Figure 3.17 Betelgeuse is one of the few stars which is large enough and close enough for structure to be imaged on its surface. It is nearly a thousand times larger than the Sun and 20 times more massive, and is a red supergiant. It is 131 pc (430 light-years) away. Structure is visible on the surface of Betelgeuse when imaged using the Cambridge Optical Aperture Synthesis Telescope consisting of five small telescopes (b). In fact Betelgeuse is so large that it can be imaged using parts of a single telescope as an interferometer (as in (a), taken using the 4.2 m William Herschel Telescope) or, at lower resolution, using the Hubble Space Telescope (c). The bright regions, probably due to convection, have changed between the observations. ((a) P. Warner/MRAO/William Herschel Telescope; (b) courtesy COAST Group; (c) NASA)

These measured values of α, multiplied by d as in Equation 3.8, have yielded stellar radii ranging from rather less than that of the Sun, to about 1000 times greater. Thus there is a great range of stellar radii, and the Sun is a rather small star.

A few hundred stars are rather a small sample. However, there are other ways of obtaining stellar radii, and we shall meet one of these later in this chapter.

QUESTION 3.2

Betelgeuse is 131 pc away. Calculate its radius, in metres and in solar radii.

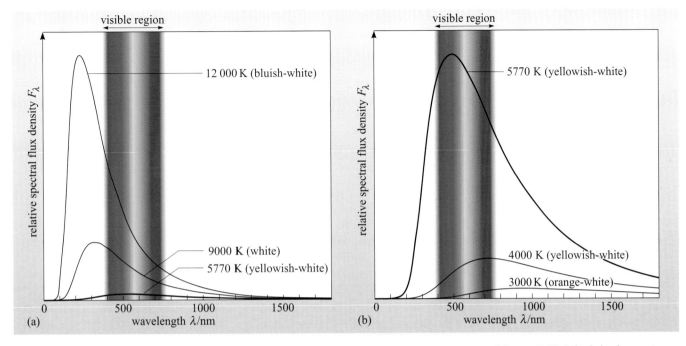

(a)

(b)

Figure 3.18 Black-body spectra at different temperatures. Each source is of the same size, and at the same distance from the detector that measures the flux density. Note that the vertical scale for the set of spectra in (a) is greatly elongated compared with that for the set in (b).

3.3.2 How hot are the stars?

You have seen in Chapter 1 that the radiation that we observe from the Sun originates largely from the photosphere, and that its distribution over wavelength – its spectrum – is close to that of a black body. If the distribution were *exactly* that of a black body then there would be a single temperature for the emitting region. Figure 3.18 shows some black-body spectra at different temperatures. The Sun's spectrum is not exactly that of a black body, and so we cannot give the photosphere a unique temperature. However, the spectrum at 5770 K in Figure 3.18 is a good fit to the solar spectrum, which means that the temperature of the source of the photospheric radiation is nowhere enormously different from 5770 K. Thus the radiation from the Sun comes largely from a region with a temperature around 5770 K. This is the Sun's 'surface' temperature.

The other stars, too, have spectra that are not very different from black-body spectra. Therefore we can obtain meaningful photospheric temperatures from well-fitting black-body spectra. A crude way of doing this is from the star colour that we perceive: from Figure 3.18 we see that white and bluish-white stars are hotter than the yellowish-white Sun, and orange-white stars are cooler than the Sun. A better way is to compare the radiation that the star emits at two different wavelengths, as you saw for the Sun in Question 1.10, or over two different wavelength *ranges*, as illustrated in Figure 3.19. From such a comparison, we can obtain the temperature of the photosphere. This is the basis of the **photometric method** of temperature determination. Note that we do not need to measure the complete spectrum: one ratio suffices.

However, for many individual stars, more accurate values of photospheric temperatures are obtained by the **spectrometric method** which is based on examination of the spectral absorption lines in starlight. The lines of interest are those formed, as in the Sun, by absorption in the star's upper photosphere and

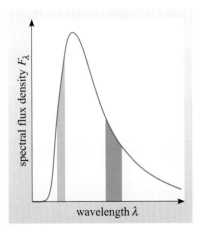

Figure 3.19 The photometric method of obtaining photospheric temperatures. The ratio of the amount of energy measured in two different wavelength regions (shaded) is uniquely defined by the temperature if the object emits like a black body.

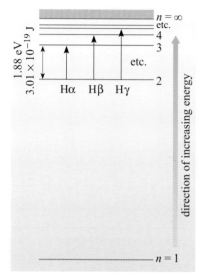

Figure 3.20 Electron transitions for the hydrogen Balmer absorption lines.

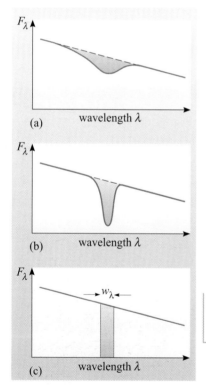

Figure 3.21 The definition of equivalent width w_λ. The shaded area between the vertical lines in (c) is the same as that bounded by each of the two spectral lines in (a) and (b). Therefore all three lines have the same equivalent width w_λ.

in the region just above it. For example, consider the hydrogen **Balmer absorption lines**. This is the name of the group of lines that you met in Chapter 1, with the individual labels Hα, Hβ, Hγ, and so on. They correspond to the electron transitions in the hydrogen atom as shown in Figure 3.20. For all the Balmer lines the lower energy level is that for which $n = 2$. If the temperature is very low, then nearly all the electrons are in the $n = 1$ level, and the Balmer lines are very weak. If, on the other hand, the temperature is very high, then nearly all the hydrogen atoms are ionized, and the lines are again weak. Between these two extremes, there will be a range of temperatures at which a larger proportion of the electrons are in the level $n = 2$ and transitions from this level will be common and hence the Balmer lines will be strong.

The strength of a spectral line is defined by the amount of radiation (as measured by the flux density) removed from the spectrum by the absorbing material (or added in the case of an emission line). Spectral lines may be broad and shallow or deep and narrow but have the same strength (in fact the shape of the line tells us a lot about the conditions in the gas which caused it as you will see later). The **equivalent width** of a line provides a quantitative measure of the strength of a spectral line. It is the width of a section of nearby continuum which has the same area as that between the spectral line and the continuum, shown in Figure 3.21. The variation of line strength with temperature is shown schematically in Figure 3.22.

Clearly there is a basis here for measuring temperature. However, there are two difficulties. Figure 3.22 indicates one of them.

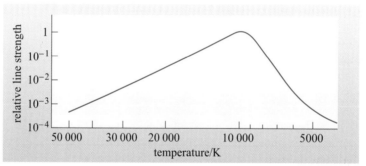

Figure 3.22 The strength of the hydrogen Balmer absorption lines versus photospheric temperature. Note the temperature scale is logarithmic (this figure is often plotted with a scale like that of Figure 3.23).

◼ What is this first difficulty?

❑ A given line strength corresponds to two temperatures.

Thus, a very hot and a very cool star will have Balmer lines of similar strength. We overcome this difficulty by observing other absorption lines that, like the hydrogen Balmer lines, are also particularly sensitive to temperature, but which have a different variation of strength with temperature from that of the hydrogen Balmer lines. Figure 3.23 shows schematically a collection of such line strengths.

◼ On the basis of Figure 3.23, what is the temperature of a star for which the hydrogen Balmer and (unspecified) helium lines have equal strength?

❑ About 20 000 K.

The second difficulty is that the strengths of the various lines are also sensitive to the elemental abundances. Clearly, you would not see calcium lines at *any* temperature if there is no calcium present. We overcome this difficulty from spectral studies that reveal the composition of the region from which the absorption lines originate (Section 3.3.6). The elements chosen for temperature measurements are those that do not exhibit large variations in abundance from star to star.

With these and other difficulties overcome, the spectrometric method of obtaining photospheric temperatures was well established by the 1920s, mainly through the efforts of the US astronomer Annie Jump Cannon at Harvard University. On the basis of the strengths of their spectral lines, stellar spectra were classified by letter in a scheme called the **Harvard Spectral Classification**. Originally, spectra were sorted into groups labelled from A (strong hydrogen lines) to O and beyond (hydrogen lines weak or missing). Later, as work progressed, the underlying physical principles represented by Figures 3.22 and 3.23 exposed the inadequacy of this progression and the scheme was modified. Some of the groups were dropped, some were merged with others, and the remainder were ordered according to temperature, to give the scheme widely used today. In order of descending temperature the spectral classes are labelled O B A F G K M, which you should remember: one useful mnemonic is 'Oh Bother, An F Grade Kills Me!' (there are many alternatives!). The classes B to M are subdivided into spectral types 0, 1 … 9, with 0 at the hotter end. For example, F9 and G0 are spectral types of stars that are not very different in temperature. Class O is (perversely) subdivided 5, 6, 7, 8, 9, 9.5(!), with 5 at the hotter end. Table 3.2 gives the temperatures around the beginning and the middle of each class. There is one further complication, the temperature for a particular spectral type also depends somewhat on luminosity (as you will discover in Section 3.3.4).

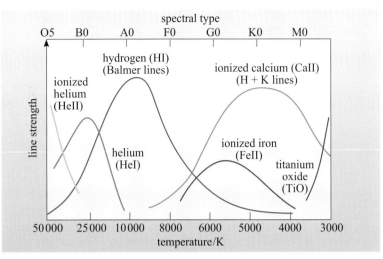

Figure 3.23 The strengths of various absorption lines versus photospheric temperature (note the temperature scale is not linear or logarithmic but in roughly equal steps of spectral type, introduced below).

ANNIE JUMP CANNON (1863–1941)

Annie Jump Cannon (Figure 3.24) was one of the first women from the state of Delaware to attend university; she graduated in Physics from Wellesley College in 1884. She then spent a decade at home with her parents, unfortunately becoming almost totally deaf following scarlet fever. In 1894, after her mother's death she returned to Wellesley and in 1896 joined Harvard College Observatory. She was one of a group of women, paid 50 cents per hour, employed to classify stars and carry out calculations – they were known as 'computers'. She is responsible for the O, B, A, F, G, K, M, spectral classification scheme and during her lifetime classified some 400 000 stars this way. Her work was published in the nine volume Henry Draper Catalogue, 1918–1924. In 1925 she became the first woman to be given an honorary degree by Oxford University. Although recognized worldwide, she was not given an official position at Harvard until age 75.

Figure 3.24 Annie Jump Cannon. (Harvard College Observatory/ Science Photo Library)

Since the Harvard Classification Scheme was devised, a new spectral class of star has been defined, class L. These are cooler than spectral class M. You will meet these again in Section 6.4.2.

■ On the basis of the strengths of its spectral lines, the Sun is spectral type G2. Is the corresponding temperature consistent with the value given earlier?

❑ From Table 3.2, G2 corresponds to a temperature of just under 6000 K, which is consistent with the value of 5770 K given earlier.

Stars are found across the full range of spectral types in Table 3.2, and so the Sun is not a particularly hot star. Far hotter is the bluish-white star Rigel A, which has a spectral type B8. (Rigel B and C are faint companion stars.) Somewhat cooler than the Sun is Betelgeuse (Figure 3.17), which has a spectral type M2, and looks orange-white. The colours of Rigel and Betelgeuse are discernible to the unaided eye and are apparent in Figure 3.1.

Table 3.2 Stellar spectral types, photosphere temperatures and prominent lines.

Spectral type	Temperature[a] /K	Most prominent lines (see Figure 3.23)
O5	40 000	ionized helium and other ionized atoms
B0	28 000	neutral helium, hydrogen
B5	15 500	
A0	9900	hydrogen, some ionized metals
A5	8500	
F0	7400	hydrogen, ionized calcium, iron and other metals
F5	6600	
G0	6000	ionized and neutral calcium, iron, and other metals,
G5	5500	hydrogen
K0	4900	neutral iron, calcium and other metals
K5	4100	
M0	3500	titanium oxide, neutral calcium
M5	2800	
M8	2400	

[a] The exact temperatures of each spectral type also depend on luminosity to a lesser extent (see Section 3.3.4) so a range of values are possible for a given spectral type.

■ Make a rough estimate of the temperatures of Rigel A and Betelgeuse.

❑ The approximate temperatures are 12 000 K for Rigel A and 3200 K for Betelgeuse.

Figures 3.25 and 3.26 show stellar absorption spectra belonging to various spectral types, in the form of graphs of relative spectral flux density (Section 1.3.2) versus wavelength: these are the sort of quantitative spectra now used to establish spectral type, and hence temperature. If we can establish a star's spectral type, then we can determine its average photospheric temperature with an uncertainty of only about 5%.

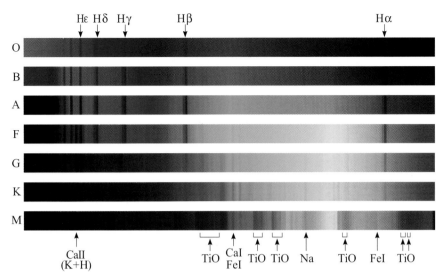

Figure 3.25 Stellar absorption spectra for different spectral types showing prominent absorption lines. (Kaufmann and Freedman, 1998)

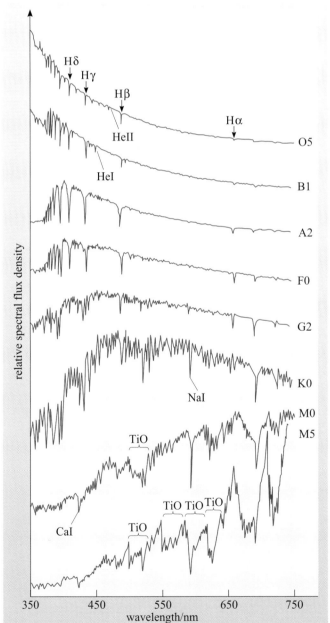

Figure 3.26 The stellar absorption spectra given in Figure 3.25 are more usually presented as graphs of relative flux density versus wavelength for ease of identification of the prominent absorption lines. The spectra have been plotted without spectral flux density scales and displaced vertically for clarity. (Kaufmann and Freedman, 1998)

QUESTION 3.3

From Figure 3.23 and Table 3.2, which spectral classes have weak hydrogen Balmer lines, and what are the corresponding temperatures? Compare your answer with the spectra in Figures 3.25 and 3.26.

3.3.3 How bright are the stars?

When we observe stars to determine their brightness we actually measure the amount of light from the star which is reaching us – its *apparent* brightness. If we want to study and compare the properties of the stars we really want to know their *intrinsic* brightness.

By intrinsic stellar brightness we mean the total amount of power a star radiates into space, over all wavelengths. This is called the **luminosity**, L, which in SI units is measured in watts. In Chapter 1 you saw that the Sun's luminosity is 3.84×10^{26} W, enormous by terrestrial standards, but how does it compare with the other stars?

The luminosity of a star depends on two of its properties that you have already met.

■ What do you think these properties are?

❏ The two properties are radius and temperature.

On the basis of everyday experience this is reasonable: as a ball of steel goes from red-hot to yellow-hot to white-hot it glows more brightly, and a pinhead of hot steel would radiate less power than a cannon ball of steel at the same temperature. We can readily develop a quantitative relationship between luminosity, radius, and temperature, because, just like the Sun (Chapter 1), any star radiates rather like a black body.

The power l radiated by unit area of a black body at an absolute temperature T (such as the kelvin scale) can be shown to be given by the simple equation $l = \sigma T^4$, where σ is called the **Stefan–Boltzmann constant**: in SI units, σ has the value 5.67×10^{-8} W m^{-2} K^{-4}. For a star, we have the approximation $l \approx \sigma T^4$, where T is the average photospheric temperature. We can take a star to be spherical, and therefore its surface area is $4\pi R^2$, where R is the radius of the star. Thus the power emitted by the whole surface – the luminosity – is given by

$$L \approx 4\pi R^2 \sigma T^4 \tag{3.9}$$

Therefore, if we know a star's radius and temperature then we can obtain its luminosity.

■ Can you see any problem with this approach?

❏ In Section 3.3.1 you saw that only a few hundred stars have had their radii measured. Therefore, this approach can yield the luminosities of only a few stars.

In reality, Equation 3.9 is often used to *determine* the radius when the luminosity is known from some independent method.

The luminosity of a star is an intrinsic property of the star, but how is the *apparent* brightness as seen by an observer affected by the distance of the observer? You will know that if you want to read a book at night, then the closer you hold the book to a source of light the greater the illumination. In more scientific terms, for a source of given luminosity, the closer the source to a surface facing it, the greater the flux density on the surface. **Flux density**, F, is the rate at which energy from a source crosses a unit area facing the source.

■ How does it differ from spectral flux density, defined in Chapter 1?

❑ Spectral flux density is the rate at which energy from a source crosses unit area facing the source within a narrow wavelength range, divided by the width of the range.

The physical units of flux density are power per unit area, which in SI units is watts per square metre.

If we know the distance d from a star to the Earth then we can work out the luminosity by measuring the flux density. At a distance d from the star, its luminosity L is spread over a sphere of area $4\pi d^2$, as in Figure 3.27. For all practical purposes a star can be considered to radiate uniformly in all directions. Thus, at any point on the sphere the flux density is given by

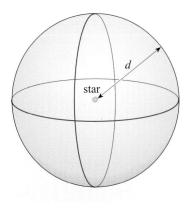

Figure 3.27 A spherical surface, radius d, centred on a star.

$$F = L/(4\pi d^2) \tag{3.10}$$

We have assumed that the effects of interstellar matter between the star and the point where F is measured are negligible. Thus, if we measure F and we know d then we can obtain L, from a straightforward rearrangement of Equation 3.10, namely

$$L = (4\pi d^2)F \tag{3.11}$$

Conversely, by measuring the flux density received from a star of known luminosity L, we can obtain a value for the distance. From a further rearrangement of Equation 3.10

$$d = [L/(4\pi F)]^{1/2} \tag{3.12}$$

This method works for distances that extend well beyond the limit for trigonometric parallax; it merely (!) requires us to determine the luminosity by an independent method. There is a range of such methods, which usually involve determining the luminosity of a nearby object and making the assumption that all other objects with the same properties have the same luminosity. These so-called **standard candles** play a vital role in determining distances in astronomy and we will discover some of them later. The luminosity of stars can in fact be determined from their spectra alone if they are bright enough for their spectra to be observed in detail. This spectrometric method is described in Section 3.3.4.

In practice, it is very difficult to measure F. One reason for this is the limited spectral range over which any one flux detector will respond. For example, instruments which can detect radiation at visual wavelengths are not suitable for detecting infrared light, and the designs of the telescopes themselves need to be completely different if we wish to detect X-rays or radio waves. We therefore require a great range of detectors, coupled to suitable telescopes, to measure F over the full wavelength range. There is a further difficulty in measuring F if we make measurements from the Earth's surface: the Earth's atmosphere is not perfectly transparent to any wavelengths, and is opaque to many, as you saw in Figure 1.38.

The best we can do is to make measurements over the more transparent wavebands, and apply corrections to obtain the flux density values that we would have obtained at the top of the atmosphere.

QUESTION 3.4

List the more transparent wavebands in the Earth's atmosphere.

We can avoid these difficulties if we can make do with flux density measurements over limited wavelength ranges. We can indeed do so: moreover, a single waveband suffices. Let's illustrate the approach by selecting one often-used waveband – the V band (or V channel). Any V-band detector has the well-defined spectral response shown in Figure 3.28: it approximates the spectral response of human vision, hence the label 'V', for visual. All stars emit a significant fraction of their luminosity in the V band, and the Earth's atmosphere can be very transparent at these wavelengths. Thus we can readily obtain the flux density F_V, where the subscript V indicates that the flux density has been measured with a detector that has the spectral response of the V band.

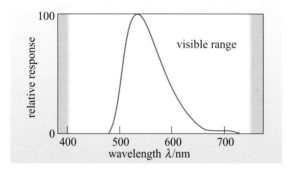

Figure 3.28 The spectral response of a V-band detector.

To obtain the luminosity in the V band, we make use of Equation 3.11, which applies not only to the whole output of a star, but also to any part of it. In the particular case of the V band, Equation 3.11 becomes

$$L_V = (4\pi d^2)F_V \tag{3.13}$$

and Equation 3.12 becomes

$$d = [L_V/(4\pi F_V)]^{1/2} \tag{3.14}$$

L_V is therefore a measure of the total power output of the star in the V waveband. We can then estimate the star's total luminosity L if we know the star's surface temperature. This is because the ratio of power output in the V band to the total power output is uniquely defined by the temperature for a black body, and stellar spectra are similar to black bodies (see Figure 3.18). For the Sun, $L_V = 4.44 \times 10^{25}$ W: you can check that this is reasonable by comparing Figure 3.28 with the 5770 K curve in Figure 3.18b, remembering that $L_\odot = 3.84 \times 10^{26}$ W.

Magnitudes

If you were to look up a table of stellar properties, you would not find F_V and L_V, but closely related quantities called, respectively, apparent and absolute visual magnitudes. Why should astronomers not simply use F_V? It is part of the burden of history, dating back to antiquity, and particularly to the Greek astronomer Hipparchus. He classified the stars according to their apparent visual brightness in six groups or *magnitudes*, the brightest stars being of first magnitude, the faintest of sixth magnitude, the ones between being placed according to their brightness into an appropriate intermediate magnitude. European astronomers used and refined this system right through to the appearance of detectors that could measure F_V. Such measurements showed that a first magnitude star is about 100 times brighter than a sixth magnitude star. Since the human perception of brightness, as measured by the eye, is logarithmic in nature (i.e. equal steps of perceived brightness correspond to equal *ratios* of flux density – see Box 3.2 if you are not familiar with logarithms) the magnitude scale is a logarithmic scale.

BOX 3.2 LOGARITHMS

A logarithm (or 'log' as it is usually abbreviated) is a mathematical formulation which was widely used to perform long calculations before the advent of the electronic calculator. We will not discuss the theory of logs here but merely their properties, which are relevant to the magnitude system. If p is the log of the number x, this is written

$$p = \log_{10} x$$

and means 10 to the power of p is equal to x, i.e.

$$x = 10^p$$

The number 10 is called the 'base' and p is correctly defined as 'the logarithm to the base 10 of x', but is often shortened to 'p is the log of x' and written

$$p = \log x$$

You can determine the value of p, for any number, x, using the 'log' button on a calculator. Conversely, if you have a value of p, then x can be determined using the 10^x button (sometimes labelled 'antilog'). In the past, books of tabulated conversions, 'log tables', were required.

Let's look at some values: if $x = 10$ (i.e. 10^1) then $p = 1$; if $x = 100$ (i.e. 10^2) then $p = 2$; if $x = 1000$ (i.e. 10^3) then $p = 3$; and so on.

Note that as each value of the number, x, increases by a *factor* of 10, the value of p increases by the *addition* of 1.

You can see that p can have negative values: If $x = 1$, (i.e. 10^0) then $p = 0$, if $x = 0.1$ (i.e. 10^{-1}) then $p = -1$, if $x = 0.01$ (i.e. 10^{-2}) then $p = -2$. Note that no value of p exists for $x = 0$ or a negative value of x.

■ If the log of 8 is 0.9, what is the log of (a) 80, (b) 8000 000, (c) 0.8?

❏ You can use your calculator but it is not necessary:
(a) 80 is 10 times larger than 8 so its log is $0.9 + 1 = 1.9$.
(b) 8000 000 is 10^6 times bigger than 8 (i.e. $10 \times 10 \times 10 \times 10 \times 10 \times 10$) so its log is $0.9 + 6 = 6.9$.
(c) 0.8 is 10 times smaller than 8 so its log is $0.9 - 1 = -0.1$.

This illustrates a very important property of logs. The range of values of p is small compared with the range of values of x. If you wish to display data with a very wide range of values on a graph, the use of logarithmic scales is recommended. An example is the horizontal axis of Figure 3.29.

QUESTION 3.5

What are the values of the logs of: (a) 5, (b) 50, (c) 5000 000, (d) 0.5, (e) 5×10^{-7}?

When performing calculations using logs, you should not confuse logs to the base 10 with *natural* logarithms, which have a different 'base'. In this case $x = e^q$ where the 'base', e, is a number $\approx 2.718 \dots$ and the natural log is written $q = \ln x$ (but is sometimes confusingly written $q = \log x$). On calculators the natural log is usually obtained from the button marked 'ln'.

Each difference of 1 magnitude corresponds approximately to a ratio of 2.5 in brightness. This was precisely quantified so that a difference of 5 magnitudes was set to be precisely a factor of 100. This meant that a difference of 1 magnitude corresponds to a factor of 2.512 in brightness since $(2.512 \times 2.512 \times 2.512 \times 2.512 \times 2.512) = (2.512)^5 = 100$. The difference in magnitudes of two stars may be expressed in the form of an equation

$$(m_1 - m_2) = -2.5 \log(b_1/b_2) \tag{3.15}$$

where m_1 and m_2 are the apparent magnitudes of stars 1 and 2, and b_1 and b_2 are the apparent brightnesses of stars 1 and 2. The brightnesses are defined by the flux density F, although they can, in this equation, be measured in any units (e.g. output signal from a measuring device) because the units cancel in the ratio b_1/b_2.

If the flux densities of the stars are measured in the V band, then b_1 and b_2 correspond to measurements of F_V for the two stars. m_1 and m_2 are then called **apparent visual magnitudes** and denoted m_V or simply V.

Note that the magnitude scale is 'upside down'. The greater the apparent magnitude, the fainter the star. Figure 3.29 shows the relationship between F_V and V. The use of telescopes meant that objects fainter than magnitude 6 became observable and so the scale was extended to larger values of V. Also, a few stars were found to be brighter than first magnitude so negative values of V are possible. It is apparent from Figure 3.29 that the magnitude scale is rather a good method for quoting stellar brightnesses. While F_V varies from 10^{-8} W m^{-2} for Sirius, the brightest star in the night sky, to 10^{-20} W m^{-2} for the faintest detectable objects, the V magnitude ranges only from -1.5 to about 27.

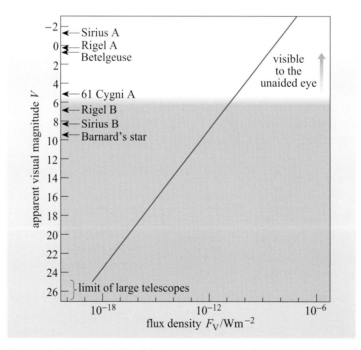

Figure 3.29 The relationship between F_V, the flux density in the V band, and apparent visual magnitude V. Approximate values of V are indicated for a number of objects. Note the log scale for F_V.

EXAMPLE 3.1

Two stars are observed with a telescope and photometer (an instrument which is sensitive to visible light and produces a signal proportional to the star's flux density). Star X, which has apparent magnitude 5.5, produces a signal 40 times stronger than star Y. What is the apparent magnitude of star Y?

SOLUTION

Using Equation 3.15, set star 1 as star X and star 2 as star Y

$$(m_{\text{star X}} - m_{\text{star Y}}) = -2.5 \log(b_{\text{star X}}/b_{\text{star Y}})$$

The ratio $b_{\text{star X}}/b_{\text{star Y}} = 40/1 = 40$, so

$$(m_{\text{star X}} - m_{\text{star Y}}) = -2.5 \log(40) = -2.5 \times 1.6 = -4.0$$

The apparent magnitude of star X, $m_{\text{star X}} = 5.5$, so

$$(5.5 - m_{\text{star Y}}) = -4.0$$

$$- m_{\text{star Y}} = -4.0 - 5.5$$

$$m_{\text{star Y}} = 9.5$$

So the apparent magnitude of the star Y is 9.5.

QUESTION 3.6

The closest star system to the Sun, α Centauri, is a triple system, with stars denoted A, B and C. The brightest, α Centauri A, has a flux density approximately 2.5 times larger than α Centauri B and 25 000 times larger than α Centauri C (also known as Proxima Centauri). If the apparent visual magnitude of α Centauri A is 0, what are the apparent visual magnitudes of α Centauri B and C?

Appendix A4 provides a list of the brightest stars visible from the Earth (i.e. those with the lowest values of apparent magnitude).

As we have seen, the value of F_V, and therefore of V, is not an intrinsic property of a star, but depends on the distance to the star (and on the effects of any intervening interstellar material). By contrast, L_V is an intrinsic property. Likewise, the **absolute visual magnitude**, M_V, is also an intrinsic property: it is the value of V that would be obtained if the star was placed at a standard distance of 10 pc (in the absence of any interstellar matter).

In the same way that the observed flux density depends on the luminosity and distance of a star, the apparent magnitude depends on the absolute magnitude and distance. The absolute magnitude M is given by

$$M = m - 5 \log d + 5 \tag{3.16}$$

where m is the apparent magnitude and d is the distance in parsecs. Most stars have absolute visual magnitudes within the range $-6 < M_V < 16$. (Remember a star with $M_V = -6$ is much brighter than a star with $M_V = 16$. Using Equation 3.15 shows that it is over 6×10^8 times more luminous.) The absolute visual magnitude of the Sun is 4.8 indicating that it is a very average star.

Magnitudes are also used to express colours of stars. Flux density can be measured using detectors and filters which are sensitive to different wavelength ranges. The best known is the **'UBV' system**; in addition to the V (visual) passband (centred at wavelength 550 nm), U ('Ultraviolet' centred at wavelength 365 nm) and B ('Blue', centred at wavelength 440 nm) are defined. Many other passbands extending into the red (R) and infrared (I, J, H, K …) have been defined but here we will concentrate only on B and V. As you saw in Section 3.3.2, the photometric method employs measurements at two different wavelengths to define the temperature of a star. The *difference* in magnitudes measured in the B and V bands, $B - V$, is called the **colour index** and is a measure of the *ratio* of flux densities in the B and V bands (see Equation 3.15) and is therefore an indicator of temperature. $B - V$ values typically range from −0.4 for very hot stars (spectral type O) to +1.5 or more for very cool stars (spectral type M).

QUESTION 3.7

The brightest stars in the southern constellation Centaurus, α Centauri and β Centauri have similar apparent visual magnitudes, $V = 0.01$ and 0.61, and lie at distances of 1.35 and 161 pc respectively. What are their absolute visual magnitudes? How do the luminosities of these stars compare with that of the Sun?

3.3.4 The power of spectroscopy – luminosities, radii and distances

In the early years of the 20th century, astronomers compared the spectra of stars with similar temperatures but different luminosities, the luminosities having been obtained non-spectrometrically. They found that, *at a given temperature*, the more luminous the star the narrower its spectral absorption lines, and the stronger the absorption lines due to certain ionized atoms.

An absorption line is *not infinitely thin*, but covers a narrow range of wavelengths. You should remember the difference between the *strength* and the *width* of an absorption line (Figure 3.21). In Figure 3.30 the two lines have the same strengths (areas), but different widths. Real examples are shown on the right.

Figure 3.30 The effect of stellar luminosity on spectral line width showing a graphical representation and a real example. (a) A small star and (b) a large star at the same temperature; note that these are *negatives* of photographs, so that the absorption lines appear bright on a dark background. (Spectra from Abt *et al.*, 1968)

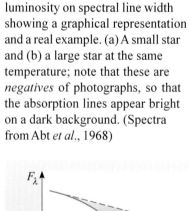

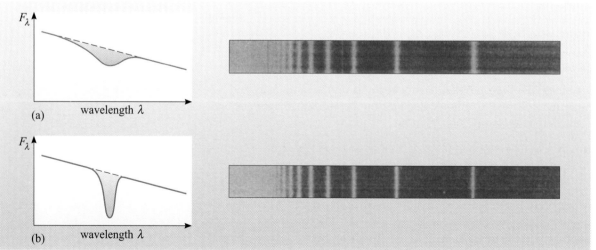

The reason for these effects of luminosity on linewidth, and on the strengths of certain ion lines, is the difference in conditions present in the atmospheres of the stars where the lines originate. The outer layers (where spectral ions are formed) of a large star has a density lower than those of a small star of the same temperature because the mass of these layers is spread over a larger volume. Atoms and ions are therefore spread over a larger volume and collisions between them are relatively infrequent. In small stars however, the density and pressure are higher and the atoms and ions collide and interact with each other. These interactions cause distortions in the energy levels of the atoms and therefore cause small random differences in the energy change (and hence wavelength) when a photon is absorbed. The spectral lines are therefore broader for small (i.e. lower luminosity) stars. This process is called **pressure broadening**. In addition, the probability of an ion interacting with an electron and recombining is higher in a small star so the spectral lines produced by that ion are weaker than for a large star, because there are fewer ions present to absorb photons.

As soon as astronomers had correlated luminosity with such spectral features, then for stars of unknown luminosity these spectral features could be used to determine their luminosity. For example, the widths of hydrogen lines are used to define luminosity for spectral types B to F, whereas the relative strengths of the ionized strontium (SrII) and neutral iron (FeI) lines are used as a luminosity indicator for spectral classes F to G. This method does not provide precise luminosities, but is used to define *luminosity classes* which distinguish different groups of stars. You will learn more about luminosity classes in Section 4.2.2.

Using this spectrometric method, many stellar luminosities have been measured, and they cover an *enormous* range, from less than 10^{-4} times that of the Sun, to over 10^6 times the solar value. Thus, the Sun is a fairly average star. Most of the stars that we see with the unaided eye are more luminous than the Sun: for example, Rigel A and Betelgeuse are over 10^5 times as luminous.

This is not because bright stars are common – in fact they are not – but because if dim stars, such as the Sun, are at distances beyond about 20 pc, they are visible only through a telescope!

Note that the effect of luminosity on the strength of lines of some ionized atoms means that spectral classification can be influenced by luminosity as well as temperature: remember that spectral classification depends on the strengths of various spectral lines, including lines from ionized atoms. Fortunately, over a wide range of luminosities, the effect of luminosity is a good deal less than the effect of temperature. Moreover, the effect of luminosity on spectral type can be allowed for. Thus the evaluation of temperature on the basis of spectral line strengths is not undermined by the effect of luminosity.

EXAMPLE 3.2

You have seen that the temperature and luminosity of a star can both be obtained spectrometrically. This gives us a new way of obtaining stellar radius.

(a) Explain what this new way is and derive an expression for the radius in units of solar radius.

(b) Aldebaran B is the fainter of the two stars in the binary system Aldebaran (α Tau; the brighter is Aldebaran A). For Aldebaran B, $T = 3400$ K and $L = 0.06 L_{\odot}$ where $L_{\odot}$ is the solar luminosity. Calculate its radius giving your answer in solar radii.

SOLUTION

(a) Stellar radius can be obtained from Equation 3.9,

$$L \approx 4\pi R^2 \sigma T^4$$

Since both L and T are defined spectroscopically the only unknown quantity is R. Rearranging Equation 3.9 gives

$$R^2 \approx L/(4\pi\sigma T^4)$$

so $\quad R \approx [L/(4\pi\sigma T^4)]^{1/2}$

A convenient form of this equation is obtained by putting the luminosity L, temperature T and radius R in solar units. Equation 3.9 can be expressed for the Sun,

$$L_\odot \approx 4\pi R_\odot^2 \sigma T_\odot^4$$

Dividing the standard form of Equation 3.9 by the solar version (i.e. dividing the left-hand side by $L_\odot$, and the right-hand side by the equivalent quantity $4\pi R_\odot^2 \sigma T_\odot^4$). We thus get

$$L/L_\odot \approx (R/R_\odot)^2 \times (T/T_\odot)^4$$

Which can then be rearranged to give

$$R/R_\odot \approx (L/L_\odot)^{1/2} \times (T_\odot/T)^2$$

This equation is useful because we can use it to compare properties of stars in solar units.

(b) For Aldebaran B

$$R/R_\odot \approx (0.06)^{1/2} \times (5770/3400)^2 \approx 0.7$$

and so $R \approx 0.7 R_\odot$.

The radius of Aldebaran B is 0.7 times the radius of the Sun.

Not only do we now have a new way of obtaining radius, we also have a new, and very important way of obtaining distance. We can obtain L and T spectrometrically and hence use Equation 3.14 to obtain the distance to a star from the observed flux density F_V. This is called the method of **spectroscopic parallax** for obtaining stellar distances, a perverse name, given that it has nothing to do with the method of trigonometric parallax outlined earlier!

The method is subject to error, in that F_V can be reduced by interstellar matter. Even in the absence of such a reduction, or if a correction is applied for it, the distance is rather uncertain, because the luminosity obtained from spectral lines is itself uncertain. However, the method can provide reasonable estimates of the distances to very bright stars well beyond those that we can obtain at present from trigonometric parallax, and it is therefore an important method of distance determination.

QUESTION 3.8

The star ι^1 Sco (Greek iota with the '1' distinguishing it from a nearby star ι^2) has $F_V = 4.4 \times 10^{-10}$ W m^{-2} (after correction for reduction by interstellar matter) and $L_V = 6.1 \times 10^{30}$ W. Calculate the distance to this star, expressing your answer in parsecs. To the eye, ι^1 Sco does not seem very bright. Given that its apparent brightness is not much reduced by interstellar matter, is it in fact a bright star at a large distance?

3.3.5 Are these properties constant?

Most stars appear to have constant properties. Over long timescales (millions to billions of years) stars evolve and their characteristics change but this process cannot generally be observed directly. Stellar evolution is the subject of Chapters 5 to 9.

However, some stars do change in brightness on measurable timescales from seconds to years and these are collectively called **variable stars**.

Extrinsic variables change as a result of geometrical effects. The majority of this type are eclipsing binaries (which you have already met in Section 3.2.3). **Intrinsic variables** change in brightness as a result of physical changes in the stars themselves.

Regular variables exhibit variations that are regular in time, and so the variations have fairly well defined periods. **Irregular variables**, as their name implies, exhibit variations that are irregular in time; some types change continuously, others stay virtually constant for long periods and then brighten (or dim) suddenly.

There are many different types of variable stars with a wide range of properties, some of which you will meet as we discover more about stellar properties and evolution. Examples of their **light curves** (i.e. how the magnitude changes with time) are shown in Figure 3.31.

About two-thirds of all known variable stars are **pulsating variables**. They are intrinsic variables with luminosity, temperature and radius all changing with time. **Cepheids** ('sefeeds') are named after δ Cephei, the first to be discovered, in 1784, by the English astronomer John Goodricke (1764–1786) and they are one of the most important classes of pulsating variables. They are regular variables with periods of anything from about a day to about 100 days, and their luminosity changes can be up to a factor of 10. Cepheids help us to understand some of the processes that drive evolution at certain stages in a star's life (Section 7.3) and provide us with yet another way to measure distance.

Figure 3.31b shows a light curve of a typical Cepheid variable star. The possibility of using Cepheids for measuring distance was discovered by accident by the American astronomer Henrietta Leavitt (Figure 3.32) from searches for Cepheids in the Small Magellanic Cloud (SMC). The Small and Large Magellanic Clouds are companion galaxies to our own (although this was not known at the time), and visible to the unaided eye from the Southern Hemisphere as apparently disconnected pieces of the Milky Way. She discovered that the longest periods corresponded to the Cepheids with the brightest apparent magnitudes. However, since the Small

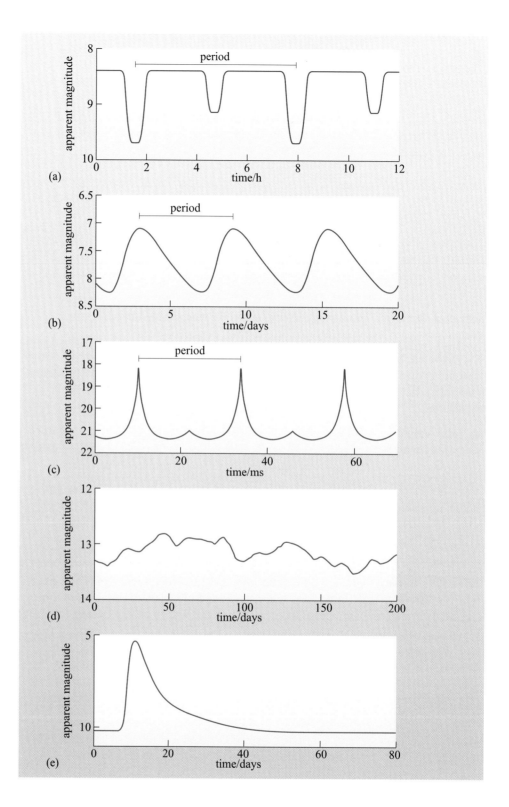

Figure 3.31 Light curves of examples of different types of variable star: (a) eclipsing binary, (b) Cepheid variable, (c) Pulsar – discussed in Chapter 9, (d) T Tauri star, (e) Nova.

Magellanic Cloud is so far away, all the stars could be assumed to be at approximately the same distance. There is therefore also a relationship between period and *absolute* magnitude (see Equation 3.16) and hence between period and luminosity. Once actual absolute magnitudes were determined from independently

HENRIETTA LEAVITT (1868–1921)

Henrietta Leavitt (Figure 3.32) developed an interest in astronomy while an undergraduate at Radcliffe College (now part of Harvard University). She graduated in 1892, and like her future colleague Annie Jump Cannon (Figure 3.24) became partially deaf following illness in the years immediately after graduation. From 1895–1900 she was a volunteer at the Harvard College Observatory, becoming a staff member in 1900, where she worked until her early death from cancer. Best known for her work on variable stars she identified over 2400, doubling the number known in her day. Her discovery of the Cepheid period–luminosity relationship gave the first means of measuring extragalactic distances and led to the recognition that the Magellanic Clouds are actually two companion galaxies to the Milky Way at (for that time) unprecedented distances from the Sun.

Figure 3.32 Henrietta Leavitt. (Royal Astronomical Society)

measured distances for some Cepheids, the **period–luminosity relationship** shown in Figure 3.33 was derived. This simple diagram provides an extremely powerful method for determining distances, since Cepheids are sufficiently luminous to be identifiable anywhere in our Galaxy and in other nearby galaxies. As soon as a Cepheid is identified from its light curve, the absolute magnitude can be deduced from the period and the distance derived using Equation 3.16.

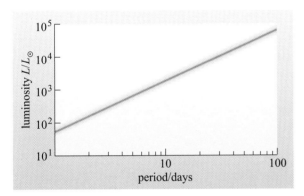

Figure 3.33 The period–luminosity relationship for Cepheid variable stars.

More spectacular irregular variables are the novae, a name that means 'new stars'. A **nova** is a star that brightens by up to 10 000 times (10 magnitudes) in a few days, followed by a slower decline to about its original luminosity – the star is not destroyed. For some novae, only one outburst has been observed. Other novae are known to repeat the performance, though irregularly, with the average interval between outbursts varying from tens of days for some stars to tens of years for others. You will learn more about the reasons for these outbursts in Section 9.5.2.

Supernovae, (e.g. Tycho's new star referred to Section 3.1) exhibit even greater outbursts in luminosity of up to 100 million times (20 magnitudes). However, these are all single events which effectively end the life of the star. The different types of supernovae are discussed in Chapter 8.

There are many other types of irregular variable, but one type that is of particular importance to the study of stellar evolution is the **T Tauri stars** ('tee tory'), named after one star of this kind. These stars exhibit variations in luminosity by factors of two or three over intervals of the order of a few days. T Tauri stars are discussed in Sections 4.2.3 and 5.3.4.

3.3.6 What are the stars made of?

The familiar world around us is dominated by certain chemical elements. Thus, the atmosphere consists mainly of oxygen and nitrogen, rocks and soil consist mainly of compounds of silicon and oxygen with metals such as calcium, aluminium, magnesium and iron, and the oceans consist almost entirely of water, a compound of two atoms of hydrogen with one atom of oxygen. The Earth as a whole is made up largely of the elements iron, silicon, oxygen and magnesium, all but iron being present mostly in compounds.

Stars represent a much larger sample of the cosmos than that provided by the Earth, and they are very different from the Earth in their composition. You have already seen in Section 2.2.3, that although stars do contain the sort of elements that dominate the Earth, stellar compositions are dominated by two elements that are but *minor* constituents of our planet, namely hydrogen and helium. Before we look more closely at stellar composition, let's consider briefly how it is measured.

Stellar compositions are obtained from studies of the spectral absorption lines in stellar atmospheres. If we can see the lines of a certain element then we can conclude that the element is present. But how much of it is there? The strength of a single absorption line is *not* a good guide.

■ Why not?

❑ Line strength is sensitive to temperature – see Section 3.3.2.

However, we can obtain the amount of an element present by comparing the absorption line strengths of different lines from the element, and it helps if lines are observed not only from the neutral element, but also from the ionized element. The great range of wavelengths covered by the various elements, particularly when we include their ionized forms, led to an enormous flood of data in the second half of the 20th century with the advent of infrared and ultraviolet astronomy. Thus, today we have very extensive knowledge of the compositions of the atmospheres of many stars.

QUESTION 3.9

Use is made of more than one absorption line in obtaining *temperature*. In what way does this procedure differ from the one to obtain *composition*, where more than one line is again used?

We shall not go into further details of how stellar compositions are obtained, except to note that spectrometry is used to obtain composition, temperature and luminosity in not quite such a separate manner as implied above. Instead, a model is constructed of a stellar atmosphere, and the composition, temperature and luminosity in the model are adjusted simultaneously until the model reproduces the great range of observed line strengths and line shapes of all the various elements.

Stellar atmospheres are largely made up of hydrogen, with remarkably small variations from star to star. For a star of average composition, like the Sun, 73% of the mass of the atmosphere consists of hydrogen. This is the lightest element, and so its dominance is even more impressive when expressed in terms of the percentage of nuclei, rather than of mass: 92% of the nuclei are hydrogen. Next comes helium, the next lightest element, at 25% by mass and 7.8% by number. If you add the hydrogen and helium figures together you will see that there is not a whole lot left for the 90 or so 'heavy' elements (atomic number more than 2) – only about 2% by mass and 0.2% by number!

It is important to realize that these measurements are for stellar atmospheres, and so the question arises of whether they are typical for whole stars. Here we have to appeal to stellar modelling, where the model has to reproduce the observed properties: the modelling procedure is similar to that for the Sun, outlined in Section 2.2.2. We find that, except for the cores of stars where considerable nuclear fusion has presumably occurred, stellar interiors are indeed dominated by hydrogen and helium, the above values being typical for most stars. Among the heavy elements considerable variations do occur, though among many stars there are still remarkable similarities.

A standard composition, the *solar system abundance*, has been defined, based on the Sun outside its core and on the composition of a particular class of meteorites (small bodies that reach the Earth's surface from interplanetary space) – carbonaceous chondrites, which are thought to have changed little since the origin of the Solar System. The standard composition is thus that of the material from which the Solar System formed. (Figure 3.34 shows the broad features of this composition, and the full details are given in Appendix A5.) Historically it has been called, rather grandly, the *cosmic relative abundance* of the elements. However, as you will discover in later chapters, this concept is flawed since the detailed abundances in stars and different regions of the Galaxy vary due to the continual processing of the elements within stars and during their death throes.

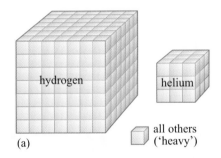

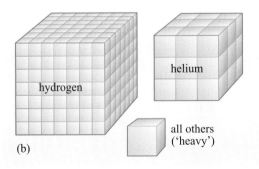

Figure 3.34 The large cubes represent the relative abundances of the elements in the solar neighbourhood (a) by relative numbers of atomic nuclei and (b) by relative mass. The *number* of particles in each small cube is the same.

QUESTION 3.10

From Appendix A5, what are the ten most abundant elements by number of nuclei, and by mass?

3.3.7 Measuring stellar masses

That stellar masses *can* be measured probably comes as no surprise to you in the light of all the measured stellar properties that you have already met. So, how is it done? The basis of the method is to observe how a star moves when a force is applied to it: this is a direct application of Newton's second law of motion, which states that the force applied to a body is equal to its mass times its acceleration. Therefore we must observe a star being accelerated.

Any star is accelerated by any other mass in the Universe through gravitational attraction, but the effects are only large enough to measure if the other mass is large, and relatively close to the star. Of particular importance therefore are systems in which just two stars are sufficiently close together to be in orbit around each other – these binary systems were introduced in Section 3.2.3.

Stellar masses are most readily obtained from those binary systems in which both stars can be *seen* to be orbiting each other, i.e. visual binary systems. The procedure, described below, is of wide applicability: for example, it can be used to obtain planetary and satellite masses in the Solar System.

Obtaining the sum of masses, *M* + *m*

Note that the symbols *M* and *m* have also been used for apparent magnitudes. This should not cause any confusion as the context and units should indicate which is meant.

Suppose that two stars are in circular orbits around each other, as in Figure 3.35a. This is a view from a point of observation that is not accelerating. One special point in the system itself is also not accelerating, and it lies on the straight line joining the two stars. This is called the **centre of mass** of the system, marked with a cross in Figure 3.35a. It lies at the centre of each of the circular orbits of the two stars, and thus the orbits shown are with respect to this centre of mass. The stars, however, *are* accelerating: their speeds are constant, but their directions of motion are constantly changing. Therefore, there must be a force acting on each of them.

■   What are these forces?

❑   The force on the star of mass *m* is the gravitational force exerted on it by the star of mass *M*, and vice versa. (The *speeds* are constant because the force on each star is always perpendicular to its direction of motion.)

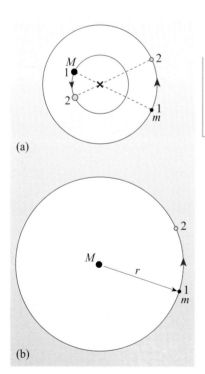

(a)

(b)

Figure 3.35 Circular orbits in a binary system: (a) relative to the centre of mass (×) of the system; and (b) of *m* relative to *M*.

Suppose now that, by some extraordinary feat, we were to sit on *M* and observe *m*. It would appear to move around us in an orbit that is called the orbit of *m* relative to *M*. If, as here, the orbits with respect to the centre of mass are circular (Figure 3.35a), then this relative orbit is also circular, as shown in Figure 3.35b, with *m* moving around its relative orbit at constant speed. Furthermore, the orbital periods are also the same from both points of view.

The relative orbit is important because its radius appears in the equation used to describe the gravitational force between the two stars. This force has a magnitude, *F*, given by Newton's law of gravitation

$$F = GMm/r^2 \tag{3.17}$$

where *G* is the gravitational constant (a universal constant), and *r* is the distance between the centres of the two stars – *this is also the radius of the relative orbit.*

Suppose that the orbital period of *m* is *P*. How do we expect *P* to depend on *M*, *m* and *r*? First, if *r* is kept fixed, then from Equation 3.17 we see that (reasonably enough) as M or m increases, F increases. Thus there will be a more rapid change of direction, and so P should decrease. Second, if M and m are kept fixed, then from Equation 3.17 we see that (again reasonable) as r increases, F decreases, and moreover the distance around the orbit, $2\pi r$, increases. Therefore, P should increase. Summarizing:

- as M or m increases, we expect P to decrease
- as r increases, we expect P to increase.

These expectations are borne out by the mathematical details, which won't concern us, but which lead to

$$P = 2\pi \left(\frac{r^3}{G(M + m)} \right)^{1/2} \tag{3.18}$$

We have thus reached the point where, for a circular orbit, if we measure P and r, then we can obtain the sum of the two masses.

QUESTION 3.11

Rearrange Equation 3.18 to obtain the explicit form for $M + m$ (i.e. $M + m = \ldots$).

Elliptical orbits

A circular orbit is rather a special case. In general, the orbit of m relative to M is an ellipse, with M at one of two special points called the **foci** (plural of **focus**) of the ellipse. Various **elliptical orbits** are shown in Figure 3.36. The size of an ellipse is given by the **semimajor axis** a, which is half of the long dimension, and which is the average separation between the two masses. The extent of the departure from circular form is called the **eccentricity** of the ellipse, and is defined as half the distance between the two foci of the ellipse (Figure 3.36) divided by the semimajor axis. In a circle, both foci coincide at the centre, and so the eccentricity of a circle is zero.

The generalization of Equation 3.18 to elliptical orbits is remarkably straightforward: we simply replace the constant separation r in a circular orbit by the average separation a in an elliptical orbit. Thus

$$P = 2\pi \left(\frac{a^3}{G(M + m)} \right)^{1/2} \tag{3.19}$$

This equation can be rearranged (as in Question 3.11) to give

$$M + m = \frac{4\pi^2 a^3}{GP^2} \tag{3.20}$$

We would have obtained the same result had we considered the orbit of M relative to m, rather than m relative to M.

Thus by measuring P and a we can again get the sum of the two masses, $(M + m)$.

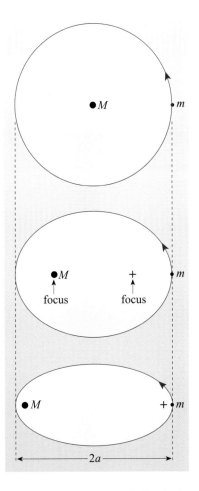

Figure 3.36 A variety of elliptical orbits.

Obtaining M and m separately

To obtain M and m separately we need to identify the position of the centre of mass of the binary system. Recall from Figure 3.35a that this lies on the straight line connecting the stars. We obtain its position on this line by making at least two observations, well separated in time. The centre of mass lies on both straight lines, and so must lie where they intersect: Figure 3.37 illustrates this procedure for the case of Sirius, where the orbits about the centre of mass are shown (elliptical in this case). These orbits are obtained by observing the motion of each star with respect to a fixed frame of reference, which for practical purposes is provided by stars that are much further away than the binary.

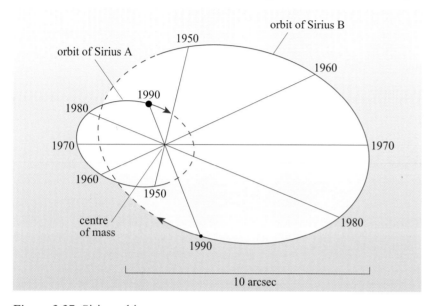

Figure 3.37 Sirius, a binary system, showing the position of the centre of mass at the intersection of lines joining stars A and B at specific times. (Prepared with the assistance of M. A. Seeds)

Consider an instant when the distances of the two stars from the centre of mass are d_m and d_M. It can be shown that the ratio d_m/d_M is constant throughout the orbital motion, and is related to the masses via

$$M/m = d_m/d_M \tag{3.21}$$

This equation is reasonable: if $M \gg m$, then $d_M \ll d_m$; that is, the centre of mass is close to the larger mass, as we might expect.

We have now achieved our goal: from Equations 3.20 and 3.21 we can obtain the masses M and m.

QUESTION 3.12

Suppose that observations have revealed that, for a binary system, $M + m = 12M_\odot$, and $M/m = 3$. Calculate the values of M and m.

Note that Equation 3.21 requires that stars behave as rigid bodies, so that they don't deform each other. In fact, stars are not at all rigid! However, provided that they are small compared with their separation then the deformations are negligible. For all visual binaries, this condition is met. This wide separation also means that the stars do not interfere with each other's evolution.

There are three important observational details to point out when deriving stellar masses from observations of binary stars:

First, it is important to realize that in Figures 3.12 and 3.37, as for any visual binary, we are observing the relative orbit as a projection onto a plane perpendicular to our line of sight, as illustrated in Figure 3.38. We thus observe the true shape of the relative orbit only if the actual relative orbit happens to lie in this plane. Furthermore, the semimajor axis a appears shorter in the projection plane than it really is. Fortunately, the true value of a, for use in Equation 3.19, can be deduced from the observational data: in essence, we are able to 'deproject' the relative orbit,

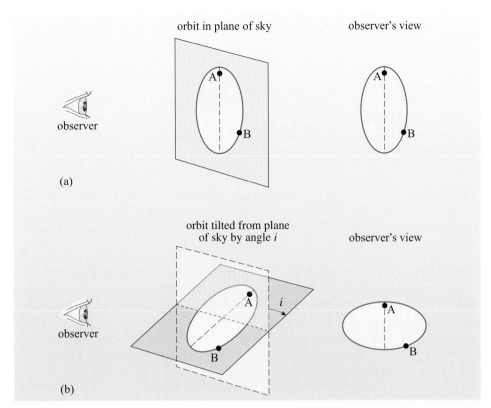

orbit in plane of sky

observer's view

observer

(a)

orbit tilted from plane of sky by angle *i*

observer's view

observer

(b)

Figure 3.38 The change in appearance of an orbit due to its projection onto the plane of the sky. The dashed line shows the major axis of the orbit. In case (b) the true major axis is not the longest axis when projected onto the plane of the sky.

though the details will not concern us. Note that the ratio d_m/d_M in Equation 3.21 is unaffected by the projection; d_m and d_M are each shortened by the same factor. The orbital period P is also unaffected.

Second, in order to obtain a we also need to know the distance to the binary: what we actually measure is the *angular* separation of the two stars, which gives us the projected value of the semimajor axis, a', using

$$a' = (\alpha/\text{radians}) \times d$$

(obtained by substituting the a', for $2R$ in Equation 3.8). The semimajor axis, a, is then determined from a' and the tilt of the orbital plane. The distance can be obtained by one of the methods that you met earlier in this chapter.

Third, note that the method of obtaining masses from binary systems outlined above requires that both stars be seen as distinct points of light, that is, we must be dealing with a *visual* binary system. Unfortunately, most binary systems are not of the visual kind. In some cases, we can detect the radiation from only one of the two stars. However, the second star makes its presence felt through its effect on the proper motion of the other star: the proper motion is not a smooth line but displays orbital wiggles (see Figure 3.39). These pairs of stars are called astrometric binaries. Indeed, before Alvan Clark saw Sirius B in 1862, its existence had been inferred by Bessel in 1844 from such wiggles in the proper motion of Sirius A.

It should be apparent from the discussion of binary stars in Section 3.2.3 that only nearby or well separated stars can be observed as visual binaries. This limits the number of stars for which we are able to obtain masses by this technique. However, it is possible to derive information on stellar masses from spectroscopic binaries (Section 3.2.3). If we assume the two stars are in circular orbits about the

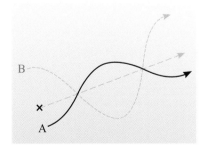

Figure 3.39 The apparent motion of an astrometric binary star. The solid track is the proper motion of a visible star against the background of distant stars. The invisible companion, B, follows the curved dashed line and the centre of mass (marked with a cross), defines the proper motion of the binary star system as a whole.

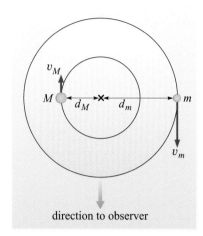

direction to observer

Figure 3.40 Circular orbits in a spectroscopic binary system relative to the centre of mass (×) of the system. The stars are at a position where the maximum separation of their spectral lines will be observed due to the Doppler shift.

centre of mass, with the orbital plane perpendicular to the plane of the sky (i.e. in the line of sight as shown in Figure 3.40), then we can derive the orbital speeds from the maximum Doppler shifts of spectral lines from the two stars (see Figure 3.9). The orbital speed of each star equals the distance around its orbit divided by the time taken for one orbit, i.e. the period P:

$$v_M = 2\pi d_M/P, \quad v_m = 2\pi d_m/P \tag{3.22}$$

Since the ratio of masses is related to the ratio of distances from the centre of mass (Equation 3.21) then from Equation 3.22 we obtain

$$M/m = d_m/d_M = v_m/v_M \tag{3.23}$$

P is obtained from the period of oscillation of the spectral lines so d_M and d_m can be calculated *individually*. The distance between the two stars, r, which in this case is also the semimajor axis, a, is simply (Figure 3.40)

$$r = a = d_M + d_m \tag{3.24}$$

We therefore can use Equations 3.20 and 3.23 to derive the masses of the two stars. Since the identification of spectroscopic binaries does not depend on distance to the binary (it only requires that the stars are bright enough for spectra to be obtained), this method is potentially very powerful. However, in general, the angle between the orbital plane and the plane of the sky is unknown and hence the observed radial velocity oscillations will be smaller than the true orbital speeds and the stellar masses will be underestimated.

▨ Are there any types of binary star for which you can be sure that the orbital plane is in the line of sight?

❑ Yes, eclipsing binary stars (Section 3.2.3) must have their orbital planes very close to the line of sight for one star to pass in front of the other.

When the above techniques are applied, measured stellar masses are found in the range from about $0.08M_\odot$ to about $50M_\odot$. We will examine the reasons for these limits in Chapter 6.

Stellar radii from eclipsing binaries

Not only can eclipsing spectroscopic binaries be used to derive masses, but they also provide an alternative technique for determining stellar radii. Figure 3.41 illustrates the geometry of an eclipsing binary eclipse with the corresponding light curve. The radii of the stars are derived from the time intervals between brightness changes in the light curve. Between time t_1 and t_2 the small star moves a distance of twice its radius, R_S, at a speed v. If the star is in a circular orbit, then v can be derived from the maximum Doppler shift as shown in Figure 3.40. During the eclipse the small star is moving along only a small part of its orbit so we can assume constant speed and therefore the distance travelled = speed × time taken:

$$2R_S = v \times (t_2 - t_1) \tag{3.25}$$

Similarly, between time t_1 and t_4 the small star moves a distance of $R_S + R_L + R_L + R_S$ at speed v. So

$$2R_S + 2R_L = v \times (t_4 - t_1) \tag{3.26}$$

Subtracting Equation 3.25 from Equation 3.26 gives

$$2R_L = v \times (t_4 - t_2) \tag{3.27}$$

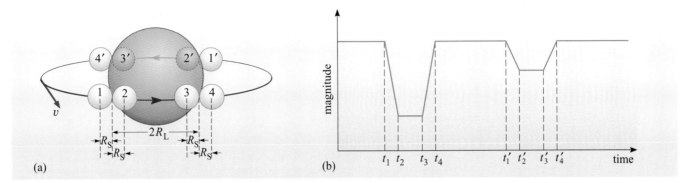

(a)

(b)

The description of spectroscopic and eclipsing binaries given above assumes circular orbits. In general the orbits are elliptical and the calculations are somewhat more complex.

3.4 Summary of Chapter 3

Stars in space

- Stars move through space with velocities that we split into transverse and radial components. The magnitude of the transverse velocity is given by

$$v_t = d \times (\mu/\text{radians}) \tag{3.1}$$

where μ is the proper motion. The magnitude of the radial velocity (determined from the Doppler shift of spectral lines) is given by

$$v_r = c \times (f - f')/f' \tag{3.3}$$

or

$$v_r = c \times (\lambda' - \lambda)/\lambda \tag{3.4}$$

- The space velocity is given by

$$v = \sqrt{(v_t^2 + v_r^2)} \tag{3.5}$$

The stars are so remote that transverse velocities produce changes to the familiar constellations that would be noticeable to the unaided eye only over intervals of thousands of years.

- Stellar distances were first measured using trigonometric parallax. The distance d is given by

$$d/\text{pc} = \frac{1}{p/\text{arcsec}} \tag{3.7}$$

where p is the stellar parallax. In this equation pc stands for parsec, an important unit of distance in astronomy.

- The distances between the stars are typically a few parsecs, of the order of a million times greater than the average distance from the Earth to the Sun (called the astronomical unit, AU). Trigonometric parallax has yielded stellar distances with useful accuracy to a range of over 100 pc.

- More than half of all stars are found in binary or multiple star systems. Only the closest and most widely separated can actually be distinguished as separate stars. Most are only identified through indirect methods such as changes in their spectra or brightness.

- Two types of star clusters (open clusters and globular clusters), with very different properties are observed in our Galaxy.

Figure 3.41 (a) Schematic of an eclipsing binary showing the different phases of eclipses. (b) Light curve indicating times of each event during the eclipses. Primary minimum occurs when the hotter star is eclipsed since the light emitted per unit area is greater for the hotter star. The example shows the case when the larger star is the hotter one.

Stellar radius

- The stars with the greatest angular diameters as seen from the Earth can have their angular diameters α directly measured. The stellar radius is then given by

$$R = [(\alpha/2)/\text{radians}] \times d \tag{3.8}$$

where d is the distance to the star.

Photospheric temperature

- Photospheric temperature can be obtained by comparing the flux density that we receive from a star in two different wavebands – this is the basis of the photometric method. However, for most stars a better method is to use the strengths of various absorption lines produced by different elements in, and just above, the photosphere. From these lines, a star is assigned to one of the spectral classes O, B, A, F, G, K, M and L and then to one of the subclasses, the spectral types. The average photospheric temperature is then obtained from the relationship between spectral type and temperature. This is the spectrometric method.

Stellar luminosity

- If we know the radius and temperature of a star, then its luminosity (its total power output over all wavelengths) can be obtained from

$$L \approx 4\pi R^2 \sigma T^4 \tag{3.9}$$

The approximation sign arises because a star's spectrum is not quite that of a black body.

- If we know the distance d from a star to the Earth then we can work out the luminosity by measuring the flux density, F

$$L = (4\pi d^2)F \tag{3.11}$$

- In general, we don't know the distance of a star, and Equation 3.11 is itself used to determine distance if the luminosity can be derived by an independent method

$$d = [L/(4\pi F)]^{1/2} \tag{3.12}$$

- F is almost impossible to measure so a restricted range of wavelengths, defined by a passband, is used. Many stars emit much of their energy in visible light where the Earth's atmosphere is transparent so the V band is commonly used

$$L_V = (4\pi d^2)F_V \tag{3.13}$$

Magnitudes

- Magnitudes are a convenient way of expressing the brightness of stars. The magnitude scale is logarithmic and 'upside-down'. The difference in magnitudes of two stars is related to the ratio of their brightnesses

$$(m_1 - m_2) = -2.5 \log(b_1/b_2) \tag{3.15}$$

- Apparent magnitudes, m, are a measure of the brightness of a star observed at the Earth. Absolute magnitudes, M, are the apparent magnitude an object would have if moved to a standard distance of 10 pc

$$M = m - 5 \log d + 5 \tag{3.16}$$

- Magnitudes measured in the visual waveband are denoted m_V (or simply V) and M_V. The difference in blue and visual magnitudes, the colour index $B - V$, is an indicator of temperature.

Stellar spectra

- The luminosity of a star can be estimated from the width of its absorption lines, and from the strength of the absorption lines of certain ions.

- We are therefore able to determine both the temperature and luminosity of a star from its spectrum and can therefore derive its radius:

$$R \approx [L/(4\pi\sigma T^4)]^{1/2}$$

- From L and T we can obtain the luminosity L_V in the V waveband. We can readily measure the flux density F_V that we receive from the star in this band, and therefore the distance is given by

$$d = [L_V/(4\pi F_V)]^{1/2} \tag{3.14}$$

Sometimes it is necessary to correct F_V for the effects of interstellar matter before applying Equation 3.14.

Variable stars

- Some stars change in brightness. These variations may be regular with periods of years or as short as fractions of a second, or irregular.

- Cepheids exhibit a relationship between period and luminosity which allows their distances to be determined by measuring their light curves.

Stellar composition

- The great majority of stars, apart from those in which considerable nuclear fusion has occurred, consist mainly of the two lightest elements, hydrogen and helium. The Sun is fairly typical, with the following composition outside its core:

  hydrogen: 73% by mass, 92% by number of nuclei

  helium: 25% by mass, 7.8% by number of nuclei

- The composition of the Sun, excluding its core, and that of a class of meteorites called carbonaceous chondrites, define a standard composition called the Solar System abundance of the elements.

Stellar masses

- Stellar masses can be obtained from observations of binary systems. This is most directly accomplished using visual binaries, the masses being given by

$$M + m = \frac{4\pi^2 a^3}{GP^2} \tag{3.20}$$

$$M/m = d_m/d_M \tag{3.21}$$

- Masses of spectroscopic binaries may be derived using the same equations but only if they are also eclipsing binaries. These allow determination of masses for much more distant stars.

- The light curves and spectra of eclipsing binaries also provide a method of determining the radii of the two stars.

Questions

QUESTION 3.13

The bright star Procyon A ('pro-sigh-on') has a proper motion of 1.259 arcsec yr^{-1}, and a stellar parallax of 0.286 arcsec. Its spectral lines are blue-shifted.

(a) Calculate how many years this star will take to travel an angular distance across the sky equal to the angular diameter of the Moon (0.5°).

(b) Calculate its transverse speed, expressing your answer in $\mathrm{km\,s^{-1}}$.

(c) Is its radial velocity directed towards us, or away from us?

(d) Why should the large proper motion of this star suggest that it should have a readily measurable parallax?

QUESTION 3.14

(a) Sirius is a binary with stars of very different apparent brightness. Sirius A appears 7600 times brighter than Sirius B. If Sirius A has an apparent visual magnitude $V = -1.46$, what is the apparent visual magnitude of Sirius B?

(b) What can you say about the difference in absolute visual magnitudes of the two stars?

(c) Sirius lies at a distance of 2.64 pc. What is the absolute visual magnitude of Sirius A?

(d) How does its luminosity compare with that of the Sun?

QUESTION 3.15

This question is about the star Rigel A, the prominent bluish-white star in the constellation Orion.

(a) From its spectral type, B8, Rigel A is given a photospheric temperature of 13 000 K. Is this reasonable? How strong are the hydrogen Balmer lines compared with the helium lines, and with the lines of ionized helium and ionized calcium?

(b) Spectroscopic studies lead to an estimate of $1.4 \times 10^5 L_\odot$ for its luminosity. How do its spectral lines differ from those of a far less luminous star of the same temperature?

(c) Calculate its radius, expressing your answer in solar radii.

(d) From its luminosity and photospheric temperature, use Figures 3.18 and 3.28 to make a *rough* estimate of its luminosity, L_V, in the V waveband. Hence estimate its distance, given that $F_V = 3.0 \times 10^{-9}\ \mathrm{W\,m^{-2}}$.

(e) Would it have been feasible to obtain its radius from its angular diameter, as measured from the Earth?

QUESTION 3.16

Using Appendix A5, plot a graph of relative elemental abundances in the Solar System by mass versus atomic number Z, for the heavy elements as far as $Z = 30$. Use a linear scale on the abundance axis extending from 0 to a suitable upper limit. (Some of the abundances will be too small to show.) Hence describe briefly the relationship between cosmic relative abundance and Z.

QUESTION 3.17

In the Sirius binary system, the orbital period is 50 years, and the semimajor axis of the relative orbit is 20 AU. From these data, and from Figure 3.37, calculate the masses of Sirius A and Sirius B, expressing your answer in solar masses.

CHAPTER 4
COMPARING STARS

4.1 Introduction

So far, we have been largely concerned with the individual properties of individual stars, in particular photospheric temperature, luminosity, radius, composition and mass. If we wish to understand more about stars and obtain some insight into their evolution, we need to look at the overall distribution of stellar properties. We would like to know the answers to such questions as 'Can stars have any combination of these properties?' and 'How many stars are there of each type?' We can potentially learn a lot more about the stars if we compare them, but what should be the basis of our comparison? We certainly want to use intrinsic properties, such as luminosity, and not properties that depend on the distance to the star, such as the flux density received on Earth. Also, as an initial step, we want to avoid properties that are well removed from what we actually observe. In this chapter we look at probably the most important diagram in stellar astronomy, the Hertzsprung–Russell diagram, and how it is used to identify the main classes of stars.

4.2 The Hertzsprung–Russell diagram

4.2.1 Constructing the H–R diagram

Three properties which are suitable for comparing stars are temperature, luminosity and radius. However, we don't need all three.

■ Why not?

❑ Since stars emit like black bodies, temperature, luminosity and radius are related, via Equation 3.9. Thus, if we know any two, we can obtain the third.

Temperature and luminosity are more directly measurable for a far greater number of stars than radius, and so it is these two properties that are used, as shown in Figure 4.1. Each point displays the temperature and luminosity of a particular star: you should check that the values given for the Sun are in accord with the values given earlier. *Note the logarithmic scales on both axes, and that temperature increases to the left.*

Such a diagram is called a **Hertzsprung–Russell diagram**, or H–R diagram, after the Danish astronomer Ejnar Hertzsprung and the US astronomer Henry Norris Russell.

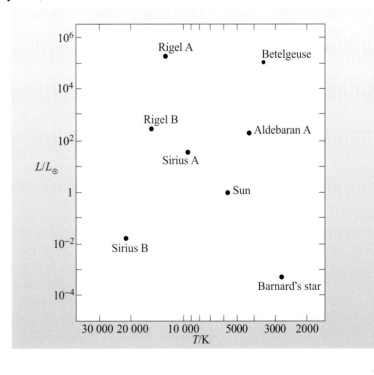

Figure 4.1 The Hertzsprung–Russell diagram for the Sun and a few nearby stars.

EJNAR HERTZSPRUNG (1873–1967) AND HENRY NORRIS RUSSELL (1877–1957)

(a)

(b)

Ejnar Hertzsprung (Figure 4.2a) born in Denmark, initially chose chemical engineering as a career because of the poor financial prospects in astronomy. However, after developing his astronomical skills as a private astronomer, he became Assistant Professor of Astronomy at Göttingen Observatory in Germany and then Professor and Director of Leiden Observatory in the Netherlands. He proposed the concept of the absolute magnitude of a star as its magnitude at a distance of 10 parsecs (Section 3.3.3). In 1906 he plotted a graph of the relationship between the absolute magnitudes and colour of stars in the Pleiades and coined the terms red giant and red dwarf. He published his work in a photographic journal without the diagrams and they were unknown to other astronomers.

In 1913 Henry Norris Russell (Figure 4.2b), then Director of the University Observatory at Princeton, plotted Annie Cannon's (Figure 3.24) spectral classification against absolute magnitude and found that most stars lay in certain regions of the diagram. The diagram, which became a fundamental tool of modern stellar astronomy, was eventually called the Hertzsprung–Russell diagram in recognition of their independent work. One of the first applications of the H–R diagram was in the development of spectroscopic parallax (Section 3.3.4) by Hertzsprung, using observations of Cepheid variable stars made by Henrietta Leavitt (Figure 3.32).

Figure 4.2 (a) Ejnar Hertzsprung and (b) Henry Norris Russell. ((a) Royal Astronomical Society; (b) Science Photo Library)

QUESTION 4.1

Where, in the H–R diagram, do the following types of star appear: hot, high luminosity stars; hot, low luminosity stars; cool, low luminosity stars; cool, high luminosity stars?

The H–R diagram in Figure 4.1 contains too few stars to give us an overall picture. Before we examine a diagram containing many more stars we can speculate on what we might find. Will we find that the stars are fairly uniformly peppered over the diagram, with, for example, as many hot, high luminosity stars as any other kind? Or will we find that certain combinations of luminosity and temperature are more common than others? In any general population there are usually more small things than big things, more faint things than bright, and more cool things than hot. Therefore we might expect there to be more stars towards the bottom of the H–R diagram and more towards the right. To some extent these explanations are borne out but with some surprises. When more data are plotted, more stars are found towards the bottom right of the H–R diagram (Figure 4.3) but there are also noticeable empty zones, and a striking locus from hot bright to cool faint stars. The shaded regions show where stars tend to concentrate: the darker the shading, the greater the concentration. Each concentration defines a particular class of stars, and we shall shortly examine each main class in more detail, but first let's add stellar radius to Figure 4.3.

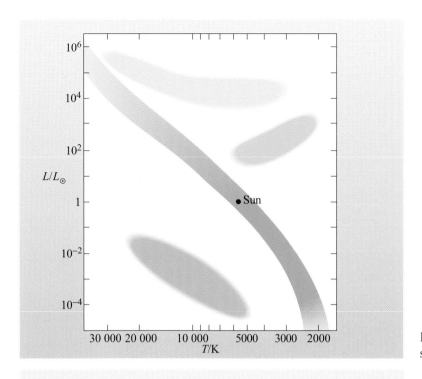

Figure 4.3 An H–R diagram, showing where stars tend to concentrate.

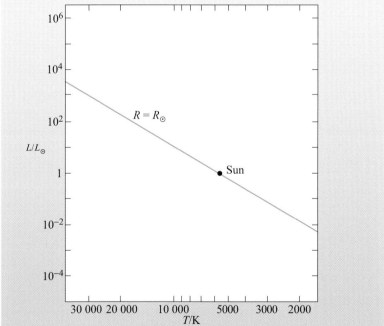

Figure 4.4 An H–R diagram, showing where stars of solar radius lie.

From the relationship between radius, temperature and luminosity in Equation 3.9, we see that at each point in the H–R diagram there is a unique stellar radius, given by $R = [L/(4\pi\sigma T^4)]^{1/2}$. Let's now add to the diagram lines of constant radius. For example, consider stars with a radius equal to that of the Sun, $R_\odot$. From Equation 3.9 we see that any other star with the same radius will have its luminosity and temperature related by $L \approx (4\pi R_\odot^2 \sigma) T^4$. Thus, as T increases, L also increases since for a given radius, the hotter the star the more power it radiates. With T increasing to the left in the H–R diagram, this gives a line sloping upwards from lower right to upper left, as in Figure 4.4. The line is straight because we are using logarithmic scales.

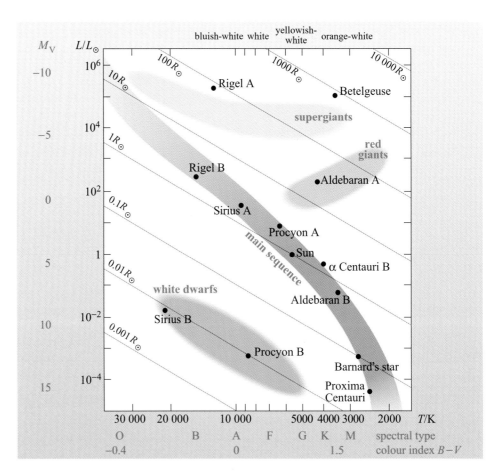

Figure 4.5 The H–R diagram in Figure 4.3, with the addition of stellar radii, and other information. (Adapted from Seeds, 1984)

Figure 4.5 is the H–R diagram in Figure 4.3 with several lines of constant radius added, and you can see that there are some classes of stars that are considerably smaller and some that are considerably larger than the Sun. These relative sizes are reflected in the names given to many of the classes, as shown in Figure 4.5; white dwarfs, red giants, supergiants. As you might expect, white dwarfs are small, red giants are large, and supergiants even larger.

QUESTION 4.2

In terms of the Earth's radius, and the Earth's distance from the Sun, how large are white dwarfs and red giants?

The class names are descriptive in ways other than size.

▨ Why *white* dwarfs, *red* giants?

▢ White dwarfs have temperatures that result in yellowish-white to bluish-white colours. Red giants have tints towards the red end of the visible spectrum, embracing orange-white and yellowish-white.

We have added to Figure 4.5 an indication of the colour associated with each temperature. However, to the unaided eye, star colours are in many cases not very striking. This is partly because too little light is being received for our colour vision to be strongly excited, and partly because the colours are, in any case, rather weak. However, stellar colours can be emphasized photographically (as you will see in Figure 4.14).

The H–R diagram can be represented in a number of different ways. As you discovered in Section 3.3.3 when we introduced the magnitude system, the colour index of a star is a measure of its temperature. The spectral classification scheme outlined in Section 3.3.2 is also a temperature sequence. Another way of representing stellar luminosity is through the absolute visual magnitude. It should be clear therefore that the luminosity axis of the H–R diagram could equally well be plotted as absolute visual magnitude and the temperature axis with spectral type or colour index. These kinds of diagrams are often used by astronomers as these are quantities that are obtained more directly from observation. A particular type of H–R diagram, called a colour-magnitude diagram, shows M_V against $B - V$. Figure 4.5 illustrates the approximate values of absolute visual magnitude and colour index as well as spectral type. The exact appearance of the H–R diagram will be slightly different when these alternative axes are used (e.g. the absolute visual magnitude M_V is directly related to the luminosity in the V band, L_V, and not the total luminosity L, as explained in Section 3.3.3). Also, spectral type depends weakly on luminosity as explained in Section 3.3.4.

Let's now look at the main classes of stars in more detail.

4.2.2 The main classes of stars

The main classes of stars are shown in Figure 4.5.

The **main sequence** is 'main' in the sense that about 90% of stars fall into this class, and 'sequence' in the sense that it is a long, thin region that trails across the H–R diagram, covering a very wide range of temperatures and luminosities. The Sun is a main sequence star, of very modest temperature and luminosity, and correspondingly modest radius. It is yellowish-white. Sirius A is a main sequence star rather hotter than the Sun, and appears bluish-white. As you learnt in Chapter 3, it has the greatest apparent visual brightness (most negative apparent visual magnitude!) of any star in the night sky. This is, as we have seen, not because it is very luminous, but because it is both fairly luminous and rather close – at 2.63 pc it's the seventh closest star after the Sun (Table 3.1).

Above the lower part of the main sequence we come first to the **red giants**. These stars are cool, hence their orange tinge, and are of order 10 to 100 times larger in radius than main sequence stars of comparable temperatures. Thus if our Sun were a large red giant, its surface would extend a considerable distance towards the Earth (as we saw in Question 4.2)!

- ▪ If you knew that a red giant was larger than a main sequence star of comparable temperature, what could you say about its luminosity?

- ❑ From Equation 3.9 we could say that its luminosity is greater than that of the main sequence star. (This conclusion is borne out by Figure 4.5.)

The bright star Aldebaran A (α Tau) is a red giant. (It's actually a visual binary, but the red giant is dominant.)

Above and to the left of the red giants we come to the **supergiants**. These are larger, and thus more luminous than red giants of comparable temperature, but they also extend to higher temperatures, where they are larger and more luminous than main sequence stars of comparable temperature. Rigel A is a hot supergiant, which appears bluish-white whereas Betelgeuse is a cooler supergiant, and it appears distinctly orange-white (see Figure 3.1).

Though we have not marked it on Figure 4.5, there is a class of stars that comprises the red giants plus the stars to their left that lie between the main sequence and the supergiants. These are the **giants**. In later chapters we shall often refer to this class, which is broader than that of *red* giants alone.

You can see from Figure 4.5 that **white dwarfs** are, as their name implies, hot and small, only about the size of the Earth (Question 4.2). Consequently their luminosities are low. Indeed, there are no white dwarfs sufficiently close to us to be visible to the unaided eye. The closest is Sirius B, the faint companion to Sirius A, but its visual magnitude is only 8.4, well outside the limit of about 6 for very good, unaided human eyes, in the very best observing conditions (Figure 3.29). Even if it were a bit brighter, its light would be swamped by Sirius A, and we would still be unable to see it.

In Section 3.3.4 you learnt how the width of spectral lines gives an indication of the luminosity of a star. We can now see how this is reflected in the H–R diagram and the description of stellar spectral types. Giant stars have narrower spectral lines than dwarf stars and stronger lines due to certain ionized atoms. These characteristics are used to define a **luminosity class**, designated by roman numerals I to V, with I being brightest. Class I is often sub-divided into Ia and Ib.

Figure 4.6 illustrates the positions of these luminosity classes on the H–R diagram.

▓ From your knowledge of the Sun's position on the H–R diagram, what is its luminosity class?

▢ The Sun is a main sequence star so its luminosity class is V.

The full designation of a star's spectral type also includes its luminosity class. The Sun is spectral type G2 V and Betelgeuse is spectral type M2 Ia.

White dwarfs are usually designated by a prefix 'D' or 'w' as in the case of Sirius B which is spectral type D A5 or w A5. Other suffixes are used for special

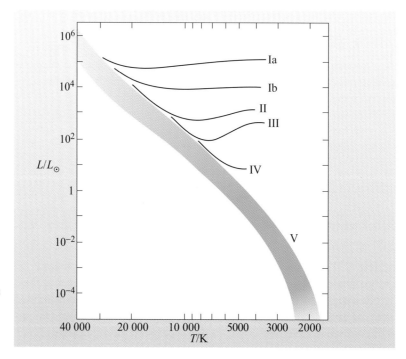

Figure 4.6 The H–R diagram indicating the approximate positions of luminosity classes I to V. Luminosity class V applies to the whole of the main sequence.

characteristics such as 'e' for emission lines or 'p' for peculiar spectrum. The spectral types of the brightest stars as seen from the Earth are listed in Appendix A4.

The tendency for stars to concentrate into certain regions of the H–R diagram is clearly meaningful. But what does it mean?

4.2.3 How can we explain the distribution of stars on the H–R diagram?

Here is a possible explanation for the concentration of stars into certain regions on the H–R diagram. It is based on the reasonable assumptions that:

- Any particular star is luminous for only a finite time;
- There are distinct stages between the star's cradle and grave, each stage being characterized by some range of temperature and luminosity; the star thus moves around the H–R diagram as it evolves;
- The stars we see today are not all at the same stage of evolution.

From these reasonable assumptions it follows that if we observe a large population of stars today, then the longer a particular stage lasts the greater will be the number of stars that are observed in that stage. Conversely, we will catch very few stars going through a short-lived stage.

We can thus explain the concentrations on the H–R diagram as those regions where the stars spend a comparatively large fraction of their lives. On this basis a star must spend most of its life on the main sequence, because this is where about 90% of the stars lie. Where it lies before it joins the main sequence, and where it goes afterwards, we cannot tell without further information, but the red giant, supergiant and white dwarf regions are where, on our assumptions, we might expect some stars to dwell for a while.

There are two other factors that influence the concentrations of stars on the H–R diagram. First, the concentration depends not only on how quickly a star passes through a region, but also on what fraction of stars pass through the region at all. Second, some regions of the H–R diagram might be bereft of stars simply because they correspond to stages in a stellar lifetime when stars tend to be shrouded in cooler material and are therefore not observable directly.

We clearly need more observational data to make further progress. Observations of individual stars actually evolving would be of enormous value. Can we see such evolution by making observations over a period of time?

Unfortunately, with very few exceptions, we can't. This is because stars evolve extremely slowly. We have good evidence (Chapter 2) that the Sun is about 4.5×10^9 years old, and that it will be about as long again before it runs out of hydrogen fuel in its core. The lifetime of an astronomer, or indeed the whole history of astronomy, are both tiny fractions of this 4.5×10^9 year timescale. Changes in the Sun and other stars in short times are usually small. No matter how obvious the changes in the Sun which are set out in Chapter 2 are to us, if we had to view the Sun as a star from a great distance they would be insignificant. However, some stars do change on short timescales – the spectacular supernovae and variable stars (Section 3.3.5).

One type of supernova, the **Type II supernova** (described further in Section 8.3), marks the end of a supergiant star. Thus, Betelgeuse and Rigel A seem fated to disappear after a final blaze of glory, their luminosity rising 10^8 times in a few days, followed by a few months of decline into oblivion, when they will vanish from the sky and from the H–R diagram.

All types of novae, which exhibit one or more short-lived outbursts, are in binary systems. In a minority of binary systems the two stars are so close together that they interfere with each other's evolution (you will learn more about this in Chapter 9). In some cases, this will lead to one of the two undergoing a nova outburst. Observations of novae thus help us to understand disturbances to the normal course of stellar evolution, and this also helps us to understand the normal course itself.

The irregular variable T Tauri stars lie just above the main sequence on the H–R diagram (Figure 4.7) in a zone that covers a wide range of temperatures, including that of the Sun, and they lie among traces of the sort of interstellar material from which stars are thought to form. These observations suggest strongly that they are very young stars, about to settle on to the main sequence. Indeed, some T Tauri stars probably have been seen to do just this. Therefore, the early phase of stellar evolution can be elucidated by the study of these stars. You will learn more about T Tauri stars in Chapter 5.

Cepheids (Section 3.3.5) and other types of regular pulsating variable stars also help us to understand some of the processes that drive evolution at certain stages in a star's life, as you will see in Chapter 7.

Have we now exhausted the main sources of observational data that help us to build models of the stars and of their evolution? No, there is one further property of enormous importance, and this is a star's mass (Section 3.3.7).

QUESTION 4.3

If most stars were to end their lives quietly, by gradually cooling at roughly constant radius, what sort of tracks would they make across the H–R diagram?

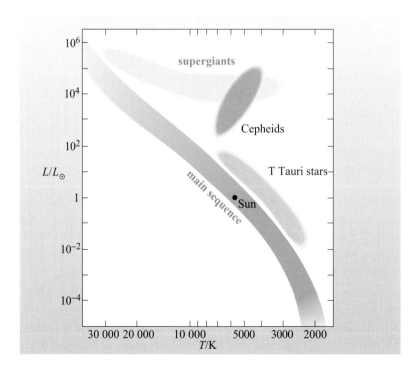

Figure 4.7 An H–R diagram, showing where the T Tauri stars and Cepheids lie.

4.2.4 Stellar masses and stellar evolution

Measured masses range from about $0.08M_\odot$ to about $50M_\odot$, a large range, with the Sun again showing up as an average sort of star. At the upper end we have some true monsters, but even at the lower end we have bodies that are still far more massive than the planets.

◼ What is the mass of a $0.08M_\odot$ star, in Earth masses?

❑ Nearly 30 000 Earth masses.

The lower the mass the greater the number of stars; the monsters are rare, and stars less massive than the Sun are more common than stars of around solar mass. These relative numbers, and the upper and lower mass limits, are all things that the stellar theories in Chapters 5 to 9 have to explain.

We can, however, throw some light on stellar evolution if we plot stellar masses on an H–R diagram. This is done in Figure 4.8, where a handful of representative stellar masses have been included. Note the following important features.

● The supergiants tend to be more massive than the red giants, which in turn tend to be more massive than the white dwarfs.

● Within each of the supergiant, red giant, and white dwarf classes, there is no correlation of mass with luminosity or photospheric temperature – the relationship is jumbled.

● Among the main sequence stars, mass correlates closely with luminosity, and hence with temperature: as mass increases, luminosity and temperature increase. (The increase in luminosity is enormous: the 500 to 1 increase in mass along the main sequence corresponds to a 10^{10} increase in luminosity.)

● In the lower part of the main sequence, the masses are comparable with the red giants, and in the upper part, with the supergiants.

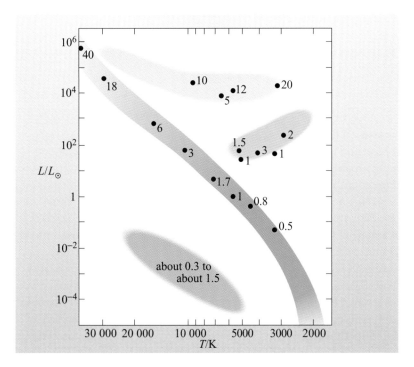

Figure 4.8 Stellar mass and the H–R diagram. Masses are given in multiples of $M_\odot$.

Before we try to construct a model of stellar evolution based on these striking features, we have to address the question 'do stars change their mass during their evolution?' There is a good deal of observational evidence to help us to answer it. We observe main sequence stars, red giants and supergiants losing mass in the form of stellar winds streaming outwards. However, the accumulated totals of mass lost by stellar winds are estimated to be only a small fraction of the initial mass of a star. A more impressive mass loss is shown in Figure 4.9, where you can see shells of material that have been flung off by the central star. Such an object is misleadingly called a planetary nebula (plural: planetary nebulae), because it looks a bit like a planetary disc when viewed under low magnification. They can account for a substantial fraction of the star's mass. In passing, we note that the central star of a planetary nebula now occupies a region in the H–R diagram somewhat hotter and more luminous than the white dwarfs, and it is plausible that it could cool to become a white dwarf.

Figure 4.9 A planetary nebula: The Helix nebula is the result of a star losing its outer layers at the end of its life. The gas is really in a shell about the remnant of the star but it appears as a ring because we see through it most easily in the direction of our line of sight to the central star. (D. Malin/AAO)

Some stars end their lives more violently than by shedding a planetary nebula.

- What stars are these, and how do they end their lives?
- Supergiants, which end their lives as Type II supernovae.

In fact, in a Type II supernova, most of the star's mass is blown away.

It thus seems to be the case that throughout most of the life of a star, severe mass loss occurs only when a planetary nebula is shed, with the resulting stellar remnant becoming a white dwarf, or when a massive star ends its life as a Type II supernova.

We are now in a position to suggest a plausible model for some of the stages of stellar evolution based on the features listed above, and on what we know about mass loss. During its main sequence phase, a star does not change its luminosity or photospheric temperature very much, otherwise it would move a good way along the main sequence, and this does not fit in with the large differences in mass along the main sequence (in fact, stars do drift very slightly above the main sequence, so it is a band rather than a narrow line on the H–R diagram). After the main sequence phase the less massive stars become red giants, and the more massive stars become supergiants: you can see that this is consistent with the masses in Figure 4.8. It is also consistent with the rarity of supergiants: there are very few main sequence precursors. Finally, red giants evolve to the point where they shed planetary nebulae, the stellar remnant evolving to become a white dwarf. Supergiants become star-destroying Type II supernovae.

We are thus continuing to unfold the story of stellar evolution. But there is one huge aspect of the story that, as yet, we have barely touched, and this is whether stars of different mass all evolve at about the same rate. Star clusters provide good observational evidence to help answer this question.

4.2.5 Star clusters and stellar evolution

Detailed observations of star clusters (introduced in Section 3.2.4) suggest that they occur because the stars in them form at about the same time. Moreover, the compositions of the stars are similar. Isolated stars (including isolated binary stars) result from the later partial or complete dispersal of a cluster.

The crucial points for us here are that all the stars in a cluster formed at about the same time, and all have similar compositions.

■ Why are these the crucial points?

❏ If the stars in a cluster have different masses, then we can discover the relative rates of evolution of stars that differ only in their mass.

These relative rates are conveniently revealed by plotting the H–R diagram of a cluster. Figure 4.10 shows two contrasting cases: the Pleiades (Figure 3.14), and a cluster that has only a catalogue number, M67 (the 67th object in a catalogue of nebulae that may be confused with comets, produced by French comet hunter Charles Messier (1730–1817)). In the case of the Pleiades, almost all the stars are on the main sequence, suggesting that this cluster is not old enough for many stars to have reached the end of this phase. The most luminous stars visible on this diagram appear to be moving away from the main sequence. The upper end of the main sequence, where the most massive stars are expected to lie (Figure 4.8), is unpopulated in this cluster. For the Pleiades, the most massive stars have already

Figure 4.10 The H–R diagrams of two star clusters: (a) The Pleiades; (b) M67.

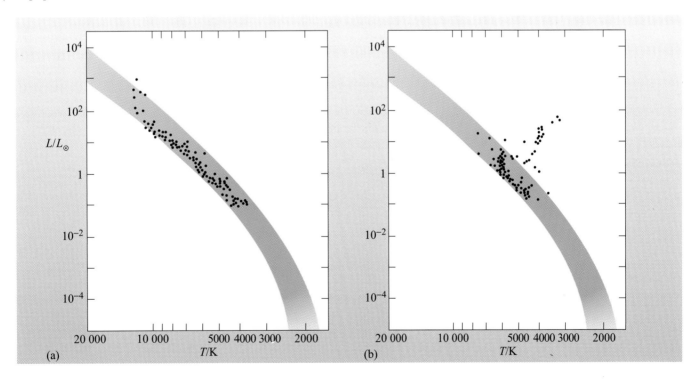

(a)

(b)

left the main sequence and therefore must have shorter main sequence lifetimes. In fact, the point at which this depopulation occurs, called the **main sequence turn-off**, is used as an indicator of the ages of clusters. The case of M67 (Figure 4.11) is the subject of Question 4.4.

Figure 4.11 This cluster, M67, is one of the oldest open clusters known, at around 3×10^9 years (almost the age of the Sun). (N. A. Sharp, M. Hanna/NOAO/AURA/NSF)

We have come a long way in constructing a plausible model of stellar evolution, and it is summarized in Figure 4.12. We have described some of the observational basis of this model; further observations not only support it, but fill in some of the missing details. However, the time has now come to move away from pure observations as the sole basis for model building, and to involve a powerful body of physical theory in the modelling process. In the following chapters we thus continue to develop the story of the stars, and of their evolution, but with considerable reliance on physical theory. This will necessarily involve us in modelling not only external events, but also stellar interiors, which will be addressed in Chapter 6.

QUESTION 4.4

Figure 4.10b shows the H–R diagram of the star cluster M67. (a) Discuss whether this is consistent with the model of stellar evolution in Figure 4.12. (b) Why is it reasonable to conclude that M67 is older than the Pleiades?

Figure 4.12 A model for stellar evolution.

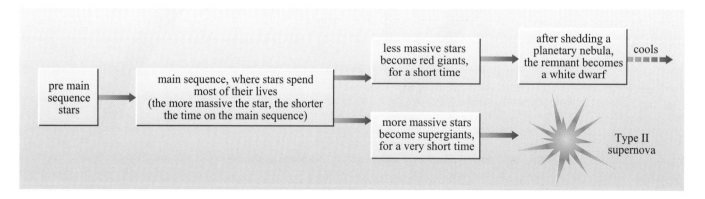

4.3 Observing through the interstellar medium

In all the analysis of stellar properties discussed so far we have made an implicit assumption – that light emitted by a star is not changed between its emission and its arrival outside the Earth's atmosphere, except by the inverse square law (i.e. it is reduced by a factor of d^2, where d is the distance to the star, Equation 3.10) and by the Doppler effect (Section 3.2.1). However, this may not be the case.

▪ What will happen to the position of a star on the H–R diagram if interstellar material causes a reduction in its brightness?

❏ If the interstellar material causes light to be absorbed then the star will appear fainter and hence be placed lower than its true position on the H–R diagram.

We will now investigate some of the properties of the interstellar material and how it affects the radiation we observe from stars.

4.3.1 Interstellar space is not empty

The difference between the apparent brightness of a star (as measured by its apparent magnitude), and its luminosity (represented by its absolute magnitude) is defined by the distance of the star. We can explicitly state this relationship as in Equations 3.11 and 3.16. However, in stating this relationship we are making the assumption that there is no intervening material that could alter the amount of light from the star that reaches the observer. In fact, interstellar space is not empty and some light is absorbed by gas and dust.

Let's imagine a star for which the flux density F_V is measured and its luminosity is derived using the method of spectroscopic parallax (Section 3.3.4).

▪ If we derive the distance of the star using Equation 3.14, $d = [L_V/(4\pi F_V)]^{1/2}$, how would the interstellar absorption affect the result?

❏ The absorption by the interstellar material would make the star appear fainter (F_V smaller) and hence the derived distance would be too large.

Conversely, if the distance to a star is known then the luminosity of the star will be underestimated if there is interstellar absorption present that is not accounted for.

In order to take account of this absorption, Equation 3.16 is written

$$M = m - 5 \log d + 5 - A \tag{4.1}$$

where A is the absorption in magnitudes. The value of A depends on the amount of material between the star and the observer and how efficiently that material absorbs the light. That efficiency depends on the composition of the material and the wavelength of light being observed.

We have used the term absorption rather loosely here. In fact, there are a range of processes which remove energy from the beam of light coming from the star in the direction of the observer (and some that add to it!).

Figure 4.13 The Orion Nebula. The gas, mainly hydrogen, is made to glow, in the main, by four very bright, massive stars that are located in the centre of the brightest region. These stars are called 'The Trapezium' and are part of a very young cluster of a few hundred stars born less than a million years ago. The dense cloud that gave birth to this cluster is apparent through the obscuration caused by the dust in it. The glowing gas is just on our side of the cloud and is material left over after star formation. (NASA)

Figure 4.14 A panorama of the Southern Skies in the direction of the centre of the Galaxy. The dark region at centre right is known as the Coal Sack. It is not a star-free tunnel but a cool dense cloud, the dust in it obscuring the light from the stars behind. The reddish glow at far right is the Carina Nebula, a glowing gas cloud lit by young stars embedded in it. Near the Coal Sack is the famous Southern Cross. Note that the different star colours have been exaggerated in this image. (Photo: Akira, Fujii, Tokyo)

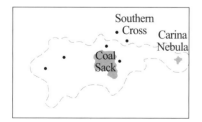

Until as recently as the 1920s, most astronomers believed that interstellar matter was confined to a handful of isolated clouds, some glowing brightly (e.g. the Orion nebula, shown in Figure 3.1 and in close-up in Figure 4.13) and some, through their obscuration of stars, appearing dark, as in Figure 4.14. The truth began to emerge from long-exposure photographs, which showed that such clouds are far more common than had previously been thought. Furthermore, by 1930 it had become clear that interstellar matter is not confined to such clouds, but is widespread in the spaces between them. There were three pieces of evidence for this.

First, it had already been observed that, in many directions in space, there are absorption lines in stellar spectra that, for various reasons, could not have originated in the stellar atmospheres, but must have originated in cool gas between us and the star. For example, the lines are very narrow, suggesting that they originate in a medium far cooler and less dense than a stellar atmosphere. In a stellar atmosphere the higher random thermal speeds of atoms or ions mean that they may be moving towards or away from an observer when they are absorbing photons. This causes a blue- or red-shift (due to the Doppler effect – Section 3.2.1) relative to the average position of the line. In addition, the higher densities in stellar atmospheres cause broadening of lines ('pressure broadening' – see Section 3.3.4).

Secondly, a characteristic type of attenuation of starlight had been observed in many directions in space, and it had been shown that this is caused by dust particles with sizes of the order of the wavelength of visible light, about 10^{-6} m. This dust attenuates starlight, partly by absorbing it and partly by scattering it. You can picture scattering as a process in which photons bounce off particles in random directions, and so some of the photons that were travelling towards us from the star do not reach us. **Scattering** plus **absorption** is called **extinction**.

Thirdly, not only did stars in distant clusters appear to be fainter than expected, they were also redder than expected. This change in colour is a result of the greater effectiveness of the dust grains at scattering shorter wavelengths (we will discuss this further in Section 4.3.3).

5 degrees

Today, the **interstellar medium** (often shortened to **ISM**) is studied at a great variety of wavelengths. These studies allow astronomers to determine the composition of the gas, and to infer the likely composition of the dust. Such studies also reveal the temperatures, densities, motions, and magnetic fields within the ISM. In later chapters we will discuss more details of the distribution of material in the ISM (Section 5.2.1) and the different sources of interstellar gas and dust (Section 8.4).

4.3.2 The effect of interstellar gas

You have seen that the ISM has been studied through the radiation that the gas and dust absorb, emit and scatter. Figure 4.15 summarizes the differences between these three phenomena.

Let's first consider the three phenomena in relation to the *gas*. The gas scatters very little light and so we need only consider absorption and emission of radiation. You have already met absorption and emission of photons by *atoms* (which we shall call **photoexcitation** and **photoemission**, respectively). Atoms can also be excited by collisions between each other as the result of their random thermal motion (**collisional excitation**). For thermal motion, the average translational kinetic energy of an atom E_k is related to the temperature T of the gas via

$$E_k = 3kT/2 \qquad (4.2)$$

where k is the Boltzmann constant. For a reasonable proportion of such collisions to be sufficiently energetic to excite an atom, E_k must be at least as large as the difference in energy between the excited and non-excited state, ε. So in order to obtain $E_k \geq \varepsilon$ we require

$$T \geq 2\varepsilon /(3k) \qquad (4.3)$$

for collisional excitation to be important.

The most prominent lines from atoms in the interstellar medium are those of ionized calcium (the 'H' and 'K' lines you met in Chapter 1). Although these lines are also prominent in the spectra of stars of spectral class G and K (see Figures 3.23 and 3.25) due to ionized calcium in their atmospheres, the interstellar lines have very different characteristics. They are much narrower than the stellar spectral lines. Also, even though they are often much fainter than the stellar lines they are usually observable because their wavelengths are Doppler shifted due to the difference in radial velocity of the interstellar gas and the star itself. In spectroscopic binaries you can see the interstellar lines remaining at a fixed wavelength while the stellar lines move due to their orbital motion (see Figure 3.13).

The processes of photoemission, photoexcitation and collisional excitation also operate in *molecules*, which are found in many parts of the ISM. Molecules are formed when atoms are bound together by chemical bonds. The electrons are 'shared' between the atoms (you could visualize this as an electron cloud surrounding the nuclei). The electrons in molecules occupy particular energy levels in a similar way to individual atoms (even though the electrons are 'shared'). Electronic transitions can take place leading to excitation, de-excitation and ionization of the molecule. The molecules also have discrete vibrational and rotational energy states and so can also undergo vibrational and rotational transitions. The vibrational energy states correspond to particular internuclear distances; when the distance becomes so large the atoms are no longer bound together, **dissociation** has occurred. A molecule can also rotate at different rates (and about different axes) resulting in discrete rotational energy states. At the molecular level, vibration and

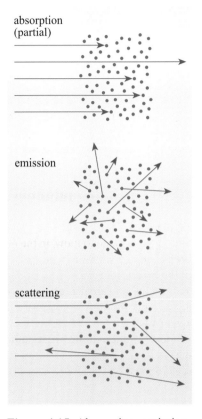

Figure 4.15 Absorption, emission and scattering of radiation.

rotation, like electron energy, is quantized, contrary to our expectations from the large-scale world where we observe an apparently continuous range of these properties. Let's look at each of these in turn. The CO (carbon monoxide) molecule is a simple case that serves to introduce the important ideas.

Figure 4.16 shows the *electronic* energy levels of the CO molecule. The levels above the lowest one correspond to the various excited states of just one of the 14 electrons that this molecule contains, in particular one of the outermost electrons, which are the least tightly bound and thus require less energy to excite them than the inner, more tightly bound electrons. For comparison, the electronic energy levels for atomic hydrogen are also shown. The excitation of a CO molecule from a lower electronic energy level to a higher one can happen through photoexcitation or through collisional excitation.

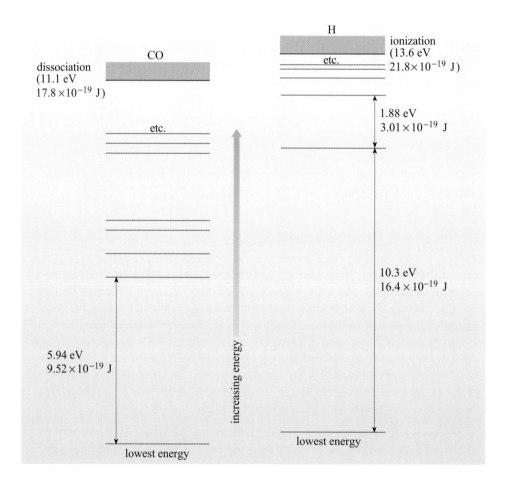

Figure 4.16 Electronic energy levels in CO and in H.

QUESTION 4.5

For the excitation of CO from its lowest electronic energy level to the one above it, calculate (a) the maximum photon wavelength for photoexcitation and (b) the minimum gas temperature for appreciable collisional excitation.

Thus, CO remains in its lowest electronic energy level unless it is exposed to photons at least as energetic as those in the near-UV region, or is at a temperature of order 10^5 K, or greater. These are the same sorts of criterion obtained for many atoms, and for many other molecules too, though in some atoms and molecules the lower electronic levels are not quite so widely spread.

Not all electronic excitations require such large energies. Thus, the higher electronic energy levels (Figure 4.16) are much more closely spaced, and excitations among them can be achieved by longer wavelength photons, and at lower temperatures.

A **vibrational transition** of CO is illustrated schematically in Figure 4.17, along with the lowest few vibrational energy levels for the case in which the molecule remains in the *electronic* state corresponding to the lowest electronic energy level. Note how much smaller are the gaps between the energy levels than is the case for the electronic transitions in Figure 4.16. This means that photoexcitation can take place at infrared (IR) wavelengths, and collisional excitation at temperatures down to the order of 10^3 K. These criteria are typical for vibrational transitions in molecules.

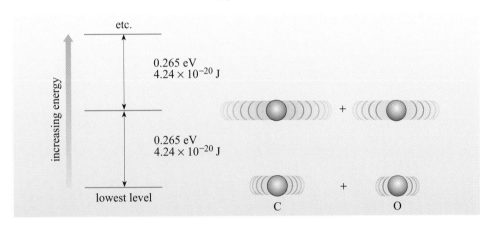

Figure 4.17 Vibrational transitions and vibrational energy levels in CO. To the right are the vibrational states corresponding to the lowest two energy levels.

A **rotational transition** of CO is illustrated schematically in Figure 4.18, along with the lowest few rotational energy levels corresponding to the lowest energy electronic and vibrational states. The energy gaps are yet smaller, and photoexcitation can now be caused by microwaves, and collisional excitation occurs at temperatures down to the order of a frigid 10 K. Again, these criteria are typical, though many molecules have even smaller rotational energy gaps, and a few have much larger gaps.

A transition from a lower to a higher energy level can also involve some combination of electronic, vibrational and rotational energy changes, necessarily so in some cases.

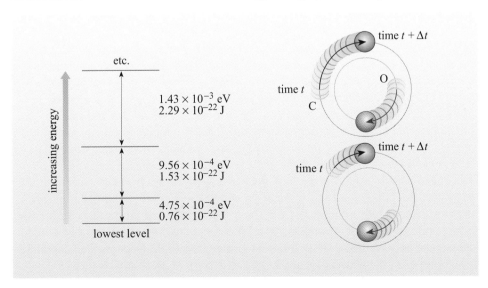

Figure 4.18 Rotational transitions, and rotational energy levels in CO. To the right are the rotational states corresponding to the lowest two energy levels.

Photoemission is the reverse process of photoexcitation, and so yields photons at wavelengths equal to those that would have caused photoexcitation between the two levels concerned.

Not all transitions involving photoexcitation and photoemission are equally probable, and so some spectral absorption and emission lines tend to be far weaker than others, and some are completely absent. Molecules consisting of two identical atoms, such as H_2, have particularly weak vibrational and rotational lines.

We have examined the processes which can cause interstellar atoms and molecules to absorb or emit radiation; let's now see what happens to starlight passing through a cloud of interstellar gas. Figure 4.19 illustrates what is seen by observers when the cloud is in the line of sight to the star and when it is out of the line of sight. Note that the prominent absorption lines in the spectrum of the star arising from the stellar photosphere (Section 3.3.2) are seen by the observer in the line of sight (Figure 4.19b) together with the superimposed, generally narrower, interstellar lines.

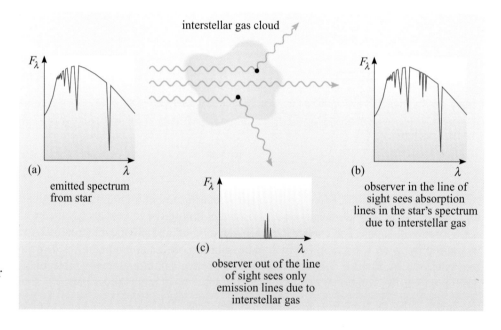

Figure 4.19 The effect of interstellar gas on radiation from a star. (a) is the spectrum emitted from the star. Spectrum (b) is seen by an observer looking at the star through the gas cloud, while spectrum (c) is seen by an observer looking at the gas cloud against a star-less background.

4.3.3 The effect of interstellar dust

Let's now consider the dust. Photoexcitation (by absorption of photons) and collisional excitation (by atoms/molecules) occur in the atoms and molecules that constitute the surface of a dust grain. Much of this energy is shared throughout the grain, raising its temperature until thermal radiation from the grain balances the energy absorbed. An alternative fate for an incident photon is to be scattered (Figure 4.15), a process that is very efficient at certain wavelengths. Figure 4.20 illustrates what is seen by observers when a cloud of interstellar dust is in the line of sight to the star and when it is out of the line of sight. The typical size of the interstellar dust grains means that they scatter short wavelengths most efficiently. This means that relatively more blue light is removed from the star's spectrum after passing through the cloud and it therefore appears redder when viewed from behind the cloud (position b in Figure 4.20). This process is called **interstellar reddening**. If the cloud is observed from out of the line of sight to a star then the dust cloud can appear as a faint blue glow from the scattered starlight (seen as the wispy blue clouds between the stars in Figure 3.14).

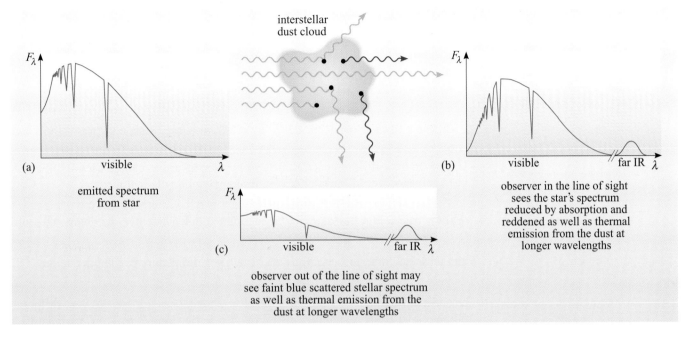

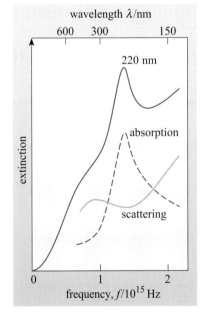

Figure 4.20 The effect of interstellar dust on radiation from a star. (a) is the spectrum emitted from the star. Spectrum (b) is seen by an observer looking at the star through the dust cloud, while spectrum (c) is seen by an observer looking at the dust cloud against a star-less background.

The combined effects of absorption and scattering (extinction) by interstellar dust are shown in Figure 4.21. Note how the extinction increases strongly through visible wavelengths, and on into the UV. Note also how broad the spectral features are, which makes it difficult to determine the composition of the dust from such spectral studies. Not much more about composition is revealed by the *emission* spectrum of the dust, which is a broad smooth thermal spectrum, depending on the dust temperature, the particle size, and only weakly on its composition. At 20 K, the dust emission lies right across the far-IR and microwave parts of the spectrum.

We have seen that absorption of starlight by interstellar dust can cause stars to appear fainter than they should and therefore cause us to overestimate their distance or underestimate their luminosity. In addition, interstellar reddening can cause stars to appear redder than they should. Since colours, as measured by the colour index (Section 3.3.3), are often used to infer temperature, the temperature can also be underestimated. If plotted on the H–R diagram a star will appear in the wrong place if the effects of interstellar absorption and reddening are not accounted for.

QUESTION 4.6

A star like our Sun is located in a star cluster at a known (large) distance and is subject to significant interstellar extinction. If its absolute visual magnitude M_V is derived from its apparent visual magnitude m_V using Equation 3.16 and its temperature determined from its observed colour index, $B - V$ (Section 3.3.3), what will be the effect on its position in the H–R diagram (Figure 4.5)? Explain how its true position can be determined if its spectrum is observed.

Figure 4.21 Extinction by interstellar dust: the top curve is the sum of the other two. The extinction is measured in magnitudes per unit distance (usually per kiloparsec).

4.3.4 Using stars to probe the interstellar medium

The effects of interstellar material on starlight can be used to probe the properties of the interstellar medium itself. A few examples are:

- The presence of particular interstellar atoms or molecules may be determined by identifying the observed spectral lines or bands.

- The temperature of the gas may be determined from the relative strengths of different lines or bands produced by different energy state changes of the same atom or molecule.

- The Doppler shift of spectral lines from interstellar gas can be used to infer the radial velocities of interstellar clouds along the line of sight to a star.

- The amount of dust along the line of sight may be inferred from the reddening if the true colour of a star can be determined independently and compared with the observed colour.

From these and other observations we have a general picture of the interstellar medium:

The chemical elements are present in relative abundances that are not very different from the Solar System abundances, introduced in Section 3.3.6. Thus, the relative percentages of atomic nuclei of hydrogen, helium and 'heavy' elements (atomic number $Z > 2$), are approximately $92\% : 7.8\% : 0.2\%$, though the proportion of an element present as ions, atoms, or combined in molecules or dust, does vary in different regions of the interstellar medium (Section 5.2.1).

Dust accounts for roughly 1% of the mass of most types of region. The particles are very small – about 10^{-7} m to 10^{-6} m in diameter – and consist of some fraction of each of the less volatile substances found in the ISM, such as carbon and silicates. In the cooler regions of the ISM, substances with greater volatility also condense to form icy coatings on the grains, so there are regional differences in the composition of the dust. The gas is always dominated by hydrogen and helium, which are abundant and very volatile.

There is a wide variety of conditions present in different regions of the interstellar medium, with temperatures ranging from a few kelvin in dense star-forming regions to 10^6 K in supernova remnants. Densities vary from $\sim10^3$ atoms per m^3 in rarefied regions of the ISM to more than 10^{10} atoms per m^3 in dense clouds.

The various types of region are far from quiescent, being racked by internal motions, and by physical and chemical transformations, often rapid compared with many astronomical changes. Each type of region is also highly structured, and far from uniform.

You will discover more about different regions of the interstellar medium when we look at the birth (Chapter 5) and death (Chapter 8) of stars.

4.4 Summary of Chapter 4

The H–R diagram

- The Hertzsprung–Russell (H–R) diagram displays the photospheric temperatures and luminosities of the stars. The corresponding radii are obtained from Equation 3.9. The H–R diagram is a very useful aid to our understanding of the stars and their evolution.

- The stars tend to concentrate into certain regions of the H–R diagram, and so some combinations of temperature and luminosity occur far more commonly than others. These concentrations define various classes of stars, the main classes being main sequence stars (about 90% of observed stars), red giants, supergiants, and white dwarfs.

- We can explain the concentrations on the H–R diagram as places where stars spend comparatively large fractions of their lives, the main sequence phase accounting for the largest fraction.

- Different types of variable stars help our understanding of stellar evolution. The supergiant phase ends in a Type II supernova – a huge explosion that destroys the star. The T Tauri stars (one type of irregular variable) seem to be on the threshold of joining the main sequence, approaching it from above on the H–R diagram. Regular variables, such as the Cepheids, give us clues about some of the processes that are of importance in stellar evolution. The novae (another type of irregular variable) help us to understand disturbances to the normal course of evolution that occur in binary systems, and this aids our understanding of the normal course itself.

Stellar masses and stellar evolution

- Measured stellar masses range from about $0.08M_\odot$ to about $50M_\odot$, with stars of lower mass being more common.

- Stars lose a rather small fraction of their masses during much of their lifetimes, but much larger fractions when they shed planetary nebulae, or when they undergo supernova explosions.

- When stellar masses are placed on an H–R diagram, and coupled with observations of mass loss, we obtain important clues to stellar evolution, leading us to a plausible model of some of the stages, as follows:

    after the main sequence phase the less massive stars become red giants, and the more massive stars become supergiants

    red giants evolve to the point where they shed planetary nebulae, the stellar remnant evolving to become a white dwarf

    supergiants end their lives as star-destroying Type II supernovae.

Star clusters and stellar evolution

- Since all the stars in a cluster formed at about the same time, and all have similar compositions, they provide a powerful tool for the study of stellar evolution.

- The lack of massive stars lying at the top of the main sequence in clusters indicates that they evolve fastest. The ages of clusters are inferred from the position of the main sequence turn-off.

Observing through the interstellar medium

- Material in the interstellar medium absorbs radiation. An extra term, A, the absorption in magnitudes, is required in Equation 3.16:

$$M = m - 5 \log d + 5 - A \qquad (4.1)$$

- Radiation is both scattered and absorbed by interstellar matter. The combined effect of scattering and absorption is called extinction.

- Atoms (and molecules) can be excited by collisions as well as by absorption of photons.

- Molecules have quantized vibrational and rotational energy states in addition to electron energy states. The energy gaps for vibrational and rotational states are generally much smaller than for electronic states so photoexcitation of (and photoemission from) vibrational states occurs at infrared wavelengths and of rotational states at microwave wavelengths.

- Interstellar dust causes greater extinction at short wavelengths. Distant stars therefore appear fainter *and redder* due to interstellar extinction.

- The properties of the interstellar medium itself can be inferred from its effects on starlight.

Questions

QUESTION 4.7

In what ways, if any, does the distance to a star influence its position on an H–R diagram?

QUESTION 4.8

The photospheric temperatures and luminosities of five stars that are visually fairly bright in the sky are given in Table 4.1.

Table 4.1

Star	T/K	L/W
Alkaid (η UMa)	17000	6.1×10^{29}
Alcyone (in the Pleiades)	12000	3.2×10^{29}
ε Eridani	4700	1.4×10^{26}
Propus (η Gem)	3000	4.2×10^{29}
Suhail (λ Vel)	2600	1.8×10^{30}

(a) Plot these stars on an H–R diagram (such as Figure 4.5), and hence try to assign each star to one of the main stellar classes described in Section 4.2.2.

(b) Suppose that we were to compare stars by preparing an H–R diagram that includes only the stars with the greatest apparent visual brightness. Discuss why such a diagram would be unrepresentative of stars as a whole.

QUESTION 4.9

In terms of photospheric temperature, luminosity and radius, compare the Sun with other main sequence stars.

QUESTION 4.10

Given that T Tauri stars become main sequence stars with little change in photospheric temperature, discuss whether this transition is accompanied by a change in stellar radius.

QUESTION 4.11

Discuss whether we can rule out the evolution of red giants to form supergiants.

CHAPTER 5
THE FORMATION OF STARS

5.1 Introduction

As with so many astronomical phenomena, the life cycle of a star takes place over a timescale that appears infinitely long in comparison with a human lifetime, or even with the entire duration of recorded human history of perhaps several thousand years. Zoologists or botanists are usually able to study the complete life cycles of the animals or plants in which they are interested, but astronomers are not afforded that luxury where the stars are concerned. The changes that occur in stars are, with a few notable exceptions, much too slow to be observed. The evolutionary pattern has to be deduced by observing a wide range of stars at different stages of their lives, and combining these observations with theory and models based on the laws of physics as determined in other environments. The fact that a credible model of the entire life cycle of most types of star has been developed can surely be regarded as one of the triumphs of 20th century astrophysics. This doesn't mean, however, that all the problems are solved – far from it. There is still much to understand.

Before we consider the physical processes and energy sources that govern the structure of stars we will consider the process of star formation.

5.2 Conception

Where should we be looking to find starbirth taking place? The fundamental idea is that stars are born from more widely dispersed gas which condenses because of the gravitational attraction of the gas on itself. The British astronomer Sir James Jeans was responsible for placing this idea on a firm mathematical footing. He showed how an extended mass of gas could be unstable because of the gravitational forces within it and we will follow his arguments in Section 5.3. We have seen (Section 3.2.4 and Figure 0.1) that the distribution of stars within the Milky Way is far from even. We have also seen, in Section 4.3, that the apparent lack of stars in certain directions arises because our view of the stars is obscured in those directions by

AN INTRODUCTION TO THE SUN AND STARS

intervening clouds of gas (consisting mainly of hydrogen) and dust (about 1% of the mass of the gas and consisting mainly of heavy elements and their compounds). Detailed study of the spectra of stars reveals that more tenuous matter is spread throughout interstellar space. Before we look at the process of star formation we will examine further the properties of different components of the interstellar medium (ISM) and the regions most likely to be the birthplace of stars.

5.2.1 The interstellar medium

The temperature and the number density are two important physical parameters of a gas, and we find that these two quantities have a wide range of values in the ISM. The number density n is the number of particles per unit volume and so has SI units of m^{-3}. Atoms can be neutral, ionized, or combined in molecules or in dust. Because of the predominance of hydrogen, to a sufficient approximation we can sometimes take n to be the number of hydrogen atoms per cubic metre, n_H. However, even when we ignore the other elements, n is not always the same as the number of hydrogen atoms per cubic metre. For example, if all the hydrogen is present in the molecular form, H_2, then the number of separate particles (molecules) per cubic metre is about $n_H/2$.

■ What is the particle number density if all the hydrogen is ionized?

❏ About $2n_H$ (the electron is a separate particle when the hydrogen is ionized so there are n_H hydrogen nuclei and n_H electrons).

Only if hydrogen is present mainly as un-ionized (neutral) atoms will the particle number density, n, be about the same as the hydrogen atom number density n_H. However, n_H always gives a good measure of the mass density ρ. Because hydrogen nuclei are the main contribution to the mass of any sample of ISM, $\rho \sim n_H m_H$, where m_H is the mass of the hydrogen atom, little changed by ionization, and unchanged by combination in molecules.

Figure 5.2 shows the typical number densities and temperatures found in the various regions of the ISM. First, note the enormous range of number densities, from a minuscule value of about $100 \, m^{-3}$, to the much greater value of about $10^{17} \, m^{-3}$ (although this is still a very low density as you will see). Second, note that the temperature range is also enormous, from less than 10 K, to several million kelvin – comparable with temperatures found in stellar interiors. However, within these wide ranges, the various temperatures and number densities are not equally represented. Thus, the names given in Figure 5.2 are not of arbitrary subdivisions, but correspond to locations on the diagram where the measured temperature and number density values tend to concentrate. It is because there are such concentrations that temperature and number density provide a useful basis on which to define some of the different types of region. Table 5.1 lists the properties of the different regions.

The intercloud media (hot and warm) account for most of the volume of the ISM, and together form a low-density, optically transparent, widespread matrix in which the other types of region are embedded. Each of these other types is present as a large number of separate objects: typical sizes are given in Table 5.1.

The **hot intercloud medium** is very widespread. You might therefore be wondering why it doesn't blaze down on us from the night sky. The reason is its extremely low density, coupled with its highly ionized state.

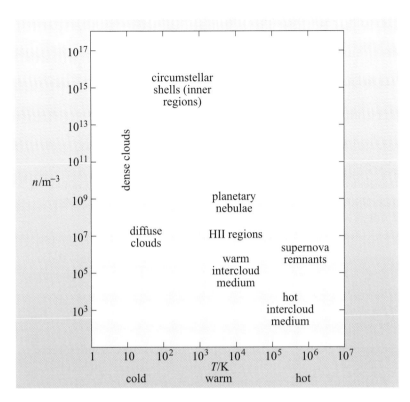

Figure 5.2 Various types of region in the ISM, distinguished on the basis of the number density n of particles, and the temperature T. The atoms can be neutral, or ionized, or combined in chemical compounds.

Table 5.1 Some features of the various types of region in the ISM.

Type of region	Fraction of the ISM (%)[a]		Typical size/pc[c]	Typical mass/$M_\odot$	Predominant form of hydrogen	Abundance of molecules
	By volume[b]	By mass				
hot intercloud medium	~60	≤0.1	–	–	H^+	very low
warm intercloud medium	~30	~20	–	–	H^+ or H	very low
diffuse clouds	~3	~30	~3 to ~100	1 to 100	H or H_2	diatomic molecules common
dense clouds	≤1	~45	~0.1 to ~20	1 to 10^4	H_2	molecules common, even large ones
HII regions	~10	~1	~1 to ~20	10 to 10^4	H^+	very low
circumstellar shells	negligible	negligible	≤1	≤1	H or H_2	diatomic and small molecules common
planetary nebulae	negligible	negligible	≤2	≤1	H^+	very low
supernova remnants	[d]	negligible	≤1000	~3	H^+ or H	very low

[a] These percentages are only rough estimates, so do not sum to 100%.

[b] The total volume of the ISM is taken to be a disc with a diameter roughly that of the spiral arms of the Milky Way, i.e. 30 000 pc, and a thickness of 300 pc.

[c] These are typical distances across a region (e.g. diffuse and dense clouds are usually irregularly shaped, and are often more sheet-like than spherical). For roughly spherical regions the size is roughly the diameter.

[d] The volume is included in the hot intercloud medium (see text).

■ Would you expect any radiation from electronic transitions (see Section 1.3.2) in a low-density, fully ionized hydrogen gas?

❏ No. Low-density gases can produce emission lines when illuminated by a hot source but hydrogen has only one electron and there are therefore no bound electrons in ionized hydrogen to change energy state and produce a spectral line.

The hot intercloud medium consists mostly of fully ionized hydrogen, and the low density means that in any case there is little material to radiate. The low density also means that it is highly transparent, and so the medium does not obscure anything lying in or beyond it. For much the same reasons the **warm intercloud medium** is also not very apparent.

By contrast, the other types of region are much more readily detected, often at a variety of wavelengths. These other types of region can be subdivided into those that are associated with individual stars at or near the end of their lives (planetary nebulae, supernova remnants and circumstellar shells that you will hear more about in Chapter 8), and those that are not (diffuse and dense clouds and HII ('aitch 2') regions described below).

Diffuse clouds are cold regions of moderate density such that stars can be seen through them at visible wavelengths, and consequently they are not apparent to the unaided eye. Hydrogen is present in both atomic (H) and molecular (H_2) forms, and a number of other simple molecules are also found. The clouds are often mapped, and otherwise investigated by various techniques. Atomic hydrogen can be detected through an emission line at a wavelength of 21 cm. Molecular hydrogen cannot be mapped directly, but the CO molecule, which usually occurs with H_2, does emit spectral lines in the microwave band. Finally, the diffuse clouds contain dust that can be mapped from its far infrared emission.

Dense clouds are as cold as or colder than diffuse clouds, and, unsurprisingly, are denser! Unless it is very thin in the direction of our line of sight, a dense cloud is opaque at visible and UV wavelengths, because of extinction by the dust in it. It is then seen by its obscuration of the stars beyond, and is often called a dark cloud. Such obscuration is apparent in Figure 4.14, which shows the Coal Sack, a dark cloud near the Southern Cross, readily seen with the unaided eye from any latitude south of about 25° N. The gas in dense clouds consists largely of molecules. In addition to H_2 there are a great variety of less abundant but nevertheless important molecules. Moreover, extinction by dust is far less severe at longer wavelengths, and so mapping, and other investigations, often rely on the (collisionally excited) microwave emissions from molecules such as HCN, OH, CS and CO. Figure 5.3 shows a map of CO emission in and around Orion: the more copious the emission, the greater the amount of material. Roughly speaking, the regions apparent in this map are dense clouds, and, though unseen here, they are fringed in most places by diffuse clouds. Typically, the two types of region are closely associated.

An **HII region** is a low-density cavity created inside a dense cloud by one or more hot, bright stars (of spectral class O or B) which emit large amounts of ultraviolet radiation. This ionizes the hydrogen gas in their proximity. For historical reasons, the term HII is used to indicate hydrogen in its ionized form.

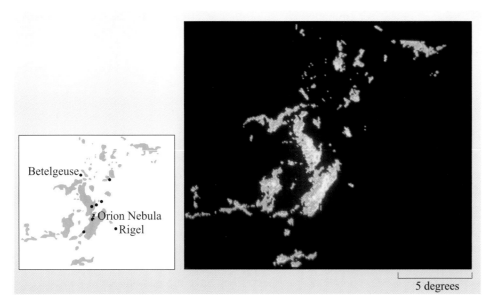

Figure 5.3 The constellation of Orion as seen at radio wavelengths emitted by CO molecules. The density of the gas is indicated by the colour, black where it is undetectable, violet where it is least dense, then through blue to white where it is most dense. Although covering a similar region of sky as Figure 3.1, the view is very different. Molecules pervade the cool regions of the interstellar medium whereas the visible image is dominated by stars or the effect of starlight on the surrounding gas and dust (see Section 4.3). (Image: R. Maddalena/NRAO)

(An alternative name for this type of region is 'diffuse nebula', but this is easily confused with the very different diffuse clouds, so we won't use it.) The hot ionized hydrogen has a greater pressure than the colder un-ionized hydrogen so it expands outwards into the colder material, creating a low-density region inside the cold dense cloud.

In many cases, the HII region has burst through the surface of the dense cloud, and is seen at visible wavelengths. This is the case for the Orion Nebula shown schematically in Figure 5.4. The nebula is lit up by the intense radiation from the young stars (four of which at the centre of the nebula form the Trapezium shown in Figure 6.14).

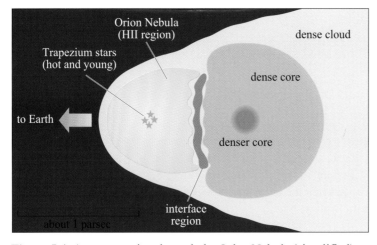

Figure 5.4 A cross-section through the Orion Nebula (simplified).

So, what produces the characteristic red colour of HII regions? The key point here is that not *all* the gas in an HII region is ionized all of the time. Occasionally a free electron and a hydrogen nucleus (proton) can reform a neutral hydrogen atom – a process called **recombination**. The electron is typically captured into a high-energy orbit and then cascades downward through the atom's energy levels emitting photons as it does so. The red coloration from HII regions is due to a transition from energy level $n = 3$ to $n = 2$ (see Figure 3.20). This transition gives rise to the Hα line at a wavelength of 656.3 nm.

Any region in which hydrogen is predominantly in molecular form is called a molecular cloud. Thus *all* dense clouds are molecular clouds, and so too are some diffuse clouds.

Many diffuse clouds have dense clouds moving around inside them (Figure 5.3). Moreover, within a dense cloud, whether inside a diffuse cloud or not, there are often even denser regions called cores and clumps, with masses in the range from $\sim 0.3 M_\odot$ to $\sim 10^3 M_\odot$. The largest of such clouds are called **giant molecular cloud complexes**, or GMC complexes for short. They have dimensions up to ~ 100 pc and masses up to $\sim 10^6 M_\odot$. They are being increasingly recognized as the fundamental cloud structure in the ISM, rather than the individual diffuse or dense clouds. Figure 5.5 illustrates the hierarchical structure of a GMC complex, and Figure 5.3 shows that the Orion Nebula (Figure 4.13), which you have already met, is just one small part of a larger complex.

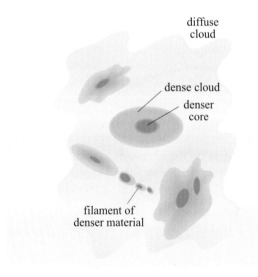

diffuse cloud

dense cloud

denser core

filament of denser material

Figure 5.5 The hierarchical structure of a giant molecular cloud complex.

At this point we should put the term 'dense' into context.

QUESTION 5.1

At the surface of the Earth, the Earth's atmosphere has a temperature of about 300 K, and a density of about 1 kg m^{-3}. Where, in Figure 5.2, do these physical conditions lie? You can assume the atmosphere is entirely made up of its major constituent, nitrogen molecules, N_2.

So, comparing a cubic metre of material from a dense interstellar cloud with a cubic metre of the Earth's atmosphere at sea-level, you have found that the sample of atmosphere has about 10^{15} (a thousand million million) times more gas molecules and atoms than the sample of the same volume from the dense cloud. Even if you took a similar sized sample from what we would describe as 'a good laboratory vacuum', you would still find that a sample from the vacuum would have a factor of about 10^6 more particles than the sample from the dense cloud. However, the clouds are large, and if you multiply the large volumes by the low densities, you do end up with a significant amount of material. The dense clouds make up perhaps 45% of the total mass in the interstellar medium. It is in the dense clouds where it appears that conditions are particularly favourable for stars to form.

5.2.2 Dense clouds – the interstellar nursery

Let's look at the evidence that supports the belief that dense clouds are the site of starbirth.

Some young star clusters seem to be surrounded by the remnants of the original cloud from which they formed. Figures 3.1 and 4.13 show the Orion Nebula. It is visible to the naked eye or through binoculars as a haze in the constellation of Orion. Observations with a telescope show this region to be one of the most visually magnificent in the sky. There is plenty of evidence to suggest that star formation took place here very recently, that is it is recent on an astronomical timescale. For example, if you look at an H–R diagram of the stars in the central region of the nebula (Figure 5.6), you see that they mostly lie above the main sequence.

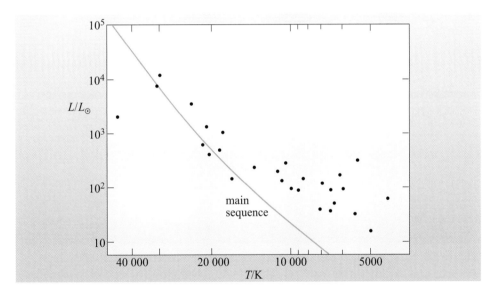

Figure 5.6 H–R diagram for stars in the Orion Nebula, showing that most of them fall above the main sequence. (Note that this H–R diagram covers only part of the range of L and T shown in Figure 4.5.)

▨ What does this suggest about the stars in this region?

❏ You saw in Chapter 4 that stars spend most of their life, maybe 90% of it, on the main sequence. The fact that the stars associated with the Orion Nebula aren't on the main sequence suggests therefore that they are at either the beginning or the end of their lifetime.

Figure 5.7 The star cluster NGC 2264 is formed of young stars illuminating the remnants of the dense cloud from which it formed. (D. Malin/AAO)

All other evidence points to these stars being very young. What interests us here, though, are the vast clouds of glowing gas and obscuring dark clouds of dust and gas. These are almost certainly the remnants of the dense cloud from which the young stars in the nebula formed.

Another example is in Figure 5.7 where the star cluster NGC 2264 (object number 2264 in the 'New General Catalogue of Nebulae and Clusters of Stars' which was an updated version of John Herschell's (see Figure 3.11) 'General Catalogue of Nebulae and Clusters' and published in 1888!) is seen to be intimately associated with the remnants of a dense cloud. Most of these stars also lie above the main sequence on the H–R diagram, though they seem to be rather older than those in the Orion Nebula. Theoretical estimates yield ages of about 5×10^6 years for the stars in NGC 2264, and about 10^6 years for those in the Orion Nebula.

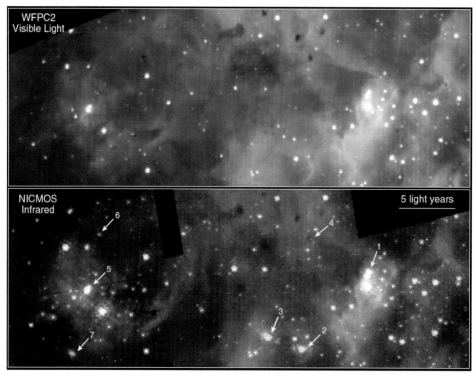

Figure 5.8 Visible light (top) and infrared (bottom) views taken by the Hubble Space Telescope of a highly active region of star birth in 30 Doradus. The orientation and scale are identical for both views. Seven very young objects are identified with numbered arrows in the infrared image. Numbers 1 and 5 are new-born, compact clusters seen in visible light as well as the infrared image. Numbers 2 and 3 are new born stars or stellar systems still immersed within their *natal* dust and can be seen only as very faint, red points in the visible-light image. Number 4 is a very red star that has just formed within one of several very compact dust clouds nearby. Numbers 6 and 7 have been interpreted as 'impact points' produced by twin jets of material (see Section 5.3.4), originating from one of the young stars in number 5, slamming into surrounding dust clouds. The jets may be rotating anti-clockwise, thus producing moving, luminous patches on the surrounding dust, like a searchlight creating spots on clouds. (N. Walborn (STScI)/R. Barba (La Plata Observatory)/NASA)

Another strand of evidence is that some dense clouds are found to contain a large number of compact infrared sources. A good example is shown in Figure 5.8, which shows visual and infrared images of part of the 30 Doradus nebula. This nebula contains some very young stars, which are unobservable in visible light but detectable in the infrared part of the spectrum. This may be due to the absorption of light at visual wavelengths by the dust surrounding a hot source or may be because the source is cool and emits only in the infrared.

QUESTION 5.2

Why should we expect dust clouds surrounding stars in the process of formation to emit in the infrared part of the spectrum? (*Hint*: think of Wien's displacement law.)

It seems that the infrared radiation comes not from the young star itself but from the cocoon of dust still surrounding it (called a **cocoon nebula**). Heated to a temperature of a few hundred kelvin by the recently formed star, the warm dust re-radiates in the infrared part of the spectrum.

The theoretical models of stellar evolution that you will look at in more detail shortly all point to regions of low temperature and high density as being the most likely sources of starbirth. The dense clouds seem to fit the bill quite well.

By way of a brief introduction to these models, recall that, according to Isaac Newton's 17th-century formulation of the theory of gravity, each piece of matter attracts every other piece by the force of gravity. The Earth attracts the Moon, and the Moon attracts the Earth. The Earth attracts an apple, making it fall down from a tree (and the apple also attracts the Earth). One apple also attracts another apple hanging on an adjacent branch, but in this case both of the masses are so small that a very sensitive device would be needed to measure the force. Just as the force between adjacent apples on a tree can be neglected for most practical purposes, so can that between molecules in the Earth's atmosphere, since these forces are much smaller than that between a molecule in the atmosphere and the Earth itself. However, when we are dealing with the gas and dust clouds spread through the vast expanses of the interstellar medium, we cannot ignore this force for two reasons. First, there are generally no other nearby large masses which dominate, and second, although the mass of each molecule or dust grain is very small, there are so many in an interstellar cloud that its total mass can be very large. Each molecule is affected by the gravitational attraction of the combined mass of all the others. James Jeans was able to show that, under appropriate conditions, a cloud (or part of one) would start to contract under the influence of the gravitational force. He derived a formula for calculating the mass and size that a cloud would have to reach, as a function of its temperature and density, before gravitational contraction could start. It is the details of this and subsequent processes, leading to the formation of stars, that are the subject of the remainder of this chapter.

5.3 Starbirth

5.3.1 Contraction of a dense cloud

All atoms, molecules and particles in a cloud are attracted to each other by gravitational forces. However, observations show that many clouds appear to be in a state of equilibrium – in other words, they don't seem to be contracting. Why is it then that the particles don't all collapse into a very small volume?

- ▆ Can you suggest a possible force to oppose gravitational contraction?

- ❑ Each gas particle (atom, molecule or free electron) is in continuous motion (with an average translational kinetic energy of $\frac{3}{2}kT$, where T is the temperature and k is the Boltzmann constant; Equation 4.2). This motion produces a gas pressure which provides an outward force to counteract the tendency of the gas to contract.

The basis of the approach used by Jeans was to consider the balance between the two forces. He proposed that if the force due to gravity was the greater, then gravitational contraction could occur. Using this simple criterion, Jeans was able to show that, for a given set of conditions of temperature and particle number density, there is a value for the mass of a uniform spherical cloud above which the force of gravitational attraction will overcome the opposing pressure due to the motion of the particles, and contraction will occur. This critical mass is known as the **Jeans mass** and is given by the following expression:

$$M_J = \frac{9}{4} \times \left(\frac{1}{2\pi n}\right)^{1/2} \times \frac{1}{m^2} \times \left(\frac{kT}{G}\right)^{3/2} \tag{5.1}$$

where n is the particle number density, m the mass of the 'average' gas particle in the cloud, and T the gas temperature. Figure 5.9 shows how the Jeans mass, M_J, varies with temperature, T, and number density, n, when the particle mass, m, equals that of a hydrogen molecule. If the actual mass of a cloud exceeds M_J, then contraction is predicted.

QUESTION 5.3

According to the Jeans criterion, Equation 5.1, (or alternatively from Figure 5.9), what conditions of temperature, T, and number density, n, are likely to promote gravitational contraction? (A qualitative answer only is required.)

You have seen that the dense clouds are the densest and coolest regions in the interstellar medium, and so, by the Jeans criterion, would contract at lower mass than other types of region. Figure 5.9 shows that dense clouds of only a few solar masses would contract. Many dense clouds are far more massive than this. Moreover, there are yet denser regions within dense clouds (called cores and clumps), with masses between about $0.3M_\odot$ and about $10^3 M_\odot$, that can therefore satisfy the Jeans criterion on their own. Thus, gravitational contraction of dense clouds is to be expected.

The picture that has been painted so far, however, is a highly simplified one.

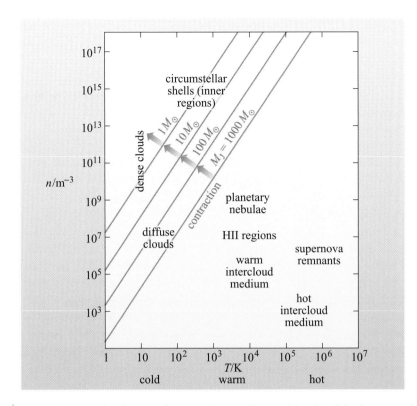

Figure 5.9 The relationship between temperature, T, particle number density, n, and the mass M_J, (the Jeans mass given by Equation 5.1) required in a spherical cloud for the gravitational force to cause contraction to occur. Any cloud or cloud fragment with a mass M_J which lies to the upper left of the diagonal lines will contract. It is assumed that the particles are hydrogen molecules, so $m = 2 \times 1.67 \times 10^{-27}$ kg.

▨ What are the factors that could complicate this simplified approach to the gravitational contraction of a dense cloud?

▭ We wouldn't necessarily expect a typical cloud to be spherical or to have the same temperature and density throughout. In addition, no account has been taken of rotation or of magnetic fields, which almost certainly play an important part. These two factors may well inhibit a cloud's tendency to contract under gravity.

Any internal energy source will also act to inhibit contraction. In stars, which are the end result of contraction of a dense cloud, an energy source is provided by nuclear fusion. In Chapter 6 you will learn more about how stars are supported against gravitational collapse.

Despite its over-simplifications, the Jeans approach is a good starting point for more sophisticated, and perhaps more realistic treatments of the early stages of star formation. Whatever approach is adopted, we can speculate as to what might be the **trigger mechanism** that is probably needed to cause a cloud, or a region inside it, to change from an equilibrium state to one in which contraction has been initiated. Diffuse clouds do not satisfy the Jeans criterion for collapse but are often close to it, so any mechanism that pushes them over the limit acts like a trigger for collapse. This generally requires an increase in density (n in Equation 5.1) without a commensurately large increase in temperature. There are several mechanisms that have been suggested, and each appears plausible:

● Supernovae, about which you will learn more in Chapter 8, produce prodigious amounts of energy, some of which is carried away by a rapidly expanding shell of gas. This fast moving gas sweeps up the local interstellar material in front of it, producing a **shock front** or shock wave, a region of compressed gas immediately in front of the expanding shell. (This is a similar process to that described for bow shocks, see Section 2.4.2 and 2.4.3.) Shock fronts alone are probably not sufficient to *produce* clouds but because they can gather matter *and* increase its density, and can also leave it in a cool state, they are particularly effective at making matter susceptible to Jeans contraction.

- Any shock will affect interstellar matter in this way: it does not have to be from a supernova. Another source of shock, though over a smaller volume of space than that from a supernova, is from regions where several O and B stars form – such groups of stars are called **OB associations**. Young O and B stars, which lie near the top of the main sequence, are massive, and highly luminous. The formation of several such stars close together produces large amounts of visible and ultraviolet light which causes a shock in the material in the surrounding ISM (due to the force of radiation pressure, see Section 5.3.4).

- The spiral structure of our Galaxy cannot be due simply to the different orbital speeds expected of material at different distances from the centre, as the spiral arms would 'wind up' in a relatively short time compared with its age. In order to maintain the spiral structure, a so-called **spiral density wave** sweeps around the Galaxy, compressing all the material that it passes, including clouds. In fact, the speed of these density waves is lower than the orbital speeds so the material is compressed as it sweeps through the density wave. Regions of star formation appear to be concentrated in spiral arms observed in other galaxies. This is consistent with the view that the density wave triggers Jeans collapse and forms stars.

- The close approach, or collision, of another cloud, or even a star, may be sufficient to produce a local gravitational disturbance that could trigger gravitational contraction.

- Star formation can also be triggered throughout a whole galaxy (called a starburst galaxy) by the interaction with another nearby galaxy.

Given that, somehow, a dense cloud, or a region inside it, starts to contract, we must now consider the question of what happens next.

5.3.2 Fragmentation

You have seen in Chapter 3 that stars are often found in groups, referred to as clusters. You also learnt that the stars in a cluster appear to have formed at about the same time. How can this be consistent with the picture of gravitational contraction that we've painted so far?

The answer is believed to lie in the phenomenon of **fragmentation**. As a dense cloud contracts then the density of the cloud (i.e. n) increases.

■ How will the Jeans mass change if n increases but all other quantities remain constant?

❑ From Equation 5.1, you can see that an increase in n will cause the Jeans mass, M_J, to decrease.

It is therefore possible that smaller parts of a massive cloud, which did not initially satisfy the Jeans criterion, can do so after the massive cloud has started to contract. Although the causes of fragmentation are unclear (it may be a result either of the initial 'clumpiness' of the cloud, or of rotation) it certainly appears consistent with the observation that star clusters are common. A cloud, initially with a mass of hundreds or even thousands of solar masses, can ultimately produce a large number of small fragments, each collapsing on its own, to yield a cluster of stars. Star clusters formed in this way are called open clusters (see Section 3.2.4), reflecting their open structure (e.g. Figure 3.14). They typically contain a few hundred stars.

5.3.3 From a fragment to a protostar

We can better understand the evolution of a contracting cloud fragment by looking at the energy balance. Consider a single gas molecule in a contracting fragment. Initially, it will possess both gravitational potential energy and thermal energy (as translational kinetic energy of the molecule). For a molecule near the surface of the fragment, its gravitational energy is given by

$$E_g = -GMm/R \tag{5.2}$$

where M is the total mass of the fragment, m is the mass of the molecule, R is the radius of the fragment and G is the universal gravitational constant. Its translational kinetic energy is given by $E_k = \frac{3}{2}kT$ (Equation 4.2) where T is the temperature and k is the Boltzmann constant.

> ■ Assuming that the contracting fragment can be treated as an isolated system, what can we say about the total energy of the fragment as contraction progresses?
>
> ❑ The law of conservation of energy tells us that the total energy of an isolated system remains constant.

What does this mean for the contracting fragment? As contraction continues, the distance R of our molecule from the centre of the fragment will decrease. This results in a decrease in gravitational potential energy (if you are surprised by this, remember that there is a minus sign in Equation 5.2 for the gravitational energy). Because energy must be conserved, this reduction in gravitational potential energy is accompanied by an increase in other types of energy. The gravitational potential energy is converted into the molecule's kinetic energy. The molecules collide, and so the increase in the individual kinetic energies can be expressed as an increase in the thermal energy, $\frac{3}{2}kT$. This means that the temperature near the cloud surface will increase. Had we considered a molecule inside the cloud the expression for its gravitational energy would have been slightly different, but we would have reached the same conclusion.

Overall, therefore, as the fragment contracts, the gravitational energy of the particles is converted into the translational kinetic energy of molecules, which in turn is converted by mutual collisions into thermal energy of the gas and the temperature rises.

However, there are various complications. One is that collisions between molecules can leave them in excited states, which can emit characteristic radiation. In this case, the radiation is most likely to be in the radio wave, microwave or infrared part of the spectrum. Initially, this radiation tends to escape from the collapsing cloud and the resultant overall rise in temperature is minimal – perhaps only from 10 K to 20 K! However, as the contraction progresses, the number density of the molecules increases and this makes it more difficult for the emitted radiation to escape; it tends to be trapped by the surrounding layers. In other words, the gas becomes opaque to the radiation and now the internal temperature can rise more rapidly.

So although during the collapse process some molecules will be excited to emit radiation, this will only temporarily slow down the inexorable rise in temperature of the cloud fragment.

Section 4.2 introduced you to the Hertzsprung–Russell diagram and the role it plays in our understanding of stars.

■ Where does a contracting fragment of a dense cloud fall on the H–R diagram at the *beginning* of its life?

❑ If the H–R diagram is plotted as in Figure 4.5, we would expect the fragment to be at the bottom right because of its low temperature and, before significant contraction has started, low luminosity.

The track of the contracting cloud fragment across the H–R diagram is far from certain and is difficult to observe directly for two reasons. One is that the process is taking place behind a shield of gas and dust which effectively screens the fragment from view. The second reason is the subject of Question 5.4 (below). For both reasons, we are forced to fall back on theoretical calculations and computer models of this phase. These seem to show that, after only a few thousand years of gravitational contraction, the surface has heated up to between 2000 and 3000 K. The fragment is still quite large at this stage and therefore the luminosity (which also depends on the surface area) can be quite high (10–100 times its eventual luminosity as a star on the main sequence). The exact track depends on the balance between the increasing surface temperature, which tends to increase the luminosity, and the decreasing surface area, which has the opposite effect.

■ Can you recall an equation that reflects this balance?

❑ Equation 3.9, $L \approx 4\pi R^2 \sigma T^4$.

At this stage, the chain of events has started that will lead the fragment, almost inevitably, to become a normal main sequence star. For this reason, we are now justified in calling the fragment a **protostar**.

5.3.4 From a protostar to the main sequence

Figure 5.10 shows the predicted tracks for protostars of various masses as they evolve towards the main sequence region on the H–R diagram. They are called **Hayashi tracks** after the Japanese astrophysicist Chushiro Hayashi (1920–) who was pre-eminent in studying the evolution of pre main sequence stars in the 1960s.

Figure 5.10 also shows the timescale for this early stage of a star's evolution (from when the fragment that becomes the protostar breaks away from the parent cloud). It's clear from this that the more massive the protostar, the quicker it reaches the main sequence. For example, a protostar of $15M_\odot$ takes only about 10^5 years to reach the main sequence, less than 1% of the time that it would take a protostar of $1M_\odot$ to reach the same stage. Virtually all fragments that ultimately become main sequence stars take less than about 10^8 years to pass through the protostar phase – rather short on the astronomical timescale.

QUESTION 5.4

What are the implications of this short timescale on our ability to observe this phase of stellar evolution?

The details of the tracks in Figure 5.10 depend on complex changes in the internal structure and the way in which energy is transported through the protostars as they

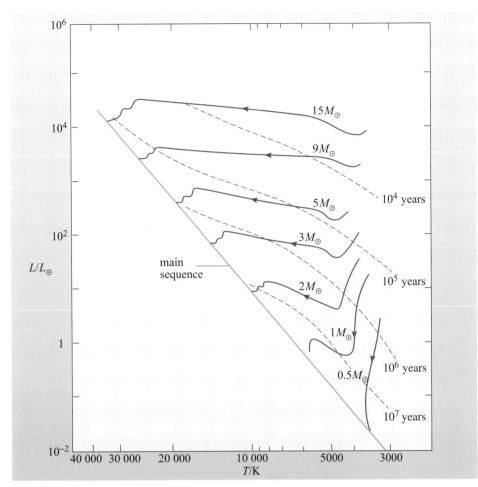

collapse. For protostars of intermediate and low mass, the early drop in luminosity is due to the effect of the increase in surface temperature T being more than offset by the effect of the decrease in radius R (remember Equation 3.9). Subsequently, for protostars more massive than about $2M_\odot$, the effects of increasing surface temperature and decreasing surface area just about balance so that the luminosity changes little as the temperature increases (indicated by an approximately horizontal track on the H–R diagram). Shortly before joining the main sequence, the tracks generally show a drop in luminosity as the effect of the contraction of the protostar tends to dominate over temperature effects.

Knowledge of the protostar phase of stellar evolution has been enhanced by observations made using radio telescopes. A large number of protostars show evidence of a phenomenon called **bipolar outflow** – that is gas flowing at high speeds, typically 50 km s^{-1}, in two streams moving in opposite directions. An example is shown in Figure 5.11, where the outflow shows up in the form of Doppler shifts in opposite directions, one a blue-shift and the other a red-shift (Section 3.2.1). Observations seem to indicate that the flows carry a significant amount of mass and require a lot of energy to sustain them – but they are believed to last for only a relatively short time, perhaps 10^4 years. As a cloud fragment contracts it spins faster and flattens into a **circumstellar disc** or torus with the protostar at the centre. If, at this stage, the protostar starts for some reason to produce a strong stellar wind – something that happens at various stages of a star's life, as we shall see later (Section 6.4.4) – then the disc will tend to channel the

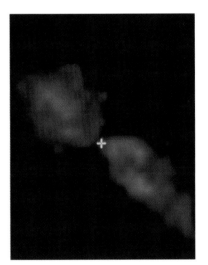

Figure 5.11 A millimetre-line image of the young stellar object IRS5, which lies in a dense cloud, catalogue number L1551. The image shows the intensity of the emission from CO molecules in the gas swept up by the material streaming away from the star (at the position marked by the cross). It has been colour-coded red where the material is moving away (i.e. a Doppler shift to longer wavelengths) and blue where the gas is approaching (i.e. a Doppler shift to shorter wavelengths). (R. Snell, University of Massachusetts)

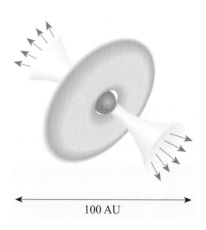

Figure 5.12 Schematic representation of bipolar outflow showing a central protostar, a circumstellar disc or torus and a strong stellar wind. The torus confines the wind to flow predominantly in two opposing directions.

100 AU

outflowing material so that it streams out preferentially along the axis perpendicular to the disc in the form of two streams. The model as shown in Figure 5.12 may explain the principal observed features in bipolar sources, although current thinking suggests that the disc alone cannot channel the outflows, and that magnetic fields probably play a part too. Figure 5.13 illustrates examples of disc and jet structures, while in Figure 5.8 the effect of jets on the surrounding nebula can be seen in the infrared image (labelled 6 and 7, see caption for details).

One method for propelling stellar winds is the phenomenon called **radiation pressure**. It is a pressure exerted by photons on any object that absorbs or reflects them (see also Section 6.4.3). Although it is a weak force it is significant for individual atoms and molecules as well as small dust grains.

Figures 5.14 and 5.15 show nebulae within which individual protostars can be observed as the surrounding gas and dust clouds are eroded by another process – **photoevaporation**. Intense ultraviolet light from nearby highly luminous stars dissociates the hydrogen molecules (H_2) in the cloud into individual hydrogen atoms. The densest regions of the cloud survive longest and 'shield' material behind them forming the observed structures.

The winds, shocks, and UV radiation from young stars, particularly from O and B stars, are the main cause of disruption of dense clouds, and of any complexes of which they may be a part, and this disruption ends the process of star formation.

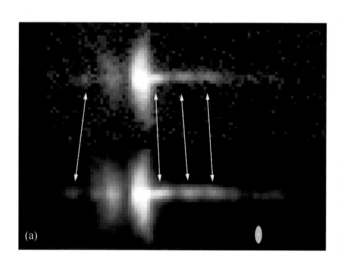

(a)

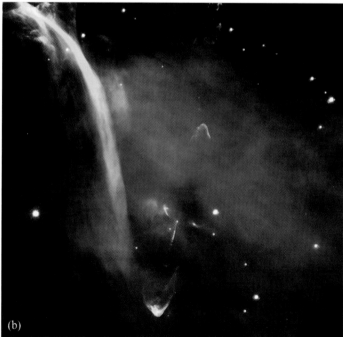

(b)

Figure 5.13 Images of protostellar objects illustrating the features shown in the schematic in Figure 5.12. (a) Protostar HH30. The densest parts of the vertical edge-on disc obscure the newly forming star which illuminates the outside surfaces of the disc. The motion of blobs of gas within the jets indicated by the arrows can be seen in images taken a year apart. The small disc to the lower right indicates the size of Pluto's orbit around the Sun.

(b) Protostar HH34. The beaded structure of the jet (the red object in the lower centre of the image) indicates episodic outbursts of dense gas ejected when chunks of material fall onto the star from the surrounding disc. The effect of the two jets (the upper jet is obscured by intervening dust) is visible as they ram into the surrounding interstellar matter. ((a) C. Burrows (STScI)/NASA; (b) ESO)

Figure 5.14 (a) One of the most famous images from the Hubble Space Telescope showing part of the Eagle Nebula in the constellation Serpens. The three pillars of dense gas and dust are being illuminated by newly formed massive stars that lie above the top of the image. The intense ultraviolet light is also eroding the nebula leaving columns behind the densest regions. The colour image is constructed from three separate images taken in the light of emission from different types of atoms; red from singly ionized sulfur atoms, green from hydrogen and blue from doubly ionized oxygen atoms. (b) This close-up shows 'fingers' of gas behind the cocoon nebulae containing protostars. In some cases the nebulae have themselves been eroded and the star inside revealed. (J. Hester and P. Scowen (Arizona State University)/ NASA)

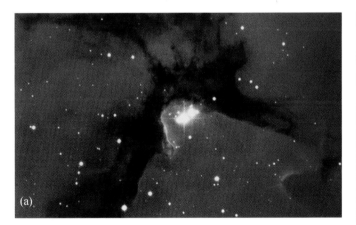

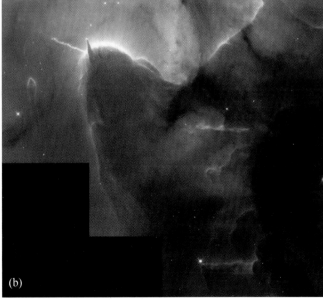

Figure 5.15 (a) The Trifid Nebula is a spectacular dense cloud in the constellation Sagittarius. The reddish colour is due to emission from hydrogen gas and the blue is due to starlight scattered by dust. (b) A Hubble Space Telescope view of a region 3 pc away from the nebula's central star, which is beyond the top of this picture. As in the case of the Eagle Nebula (Figure 5.14), this region is being eroded by the central star. A stellar jet (the thin, wispy object pointing to the upper left) protrudes from the head of a dense cloud and extends 0.2 pc into the nebula. The jet's source is a very young stellar object that lies buried within the cloud but which is likely to be exposed within the next 10 000 years. The nearby 'finger' pointing upwards shows a protostar which has already suffered this fate. ((a) NASA; (b) J. Hester (Arizona State University)/NASA)

■ Why would winds, shocks and UV radiation end the process of star formation?

❏ Extra energy sources in the collapsing cloud raise the temperature and hence raise the Jeans mass.

Only a small fraction of dense cloud mass is transformed into stars. The remnants of the dense cloud, and of any associated complex, can survive as independent clouds, perhaps to play a role in the build up of new, more massive clouds, and new giant molecular cloud complexes.

Another class of objects that is thought to be relevant to the early stages of stellar evolution is T Tauri stars, (which you have briefly met in Sections 3.3.5 and 4.2.3) named after the first to be discovered, the star designated 'T' in the constellation Taurus. They show several tell-tale signs of instability and youth, for example:

- They generally lie to the upper right of the main sequence on the H–R diagram, just where protostars are expected to lie;

- They usually appear in or near dense clouds;

- Many show an infrared excess (a higher flux in the infrared part of the spectrum than would be expected from a main sequence star at the appropriate temperature), suggesting a surrounding dust shell, probably in the process of being blown away;

- They often show irregular variability associated with strong magnetic fields that are not usually seen in older stars;

- They contain the element lithium that is normally destroyed later in the life of a star.

Other evidence indicates that they are young stars of age 10^5 to 10^8 years and that they are losing mass through stellar winds of high speed (up to 100–$200\,\text{km}\,\text{s}^{-1}$). This is consistent with the previously discussed model of bipolar outflow sources that requires a strong stellar wind. It is thought that stars can lose as much as $0.5M_\odot$ in the form of a stellar wind during the T Tauri stage. Some T Tauri stars also show evidence of thin discs of circumstellar material, again consistent with the models of bipolar outflow sources, though in T Tauri stars the outflow is in all directions. From these observed properties, it is concluded that T Tauri stars are pre main sequence stars, in the mass range from about $0.2M_\odot$ to about $2.0M_\odot$, which are in their final spasmodic stage of birth and approaching the main sequence. If this interpretation is correct, they can be used as tracers of the behaviour of pre main sequence stars of these masses.

Let's attempt to bring together some of this observational evidence on the ever-useful H–R diagram. In Figure 5.16 a series of evolutionary tracks for pre main sequence stars of various masses is plotted. The line marked as 'birthline' represents the positions on the various Hayashi tracks where, according to one of the models, stars become optically visible. You may notice, incidentally, that these tracks differ in detail from those for a similar part of a star's life shown in Figure 5.10. This emphasizes that astronomers are not sure about the detailed evolution of a protostar onto the main sequence. Although the broad features are generally agreed on, the fine detail depends on various assumptions made in different models. The dots in Figure 5.16 mark the positions of observed T Tauri stars and stars that show evidence of major outflow (mostly in the form of bipolar outflow). They do indeed fall mostly in the region between the birthline and the main sequence. In addition, there seem to be more lying on Hayashi tracks for the lower masses, consistent with

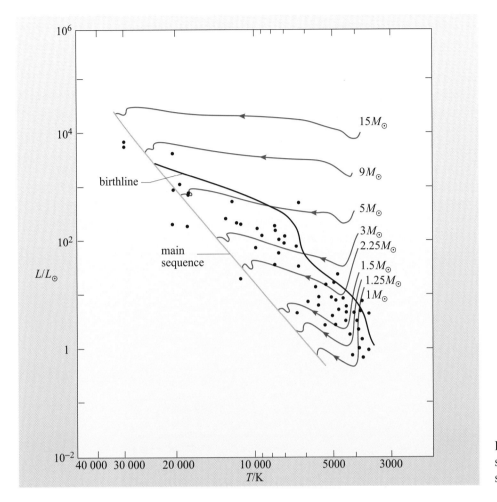

Figure 5.16 H–R diagram for a selection of T Tauri stars and stars showing evidence of major outflow.

our knowledge that low-mass stars are more common than high-mass stars. These observations confirm the general interpretation of both T Tauri stars and stars showing major outflow being associated with the early stages of stellar evolution.

QUESTION 5.5

What observational evidence from Doppler shifts can distinguish the mass outflow in a typical T Tauri star from that in bipolar outflow?

Although there are differences in the details of the various models of the later stages of collapse, they all essentially show a continued rise in temperature as the gravitational energy decreases. The critical point comes when the temperature in the centre, or core, of the protostar becomes sufficient for nuclear fusion to be triggered in the core. The energy released raises the core temperature sufficiently to halt contraction, and marks the protostar's arrival as a new main sequence star. You learned in Section 2.2.4 that it is nuclear fusion that is the power source in the Sun, so you shouldn't be surprised to hear that this is so for all main sequence stars. Before we consider the life of stars on the main sequence, we will briefly look at another formation process, (vital for our existence!) which has occurred for at least a few stars, the formation of planets.

5.3.5 Formation of planetary systems

The suggestion that discs of material may be associated with bipolar outflow sources and with some T Tauri stars posed the exciting possibility that this might also be material from which a planetary system may form, and that planetary formation might therefore be a widespread phenomenon in our Galaxy. Little is known about the formation of planets outside our own Solar System, their frequency, mass and orbital radius distributions. The current emphasis is on discovering other planetary systems and gathering evidence for their formation.

Observations of star-forming regions such as the Orion Nebula have revealed concentrations of dust around very young stars (see Figure 5.17). These discs are called protoplanetary discs, or proplyds, because it is believed that planets will eventually form from them. They are around ten times larger than our own Solar System and contain several Earth masses of dust.

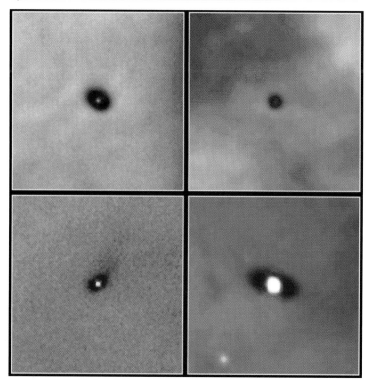

Figure 5.17 Protoplanetary discs in the Orion Nebula. The discs, surrounding the cool red central stars are seen in silhouette against the bright nebula, in these visible light images. (M. J. McCaughrean, (MPIA)/C. R. O'Dell (Vanderbilt University)/NASA)

In the last twenty years dust discs have been discovered around main sequence stars, and although it appears that these discs are not planetary systems in formation, they do provide some evidence of the presence of large objects in orbit.

Astronomers using the Infrared Astronomical Satellite (IRAS) in 1983 discovered that the spectra of various stars showed an infrared excess, probably because the stars are surrounded by a shell or disc of solid particles or dust. This material absorbs radiation from the central star and re-radiates at infrared wavelengths corresponding to the lower temperature ($T < 100\,\mathrm{K}$) of the dust. For some stars, the observations implied that the dust was in the form of a disc rather than a shell, giving support to the belief that what was being observed was the early stages of planetary system formation. The unexpected discovery by IRAS of an infrared excess from the bright star Vega ('vee-ga') was interpreted as emission from dust grains, about a thousand times larger than typical interstellar dust grains, in a shell or disc with a diameter of roughly 170 AU (about twice the diameter of our Solar System). Although Vega is a relatively

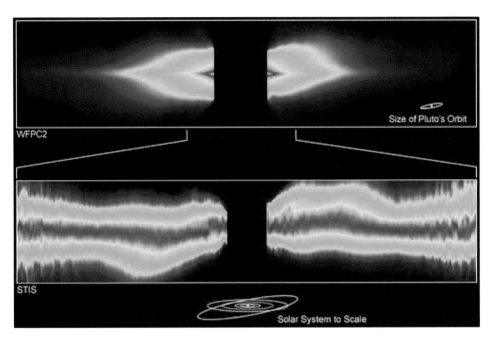

Figure 5.18 The upper image shows the central part of the disc of the young star β Pictoris, which is only a few hundred million years old, imaged at visible wavelengths using the Hubble Space Telescope. The black strip, much larger than the star, prevents the light from the star swamping the much feebler radiation from the disc. In the lower panel the intensity levels have been represented by false colours to enhance the warped shape of the disc believed to be due to the presence of one or more planets orbiting the star. (A. Schultz (Computer Sciences Corp.)/S. Heap (NASA GSFC)/NASA)

young star, perhaps only 20% of the Sun's age, it is probably still too old for dust to have survived from its formation. The disc may be a result of the break up of larger objects such as comets, but is still consistent with the suggestion that at least some of the process of planetary system formation has occurred.

Figure 5.18 shows an image of the region close to the star β Pictoris. The light from the star itself has been blocked out as it would otherwise swamp the much fainter light from the disc, which appears to extend out to several hundred AU from the star. The disc appears to be warped and has an inner edge (not seen in these images) implying the presence of one or more large unseen planets. The possible presence of planets, influencing the distribution of material in dust discs is even more apparent in the examples shown in Figure 5.19.

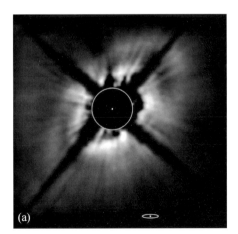

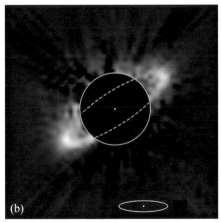

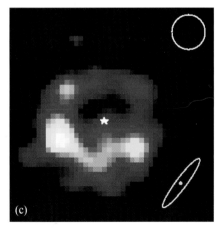

Figure 5.19 Images of circumstellar dust that suggest the presence of planets. The orbit of Pluto about the Sun is shown to the same scale. (a) The young star HD 141569 in the constellation Libra has a bright inner disc separated from a fainter outer disc by a gap which may have been carved out by an unseen planet.

(b) HD 4796A has a dust ring which can only stay intact by some mechanism confining the dust. The gravitational influence of unseen planets may be preventing the dust from dissipating due to collisions and radiation pressure. ((a), (b) B. Smith (University of Hawaii)/G. Schneider (University of Arizona)/NASA)

(c) Microwave image of a dust disc around the nearby star ε Eridani taken using the James Clark Maxwell Telescope. The irregular distribution of dust may indicate the presence of planets. The circle to the top right indicates the size of the telescope detector beam. ((c) Joint Astronomy Center, Hawaii)

Direct observation of planets is technically challenging due to their extreme faintness and close proximity to a much brighter star. However, it is possible to detect planets due to their effect on the motion of the star. In 1995 the first direct evidence of a body of planetary mass orbiting another main sequence star was obtained from observations of minute oscillations in the wavelength of spectral lines from the star 51 Pegasi with a period of 4.2 days. The Doppler shift of the lines was due to the orbital motion of the star about the common centre of mass of the star–planet system. This allowed a lower limit to the mass of the planet to be determined in a similar way to that for spectroscopic binary stars (see Section 3.3.7). The Doppler shift implied a planet of mass ≥ 0.47 times the mass of Jupiter (the exact mass is unknown because the tilt of the orbit to the line of sight cannot be determined by this method). Since then, many other stars have been found to have companions with masses comparable to, or a few times larger than, Jupiter. The majority of such objects have large masses but orbit very close to the parent star sometimes in highly eccentric orbits (unlike our Solar System where the massive gas giant planets formed in cool regions in the outer Solar System in near-circular orbits). It is possible that a similar formation process occurred as in our Solar System, with massive planets forming far from the stars, but subsequently undergoing interactions with remaining disc material or other protoplanets, with the result that they migrated inwards towards the central star. Another possibility is that some of these objects may not be planets at all but more massive brown dwarfs (see Section 6.4.2). It is certainly true that planets with large masses close to the primary star are easiest to detect.

■ Why are planets with larger masses close to a star easier to detect?

❑ Planets with larger masses and small distances exert a greater gravitational force on the star and therefore induce a greater 'wobble' of the star (corresponding to its orbit about the centre of mass of the star–planet system). This produces a greater Doppler shift in its spectral lines, which is more likely to be detectable, compared with a low-mass or distant planet.

The apparent scarcity of Jupiter-like planets in Jupiter-like orbits around other stars may therefore simply be a consequence of the fact that they are more difficult to detect combined with the short time since observations started.

The study of **extrasolar planets** is a rapidly advancing field in astronomy. One objective of such research is to determine whether the formation of planetary systems is commonplace. If our Solar System is typical, then there may be many planets capable of sustaining life elsewhere in the Galaxy. The next big step is to detect such Earth-like planets (and hence those which may harbour life) around other stars. Such low-mass planets will be difficult to detect using the spectroscopic method described above, but spectroscopy is likely to provide the solution in a different way. If a planet with an atmosphere passes in front of a star as it orbits, as viewed from the Earth, then it is possible to detect the gases present by the faint absorption lines they produce in the star's spectrum. It will therefore be possible to determine if the conditions for life, including the presence of oxygen, abundant in the Earth's atmosphere due to the respiration of plants, are present. Such techniques are already being attempted using ground-based telescopes. In the future, space observatories could provide a census of planetary systems in our neighbourhood of the Galaxy. The Gaia mission – see Section 3.2.2) will make precise measurements of the motions of the billion brightest objects in the sky, which is estimated to include 50 000 stars in our Galaxy possessing planets. Darwin, a possible future ESA mission, will make spectroscopic observations to search for Earth-like planets.

5.4 Summary of Chapter 5

- Stellar evolution occurs on such a long timescale that it can only rarely be directly observed.

The interstellar medium

- Stars form from material in the interstellar medium (ISM). The ISM contains regions with enormous variations of size, temperature and number density of particles.

- Even the densest clouds are far more tenuous than the Earth's atmosphere.

- Observational and theoretical evidence points to dense interstellar clouds as being the place where star formation begins.

Starbirth

- The Jeans criterion, despite being a simplified approach, makes useful predictions of the mass that a cloud, or part of one, must achieve before gravitational forces overcome those due to the thermal energy of the gas and contraction is able to start. The Jeans mass is given by

$$M_J = \frac{9}{4} \times \left(\frac{1}{2\pi n}\right)^{1/2} \times \frac{1}{m^2} \times \left(\frac{kT}{G}\right)^{3/2} \tag{5.1}$$

An external trigger mechanism is believed to cause a cloud to start contracting under the influence of gravitational forces. Compression of the cloud by shocks generated by supernovae and OB associations or by a density wave in spiral arms may be that trigger.

- As contraction of a dense cloud continues, the cloud fragments into smaller parts, each of which may continue to contract, as long as the Jeans criterion is satisfied for that particular fragment.

- Gravitational contraction is accompanied by a rise in the temperature throughout the fragment, though this is moderated by the escape of radiation from the fragment, particularly until the fragment becomes opaque.

- The temperature continues to increase and the size to decrease, both at a rate and in a way that depend predominantly on the mass of the fragment – now a protostar.

- Some protostars show evidence for circumstellar discs and for bipolar outflows, in which material flows out in opposite directions at high speed.

- T Tauri stars are pre main sequence stars of mass below about $2M_\odot$, showing strong stellar winds and variations in luminosity.

- When the temperature in the core of the protostar rises sufficiently, nuclear reactions are triggered. This provides the energy source to prevent further contraction, and at this stage the protostar joins the main sequence and becomes a fully fledged star. The time for a fragment to reach this stage is generally less than about 10^8 years; the more massive the fragment, the shorter the time.

- The presence of circumstellar discs around young stars led to the belief that they may be the precursers or remnants of planetary formation. The presence of planets around other stars has been inferred from the tiny *wobble* in the motion of those stars about the centre of mass of the star–planet system. The search for Earth-like planets and the conditions for life to form is currently one of the most rapidly expanding branches of astronomy.

Questions

QUESTION 5.6

A particular spherical dense cloud has a radius of 3 pc and a particle number density of $10^9 \, \mathrm{m}^{-3}$. Calculate whether this cloud contains enough material for a star or stars to form from it. (Assume that all the material in the cloud is in the form of H_2 molecules.)

QUESTION 5.7

A dense cloud is compressed, such that a denser core of 1 solar mass inside it increases in density and temperature, as in the table below. Assuming that the core consists mostly of H_2 molecules, and by considering the Jeans mass in each case, discuss whether the core will contract to form a star.

	Uncompressed	Compressed
number density/m^{-3}	5×10^9	8×10^{11}
temperature/K	10	25

CHAPTER 6
THE MAIN SEQUENCE LIFE OF STARS

6.1 Introduction

In Chapter 4 we saw that if we determine the temperature and luminosity of a large number of stars and plot their positions on a Hertzsprung–Russell diagram, we will find that they are not randomly distributed. The majority lie on the main sequence, ranging from hot, luminous stars of spectral type O and B to faint, cool stars of spectral type M. The Sun (spectral type G2) is a very ordinary star lying near the middle of this sequence. The fact that we see so many stars on the main sequence leads us to believe that any star must spend a relatively large fraction of its lifetime there. The Sun's luminosity must have remained relatively constant for most of its life, otherwise the liquid water on the Earth (which is such an essential requirement for life) would have boiled away or frozen out, whereas the geological record tells us that oceans have been present for much of the Earth's lifetime of 4.5×10^9 years.

We have considered the basic processes of star formation in Chapter 5. All the evidence tells us that young stars evolve onto the main sequence, the most massive stars being the hottest and most luminous. In this chapter we will examine the physical properties and energy sources of stars on the main sequence. We can make use of our knowledge of the star that we know best, the Sun (Chapters 1 and 2), but it would be dangerous to use only the Sun as a model, as we would have no idea how typical of the whole population of stars it might be. It is therefore important to use observations of all stars, as described in Chapters 3 and 4, to try to piece together the life stories of different types of stars. When we understand more about the processes that dictate the structure of main sequence stars and produce their immense power output, we will be better placed to investigate how they evolve.

6.2 Stellar structure

6.2.1 Stellar models

We shall start our investigation of main sequence stars by looking at the physical conditions that exist inside a typical star and the way in which these parameters vary with position within a star. If we take our Sun as an example, you saw in Chapter 1 that these parameters can be measured for the outer layers, the corona, chromosphere and photosphere. How they change deeper in the Sun, however, can't be known from direct observation, as those regions are essentially inaccessible. We have to rely on indirect information such as the results of helioseismology (Section 2.2.6) which constrain theoretical models of the variation of temperature, pressure, density and composition with depth, as shown in Figures 2.1 and 2.2.

The situation with other stars is clearly much more problematic because results from asteroseismology provide less information since the discs of stars cannot be resolved. Yet, in order to understand the way that different stars evolve we *do* need to have an understanding of these interior conditions, because it is these that dictate, to a large extent, the nature, the rate and the extent of the nuclear processes which power the stars.

The study of these conditions involves the construction of a set of equations, known as the **equations of stellar structure**, which permit predictions of the various physical properties (e.g. temperature, density, pressure) throughout a star. Although the equations are relatively simple, the associated theory, and the techniques for their solution are complex. However, by making various simplifying assumptions, it is possible to study basic properties of stars and to derive some limited quantitative information about their structure. We shall briefly look at the way in which some of the equations are derived, without in any sense attempting a rigorous determination.

Before that, however, we can make some general observations. Because energy appears to be lost from the surface of a star, in the form of radiation and perhaps particles, and is released within a star, it appears likely that the temperature generally increases with depth into a star. A temperature gradient in this direction is needed to maintain an outward flow of energy. Also, the pressure must increase with depth because of the increasing amount of overlying material.

▪ Is the assertion that temperature and pressure increase with depth true for the Sun?

❑ Yes. Figure 2.1 shows that both parameters increase quite sharply with depth.

However, properties such as pressure, temperature, density, and various of the intrinsic properties of the material that constitute a star, such as composition, thermal conductivity, opacity (the degree to which the gas absorbs radiation), and the rate of energy generation, are all intricately interrelated, and it is one of the challenges of stellar physics to determine these relationships.

6.2.2 Internal temperatures and pressures

In order to set up the equations that describe the flow of energy, the pressure balance and other properties of a star, astronomers often resort to the strategy of considering a small volume at an arbitrary position within the star. They then write down the appropriate equations for that small volume only. By applying what are called boundary conditions (e.g. the fact that the temperature of a small volume at the edge of the star is simply the observed surface temperature) and other constraints (such as the fact that the masses of all small volumes added up is the total mass of the star), they look for consistent solutions to the equations that yield, amongst other things, profiles of temperature and pressure against depth in the star.

If we make the assumption that the star is spherically symmetric – in other words, that its properties vary in the same way in any radial direction away from the centre – then the small volume that we should consider is a thin spherical shell, as shown in Figure 6.1. In this figure, Q is the flux of energy flowing outwards in the radial direction through the shell. It has the units of energy per unit cross-sectional area perpendicular to the direction of flow per unit time. In order to maintain this flow of energy, there must be a temperature difference between the faces of the imaginary shell. Referring to Figure 6.1, T_1 must be greater than T_2 in order to maintain an outward flow of energy, i.e. we expect temperature to increase with depth.

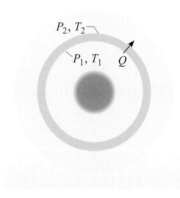

Figure 6.1 A thin spherical shell within a star to which calculations of various stellar properties are applied. The pressure and temperature at the inner and outer surfaces of the shell are calculated, as well as the heat flowing through it.

The exact difference between T_1 and T_2 is related not only to the rate of energy flow (a higher rate of energy transfer generally requiring a larger temperature difference), but also to the pressure and composition of the gas itself.

One of the simplifying assumptions made is that the stellar material, at least for main sequence stars, behaves as a so-called perfect or **ideal gas**. If you have come across the concept of an ideal gas before, you may be surprised to hear that we can apply it at the high densities that are believed to exist deep inside a main sequence star. The very high temperatures inside a star cause most of the gas to be ionized, and these ionized particles and electrons act as ideal particles. This means that we can use simple equations, such as the **ideal gas law**, which states, in one form:

$$P = k\rho T/m \tag{6.1}$$

where P is the gas pressure, ρ is the gas density, m is the average mass of the gas particles, k is the Boltzmann constant, and T is the gas temperature. Because m is dependent on the composition of the star, this form of the equation reinforces the fact that certain of the physical properties, such as pressure, do depend on composition.

How does the pressure vary with position throughout the star? To answer this question, we return to the imaginary shell in Figure 6.1. In a stable main sequence star, we assume that this shell will be stationary, i.e. there will not be significant net movement of mass on a large scale, so the forces acting on the shell must balance. Gravity acting on the shell will cause the material to be pulled towards the centre of the star. Opposing this will be the force due to the pressure in the gas. This balance is called **hydrostatic equilibrium**. If P_1 and P_2 are the pressures on the inner and outer faces of the shell, then as long as P_1 is greater than P_2, the pressure difference will result in a force acting outwards, opposing that due to gravity. The pressure in a body always increases with depth, because of the increasing mass of material above, and so P_1 must be greater than P_2. However, this gradient need not be sufficiently large to stop contraction. If there is an internal energy source generating a temperature gradient, then, as Equation 6.1 shows, this will tend to increase the pressure gradient. In a main sequence star the internal energy source is, of course, the nuclear power-house in the core. It enables the star to achieve stable equilibrium. The variation of pressure with depth can then in principle be calculated.

We have seen that a temperature difference across the shell maintains a flow of energy through that shell. If the shell is outside the region where energy is being released then the energy flowing into the shell is equal to that flowing out. However, if it is inside this region, then the rate at which energy flows out exceeds the rate at which it flows in by an amount equal to the rate of energy release within the shell.

The concepts described above can be used to define some of the equations of stellar structure. We will not attempt to derive any of these equations here as they can get quite complex and are beyond the scope of this book. However, we can, by making some very simple assumptions, make a surprisingly accurate estimate of the internal temperature of a star.

We consider a star, of mass M and radius R, composed of an ideal gas, as being divided into two hemispheres, each of mass $0.5M$, as indicated in Figure 6.2.

The two hemispheres gravitationally attract each other with a force of magnitude given approximately by

$$F = \frac{G(0.5M)(0.5M)}{R^2}$$
$$= \frac{GM^2}{4R^2}$$

We have assumed that the two hemispheres can be considered as point masses separated by a distance of R. Balancing this gravitational force on each hemisphere, which tends to make the star contract, is the outward pressure, P, of the hot gas. The pressure gradient here is provided by the difference between P inside the star,

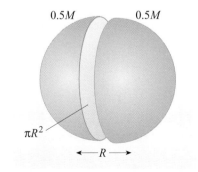

Figure 6.2 Schematic diagram of a star divided into two hemispheres for the purpose of estimating the central temperature.

and zero pressure outside it. Let us assume that P acts over the inner surface of the hemisphere, of surface area πR^2 (see Figure 6.2). So, remembering that pressure times area (i.e $P \times \pi R^2$) is force,

$$P\pi R^2 = \frac{GM^2}{4R^2} \tag{6.2}$$

We now need to use the ideal gas law:

$$P = k\rho T/m \tag{6.1}$$

The density, ρ, is equivalent to M/V, or Nm/V, where V is the volume containing N particles each of mass m. Equation 6.1 then becomes

$$P = \frac{kNmT}{mV} = \frac{NkT}{V} \tag{6.3}$$

For a sphere, $V = \frac{4}{3}\pi R^3$, so

$$P = \frac{3NkT}{4\pi R^3} \tag{6.4}$$

Thus, eliminating P from Equations 6.2 and 6.4,

$$\frac{3NkT}{4\pi R^3} \times \pi R^2 = \frac{GM^2}{4R^2}$$

which can be rearranged to give

$$T = \frac{GM}{3kR} \times \frac{M}{N}$$

The factor M/N represents the total mass of the star divided by the total number of particles – in other words it is the average mass of a particle, m_{av}, in the whole star. Thus we have the final result for the temperature of the interior of the star

$$T = \frac{Gm_{av}}{3k}\frac{M}{R} \tag{6.5}$$

where the constants are gathered into the first term. For an approximate calculation such as this one, we won't go far wrong if we assume that the star is composed entirely of ionized hydrogen, and so m_{av} is approximately equal to $\frac{1}{2}m_H$, where m_H is the mass of the hydrogen atom. (In an ionized gas a hydrogen atom of mass m_H consists of two separate particles, a proton and an electron.) If we substitute values appropriate for the Sun, we find $T \approx 4 \times 10^6$ K. We should expect this to be only an approximation to the true internal temperature, which will also vary considerably through the star, being highest in the core, and falling towards the edge. In the case of the Sun, more detailed calculations indicate a temperature in the core of 1.56×10^7 K. Our very simple calculation, therefore, which has made no assumption of the method of energy release, but has used only some simple laws of physics, has given the core temperature accurate to within a factor of four.

QUESTION 6.1

Identify some other gross assumptions that have been made in this derivation.

What this calculation has done is to show that the temperatures that are likely to be attained in a typical star are sufficiently high to trigger nuclear reactions. The types of nuclear reaction that occur are discussed in Section 6.3.

Equation 6.5 is also important because it indicates the way in which we might expect the interior temperature of a star to change for stars of different mass. From this equation we can write $T \propto M/R$. More realistic derivations show that this relationship applies to the *core* temperature of stars.

■ How does the radius change if the star had double the mass? (You may safely assume that the mean density is roughly the same.)

❏ Mass $M = \frac{4}{3}\pi R^3 \rho$ so we can rewrite this as $R = (3M/4\pi\rho)^{1/3}$. If M is larger by a factor of 2 then R is larger by a factor of $2^{1/3} = 1.26$ or by about 26%.

The mass therefore increases more rapidly than the radius, causing the ratio M/R to be larger for stars of larger mass. It therefore follows that the temperature is predicted to rise with increasing stellar mass. This is very important and is at the root of the observation that luminosity increases dramatically for more massive main sequence stars – this **mass–luminosity relationship** was established in the 1920s by Eddington (Figure 6.3), on theoretical grounds. This is the relationship that, as we will show in Section 6.2.5, leads to the result that the lifetime of massive stars is shorter than the lifetime of less massive ones.

ARTHUR STANLEY EDDINGTON (1882–1944)

Arthur Eddington (Figure 6.3) won a scholarship and started university in Manchester a few months before his 16th birthday. He got a first class degree in physics and a scholarship to Trinity College Cambridge at age 19. He completed the three-year maths course at Cambridge in two years coming top of the final year class as well – the first time a second-year student had done that! In 1907 he won a Trinity College Fellowship, on the basis of a five-page thesis, and at the age of 30 became Plumian Professor.

Figure 6.3 Arthur Stanley Eddington. (Royal Astronomical Society)

During the First World War he received, via neutral Netherlands, a paper by Einstein describing relativity theory. One of the first people to recognize the importance of this theory, Eddington also recognized that the predicted gravitational bending of light could be checked observationally during a solar eclipse. He led an eclipse expedition in 1919 which confirmed the theory. He is also known for his significant work in stellar dynamics and stellar structure, and for the disagreements he had with Chandrasekhar (Figure 9.2) and Jeans (Figure 5.1)! He had excellent physical insight; as early as 1917 he was speculating that the energy source in stars was sub-atomic, involving the transmutation of hydrogen into other elements. He discovered the fundamental role that radiation pressure plays in maintaining stellar equilibrium, argued the importance of radiative energy transfer in stars and showed that a star's equilibrium was maintained by the balance between gravity and the outward forces of gas pressure and radiation pressure. His announcement in 1924 that the luminosity of a star depends almost entirely on its mass revolutionized ideas about stellar evolution. Eddington was an accomplished writer, introducing many to the science of astronomy and was considered one of the greatest astronomers of his time. He was elected a Fellow of the Royal Society in 1914 and knighted in 1938.

6.2.3 Energy release and transport

If we accept that nuclear processes (which differ depending on a star's mass) are responsible for energy release within main sequence stars, where do these processes take place within a star? It has to be stressed that we have essentially no *direct* information on this question, since we cannot see directly into the interiors of stars (although we can probe the interior of the Sun and other stars using helioseismology/asteroseismology – see Section 2.2.6). It is possible, however, to use the results of modelling of the interiors of stars through the equations of stellar structure, together with knowledge of the relevant nuclear processes, to answer this question.

■ Why can't we assume that nuclear processes are occurring throughout the volume of most main sequence stars?

❏ The surface temperatures of stars are too low for nuclear processes to be taking place at the surface, so clearly these reactions do not take place throughout the whole star.

For stars of all masses, nuclear reactions take place where the temperatures are greatest, which is at the centre and in a surrounding region, called the core. Because the nuclear reaction rates vary so much with temperature, this means that, moving out from the centre of a star, the boundary defining the limit of nuclear reactions is fairly sharp. The size of this region varies with the mass of the star, as do the types of reaction that predominate, and the mechanism by which energy is transported to the outer layers of a star.

■ By what mechanisms can energy be transported from one place to another, and which do you think play an important part in main sequence stars?

❏ There are three mechanisms – conduction, convection and radiation. If we regard the Sun as typical of main sequence stars, then we would expect, as we saw in Section 2.2.5, that convection and radiation are the dominant mechanisms of energy transfer in these stars.

Let's consider stars of low mass (less than $\sim 1.5 M_\odot$) on the main sequence (called **lower main sequence stars**). For stars with masses $\sim 1 M_\odot$ the rate of energy release does not increase sufficiently towards the centre to set up the temperature gradient necessary to initiate convection in the core. In other words, the core is non-convective. Immediately outside the core, the temperature gradient is still too small to cause convection, and so radiation is the primary mechanism by which energy is transferred. Further out, however, there is a region in which convection does take place, in the form of a convective zone or envelope. This convective envelope starts deeper in stars of lower mass, and for masses less than $\sim 0.3 M_\odot$ the convective envelope reaches right down to the centre.

If the mass of a star exceeds about $1.5 M_\odot$ (termed **upper main sequence stars**), the temperature of the core is sufficiently high for a different set of nuclear reactions (we'll be looking at these shortly) to become predominant. The energy release in the core is then sufficiently concentrated to trigger convective instability and the centre of the star is convective. In fact, this convective zone may extend beyond the core in which nuclear reactions are taking place. The situation for both upper and lower main sequence stars is shown in Figure 6.4.

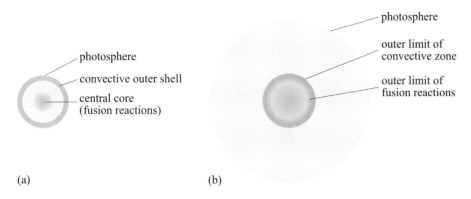

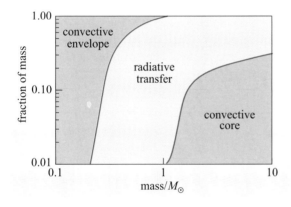

(a)   (b)

Figure 6.4 Schematic cross-section of a star of (a) $1M_\odot$ and (b) $5M_\odot$, showing the difference between the location of the convective regions.

Figure 6.5 illustrates how the situation varies with stars of different mass. It shows the extent of the **convective core** or the **convective envelope** as a function of the mass of a star. The vertical axis represents the fraction of the total mass occupied by each region.

Figure 6.5 The fraction of the mass of a star in the convective core and envelope as a function of stellar mass. Note that the mass is concentrated towards the centre of the star so the convective envelope of a $1M_\odot$ star (see Figure 6.4) contains only a small fraction of the mass.

QUESTION 6.2

In a star of $\sim 1M_\odot$, will the composition of the core change during its main sequence lifetime? If so, how?

6.2.4 Why is there a main sequence?

In trying to answer the question 'Why is there a main sequence?' we need to consider why stars have particular values of luminosity and temperature and why these values remain constant for so long.

Let's first consider the **stability** of stars on the main sequence. You have already seen that a stable star is in hydrostatic equilibrium, i.e. the inward force of gravity on each layer of a star is balanced by the net outward pressure forces.

■   What would you expect to happen to the pressure of the gas in the star if the whole star is 'squashed' a small amount, i.e. its radius is decreased at constant temperature?

❑   The density of the gas in each layer of the star would rise because the volume has decreased. Since $P = k\rho T/m$ (Equation 6.1) for each layer of the star, the pressure will also rise.

Thus the net outward force due to the variation of pressure with depth will also rise and there will no longer be a balance with the inward gravitational force. The star will therefore expand back to its original dimensions. (This is a simplification since the gravitational force will also rise a small amount if the star is 'squashed', but this will be accompanied by a rise in temperature, due to conversion of the gravitational potential energy into kinetic energy of the gas particles, and hence a rise in pressure.) This means that a star on the main sequence is very stable against any dimensional changes. (We will consider the longevity of most stars on the main sequence in the next section.)

Now we will briefly consider the fundamental matter of why there is a main sequence in the first place! In 1926, Henry Norris Russell (Figure 4.2b) and the German physicist Heinrich Vogt (pronounced 'voit') (1890–1968) derived a result, sometimes called the **Russell–Vogt theorem**, which can be stated as: 'The equilibrium structure of an ordinary star is determined uniquely by its mass and chemical composition'. To put it another way, we can say that a certain mass of stellar material of fixed composition can reach only *one stable configuration*. This stable configuration will correspond to one point on the H–R diagram. A star of different mass occupies a different point on the H–R diagram. This is exactly what we see along the main sequence, as in Figure 4.8. Thus, we have a main sequence because the stars on it are stable, with similar chemical compositions, but with different masses.

How can we reconcile the Russell–Vogt theorem with the fact that the H–R diagram, apart from the main sequence, also contains other regions, such as those of red giants, supergiants and white dwarfs? The existence of these regions would seem at first to undermine the theorem. But, as we shall see later, their existence is not contradictory, for it is the chemical composition of these stars that is different, and this allows them to occupy different regions on the H–R diagram.

6.2.5 Main sequence lifetimes

Consider the duration of a star's life on the main sequence.

- You have already come across some evidence in Chapter 4 that gives a clue as to how a star's lifetime on the main sequence varies with the star's properties. Can you remember the source of that evidence and what it tells us?

- The evidence comes from looking at the H–R diagram for star clusters (Section 4.2.5). These show that the more massive stars have shorter main sequence lifetimes.

Let's try to be slightly more quantitative about the matter of main sequence lifetime. The lifetime is given by the energy available divided by the luminosity. Masses and luminosities derived from observations (see Sections 3.3.7 and 3.3.3 respectively) imply that the main sequence luminosity depends on a high power of the mass. On average, $L \propto M^4$ (although the actual power varies between 3 and 5 depending on the mass, the exact value does not affect the basis of the discussion below). If the mass of one star is twice the mass of another, the luminosity of the first is 2^4 or 16 times more than the second. The reason for this sensitive dependence on mass is the temperature of the core, which has important consequences for the star because the relevant nuclear reaction rates are *very* sensitive to temperature.

In order to estimate the lifetime of a star on the main sequence, we shall make the assumption that the energy available is proportional to the total mass. Therefore, the main sequence lifetime is given by

$$t \propto \frac{\text{energy supply}}{\text{rate of release}} \propto \frac{M}{L} \propto \frac{M}{M^4}$$

and so $\quad t \propto M^{-3}$ \hfill (6.6)

This tells us that the lifetimes of stars on the main sequence decrease ever more rapidly for stars of higher mass. We now have an approximate relationship for the way in which the main sequence lifetime of a star depends on its mass – but the above relationship is not an equality but a proportionality. For example, a relationship like $t \propto M^{-3}$ tells us that for a star of double the mass, its main sequence lifetime is a factor of $(2)^{-3}$ or $\frac{1}{8}$ as long. In other words, the lifetime is one-eighth of that of the less massive star. However, we still need to fix, or 'calibrate', the lifetime scale. This can accomplished by a variety of means, usually consisting of a mixture of observational and theoretical approaches. Using the Sun for calibration gives a main sequence lifetime of about 1×10^{10} years for a mass of $1M_{\odot}$.

More sophisticated calculations give a dependence of main sequence lifetime on mass as shown in Table 6.1. The five most massive stars listed are upper main sequence, the remaining being lower main sequence stars.

It follows from Table 6.1 that massive stars, such as those of $15M_{\odot}$, are predicted to have relatively short main sequence lifetimes, perhaps as brief as 10 million years. This means that many of the massive upper main sequence stars currently observed must have been formed fairly recently on the astronomical timescale, and therefore provide further evidence that new stars are being formed even today.

Table 6.1 Selected properties of main sequence stars of various masses.

Mass/$M_{\odot}$	Luminosity/$L_{\odot}$	Surface temperature/K	Main sequence lifetime/yr
0.50	0.03	3800	2×10^{11}
0.75	0.3	5000	3×10^{10}
1.0	1	6000	1×10^{10}
1.5	5	7000	2×10^{9}
3	60	11000	2×10^{8}
5	600	17000	7×10^{7}
9	4000	23000	2×10^{7}
15	17000	28000	1×10^{7}
25	80000	35000	7×10^{6}

6.3 Nuclear reactions

Before we look at some of the details of nuclear fusion in main sequence stars, we should ask whether there is any other possible way in which the luminosity of these stars could be explained. In the following two questions you are asked to consider two particular alternative energy sources and to work out for how long each could keep the Sun shining at its present rate.

QUESTION 6.3

Suppose that it is chemical energy that is responsible for the luminosity of the Sun. Assuming that 1 kg of material in the Sun yields an energy output of 3.5×10^7 J (a value typical for coal-burning on Earth), determine for how long this mechanism could sustain the Sun's present luminosity.

QUESTION 6.4

Assuming that a star like the Sun contracts to a tenth of its present size, and in doing so releases gravitational energy, for how long could this power the Sun at its present luminosity? (You can assume that the gravitational energy of a sphere of radius R and mass M is approximately $-GM^2/R$.)

The answers to Questions 6.3 and 6.4 show that these two methods of releasing energy give solar lifetimes that are short compared to geological estimates of the lifetime of the Earth. It was not until the development of ideas on the structure of matter and on nuclear processes, in the early and middle part of the 20th century, that it was eventually appreciated that nuclear fusion could provide the vast amount of energy needed to power most stars. In fact, it appears to be the only possible energy source that we know of that will do the job.

In the study of the Sun and its power source in Chapter 2, you were introduced to some basic ideas concerning atoms, nuclei and nuclear processes. We shall call heavily on some of these ideas and concepts in our study of the evolution of stars both in this and subsequent chapters. However, if you need to know more about nuclear processes, this is an appropriate time to study Box 6.1.

BOX 6.1 FUSION REACTIONS AND ENERGY RELEASE

In Section 2.2.4 you met Einstein's famous equation, namely $E = mc^2$, which links mass and energy. When applied to a nucleus, of mass m, it enables us to determine the *rest energy*, mc^2, of any nucleus. This can be thought of as the energy that would be released if the nucleons (protons and neutrons) were annihilated and converted into energy alone. It is possible for us to measure the rest energy of most nuclides to high accuracy. Figure 6.6 shows a plot of the rest energy *per nucleon*, mc^2/A, as a function of the mass number A, the number of nucleons in the nucleus. The general property of the curve is

that the rest energy per nucleon initially decreases fairly rapidly with increasing mass number; there is then a broad minimum around values of A between 50 and 60. Then the rest energy per nucleon gradually increases for nuclei with higher values of A. Note that $^{56}_{26}$Fe is a nuclide with one of the *lowest* rest energies per nucleon.

We can use Figure 6.6, or more detailed tabulated data, to work out whether a reaction is **exothermic** or **endothermic** – in other words, does the reaction release energy or does it require an input of energy in order for it to take place?

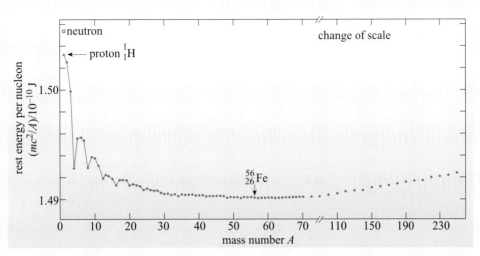

Figure 6.6 The variation of rest energy per nucleon as a function of mass number. At each mass number, the nuclide with the lowest rest energy per nucleon is shown. Beyond $A = 70$, only some mass numbers are shown.

■ Without *calculating* rest energies, use Figure 6.6 to decide whether the reaction

$$^{12}_{6}\text{C} + {}^{4}_{2}\text{He} \rightarrow {}^{16}_{8}\text{O} + \text{photon}$$

is exothermic or endothermic. (At each of the mass numbers 4, 12 and 16, these are the nuclides with the lowest rest energy per nucleon.)

❏ Figure 6.6 shows that the rest energy *per nucleon* is *lower* in the product $^{16}_{8}\text{O}$ than in either of the reactants $^{12}_{6}\text{C}$ or $^{4}_{2}\text{He}$. With the same number of nucleons (16) in the product as in the reactants, it follows that the sum of the rest energies of the reactants exceeds that of the product, and so the reaction *releases* energy, i.e. it is exothermic. (The fact that the number of nucleons in the reactants equals that in the product is in accord with the law of conservation of baryon number.)

To find the energy released in the reaction we make a quantitative calculation.

■ Calculate the energy released in the reaction. (The rest energies for the carbon, helium and oxygen nuclei are 1.7904×10^{-9} J, 5.9720×10^{-10} J and 2.3865×10^{-9} J, respectively.)

❏ The sum of the rest energies of the reactants on the left-hand side of the reaction is 2.3876×10^{-9} J. For the product, on the right-hand side, the rest energy is 2.3865×10^{-9} J. There is thus an excess of 1.1×10^{-12} J in the reactants.

The translational kinetic energy of the reactants is not very different from that of the product, and so nearly all of the energy released goes into the photon.

■ What type of photon is this?

❏ Using Equation 1.3, ($\varepsilon = hf$) we find

$f = \varepsilon/h = (1.1 \times 10^{-12} \text{ J})/(6.6 \times 10^{-34} \text{ J s}) = 1.7 \times 10^{21}$ Hz

We can see from Figure 1.36 that this frequency corresponds to a γ-ray photon.

The fusion of two light nuclei to form a single nucleus with $A < 56$ is exothermic for nearly all pairs of reactants. However, can we determine which nuclear reactions are fast enough to be significant in the interior of stars? Various factors contribute to the rate of a particular nuclear reaction. We shall look at those factors in a qualitative way only.

First, we would clearly expect the rate to depend on the concentration (the number density) of the reactants. For the simplest case of a reaction between pairs of particles, the rate, and thus the energy release per unit volume, is proportional to the *product* of the concentrations of the two interacting types of particles.

Another crucial factor is the ability of two nuclei to get close enough to each other to react.

■ What normally prevents two nuclei from approaching very close to each other?

❏ As nuclei are positively charged, it is the repulsive electrical force that tends to keep them apart.

The electric charge on a nucleus is Ze, where Z is the atomic number of the nucleus, and e is the charge on the proton (1.6×10^{-19} C). The *magnitude*, F_e, of the repulsive electrical force between nuclei with atomic numbers Z_1 and Z_2 is given by

$$F_e = A_0(Z_1 e)(Z_2 e)/r^2 \qquad (6.7)$$

and the electrical potential energy is given by

$$E_e = A_0(Z_1 e)(Z_2 e)/r \qquad (6.8)$$

where $A_0 = 8.99 \times 10^9 \, N \, m^2 \, C^{-2}$ is a universal constant, the Coulomb constant, and r is the separation of the nuclei. (Note how similar these equations are to those for gravitational force, Equation 3.17, which is also proportional to the inverse square of the distance separating two point masses.) Nuclear reactions can occur if the particles approach close enough to each other in spite of the repulsive electrical force; this occurs if the relative velocity of the two particles is high enough. Of course, the temperature is the main factor that dictates the relative velocity of particles in a gas. Although for a given temperature we can calculate the *average* velocity of a particle from its translational kinetic energy $E_k = 3kT/2$ (Equation 4.2), the particles actually exhibit a distribution of velocities with some lower and some higher than the average. However, it turns out that at the temperatures typical of the interiors of stars, the probability of one nucleus having sufficient energy to overcome the repulsive electrical force between nuclei is vanishingly small (see Question 6.5). So how then does nuclear fusion occur in stars? The answer called 'quantum mechanical tunnelling' lies in the details of quantum mechanics and the wave-like properties of atomic nuclei. The result is that for a given nucleus there is a small but non-zero probability that it can approach closely enough to another nucleus for nuclear fusion to occur despite the repulsive forces. As the temperature rises this probability increases.

QUESTION 6.5

Use Equation 6.8 to find the electrical potential energy of two hydrogen nuclei ($Z = 1$) at a separation of $r = 10^{-15}$ m (this is the distance at which the forces which hold nuclei together dominate and the two nuclei can combine). Use Equation 4.2 to determine the *average* thermal kinetic energy of a particle in the core of the Sun (assume $T = 1.6 \times 10^7$ K) to show that the temperature is too low for such particles to overcome the electrostatic forces.

We can also see from Equation 6.7 that the higher the electric charges of interacting nuclei, the greater is the repulsive electrical force between them. This generally means that nuclear reactions between light elements (which contain a small number of positively charged protons) occur at an appreciably faster rate at lower temperatures than reactions between heavy elements (which contain a large number of protons).

With this information, you should begin to see that light elements in a star can be gradually converted into heavier elements as a star evolves, and that, if the temperature within the star rises, these heavier elements can also undergo nuclear reactions. Figure 6.6 shows that most of these fusion reactions are exothermic until the product nucleus has a mass number of about 56.

6.3.1 The pp chain

You have already seen in Section 3.3.6 that the majority of stars have a remarkably similar composition to each other and to the Sun: 92% of all nuclei are hydrogen, about 7.8% are helium and the remaining 0.2% is made up of the other elements. It is clear that we should be looking to hydrogen for the source of nuclei to take part in the exothermic nuclear fusion reactions described in Box 6.1. The most important series of nuclear reactions occurring in main sequence stars are those that convert hydrogen into helium. This is termed **hydrogen burning** (although of course, the term here has nothing to do with setting fire to anything!).

There are various routes by which hydrogen can be converted into helium. However, some reactions are excluded by the conservation laws.

▪ Can you remember which conservation laws are relevant to nuclear reactions?

❑ They are the conservation of charge, baryon number and energy.

The hydrogen burning routes that are not excluded by the conservation laws each involve a chain of reactions, whose overall effect is the conversion of hydrogen nuclei into helium nuclei. Several such chains are believed to play an important part in the cores of main sequence stars. The particular temperature in the Sun's core dictates that the ppI chain (Section 2.2.4) should dominate, but that isn't necessarily the case in other stars. In stars with cores of progressively higher temperature than the Sun, we find that two other chains become important. These two are called the **ppII** and the **ppIII chains**.

■ What do the names ppII and ppIII tell you about these particular nuclear reaction chains?

❏ By analogy with the ppI chain, these reaction chains start with the interaction of two protons.

Moreover, for each of the pp chains, the net effect is the production of a helium nucleus, $^4_2\mathrm{He}$, from four protons ($^1_1\mathrm{H}$) (Figure 6.7).

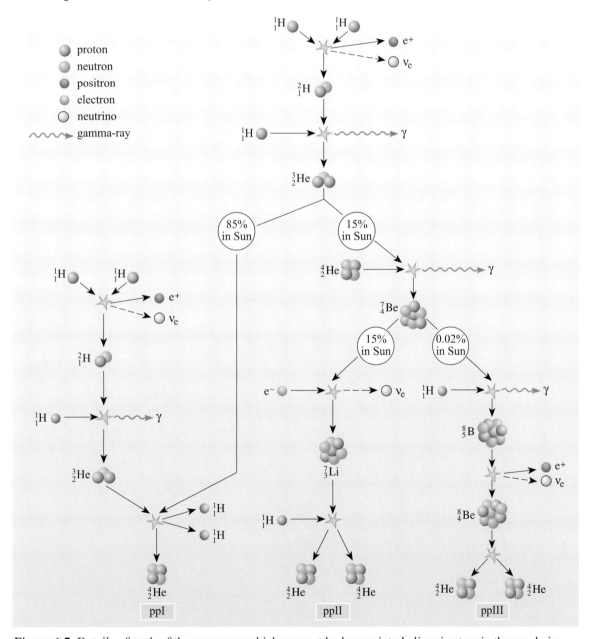

Figure 6.7 Details of each of the processes which convert hydrogen into helium in stars in the pp chains.

The neutrinos produced in the pp chains, by virtue of their very low probability of interacting with other matter, essentially all escape from the star in which they are produced. They carry off about 2% of the energy released in the formation of every helium nucleus (see Section 2.2.6 for details of the detection of neutrinos from the Sun).

6.3.2 CNO cycle

As we move to stars with yet hotter cores, we find a different set of reactions becoming important. These are reactions that involve the nuclei of carbon, nitrogen and oxygen. As in the pp chains, the net effect of this set of reactions is the production of a $_2^4$He nucleus from four protons. However C, N and O act as

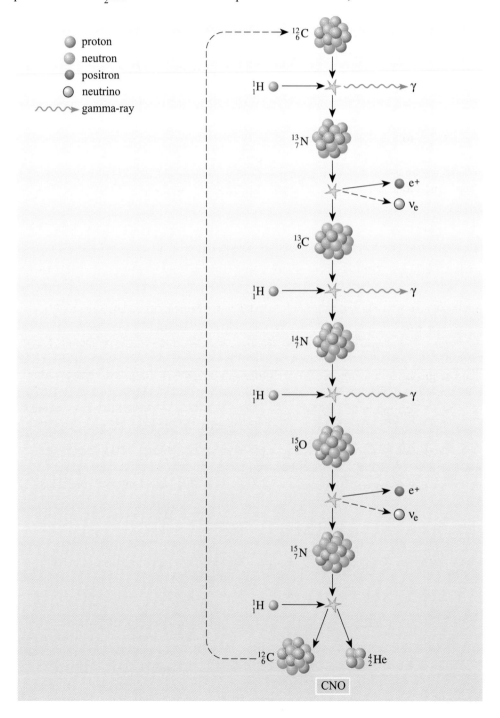

Figure 6.8 Details of each of the processes which convert hydrogen into helium in stars in the CNO cycle.

catalysts – that is, they help the reactions to take place. Although the relative abundances of the various isotopes of C, N and O may change, the combined concentration of these three elements, which is anyway low in main sequence stars, remains unchanged. This set of reactions is termed the **CNO cycle** and details are provided in Figure 6.8.

The net reactions for each of these processes are very similar:

ppI $\quad 4\,^1_1\text{H} \rightarrow \,^4_2\text{He} + 2\text{e}^+ + 2\nu_e + 2\gamma$

ppII $\quad 4\,^1_1\text{H} + \text{e}^- \rightarrow \,^4_2\text{He} + \text{e}^+ + 2\nu_e + 2\gamma$

ppIII $\quad 4\,^1_1\text{H} \rightarrow \,^4_2\text{He} + 2\text{e}^+ + 2\nu_e + 3\gamma$

CNO $\quad 4\,^1_1\text{H} \rightarrow \,^4_2\text{He} + 2\text{e}^+ + 2\nu_e + 3\gamma$

Figure 6.9 shows the rate of energy release as a function of temperature for both the pp and CNO reaction chains (the three pp chains have been added together for this purpose). As we can see, up to a temperature of about 18×10^6 K, it is the pp cycles that contribute most to energy generation within stars. Above that temperature, even though the pp cycles generate energy at an increasing rate, they are outstripped by the rate due to the CNO cycle. It is clear that the rate of energy generation depends heavily on the temperature – note the logarithmic scale. In the region where the two sets of reactions (pp and CNO) contribute equally to energy generation, the rates of energy production can be approximated by the following simple relations. For the pp chains

$$R_{\text{pp}} \propto n^2 T^4 \qquad\qquad (6.9)$$

and for the CNO cycle

$$R_{\text{CNO}} \propto n^2 T^{17} \qquad\qquad (6.10)$$

where R is the appropriate rate of energy generation per unit volume of gas, n is the number density of the reactant nuclei, and T is the temperature. The very high powers to which the temperature is raised in these equations go some way to explaining why certain stellar properties are so sensitive to a star's temperature. For example, in Section 4.2.4 you were told that for a 'mere' increase in mass along the main sequence of a factor of 500, the corresponding stellar luminosity increased by a factor of 10^{10}! This incredibly sensitive dependence of luminosity on mass can now be understood in the light of Equations 6.9 and 6.10, if you recall from Section 6.2.2 that the greater the mass of a star, the greater the temperature of its core.

The designation of *upper* and *lower* main sequence is a reflection of the division in mass (and therefore temperature) between those stars in which the pp chains dominate and those in which it is the CNO cycle that takes a dominant role. This

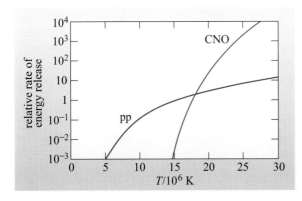

Figure 6.9 The rate of energy release for the three pp and CNO reaction chains as a function of temperature. A relative abundance of the elements as for the Sun has been assumed.

division occurs at a mass of around $1.5M_\odot$, corresponding to a peak core temperature of about 18×10^6 K. For more massive stars, in other words upper main sequence stars, the mass is greater and the temperature is higher, and the CNO cycle provides most of the energy generated. For less massive stars, lower main sequence stars, the temperature is lower and it is therefore the pp chains that are the main source of energy. The Sun falls into this latter category with a temperature nowhere higher than 16×10^6 K. In the Sun the ppI chain produces 85% of the energy, the ppII chain about 15% and the ppIII chain only 0.02%.

6.4 Stellar masses

6.4.1 Distribution of stellar masses

What can we say, if anything, about the *distribution* of the masses of stars? Are we equally as likely to find a low-mass star as to find a massive star?

■ What technique is used for the direct determination of the masses of stars?

❏ The only *direct* method of measuring the mass of a star is by studying the dynamics of a binary system, through observations either of visual binaries or eclipsing spectroscopic binaries (see Section 3.3.7). For all other stars, the techniques are less direct.

Although the measurement of mass of any one star might not be possible to high precision, if we observe a large number of stars, we can expect the overall trend to be known to a reasonably high degree of accuracy. A plot of the relative numbers of stars of different masses, called the mass distribution function, is shown in Figure 6.10. This confirms what was stated in Sections 3.3.7 and 4.2.4 – namely that star masses cover the approximate range from $0.08M_\odot$ to $50M_\odot$. The figure also indicates that far more stars have a low mass than a large mass. While we can believe that the relative rarity of massive stars is a genuine feature, at the other end of the mass range, can we be sure that the observations that contribute to Figure 6.10 are not suffering from a selection effect? If we observe stars at random, there is clearly a bias towards observing brighter stars. The less massive a star is, the less luminous it is and the harder it is to observe, so that we might be underestimating the true number of very low mass stars. The mass distribution of stars is of particular significance to understanding the total mass and therefore gravitational stability of star clusters and galaxies. If the very faintest of stars are not easily visible, a large fraction of the mass of a star system may be due to undetected faint stars. Such cool stars will be more easily detectable (and identifiable) from observations made at infrared wavelengths (remember Wien's displacement law) and recent sky surveys have revealed that they are extremely common. Results from a survey entitled the Two Micron All Sky Survey, or 2MASS for short, have led to the definition of a new spectral class of stars (L-types), which are cooler than M-types (see Figure 6.11). They appear to be twice as numerous as all other stars, but because of their low masses, they probably make up only around 15% of the total mass of stars in the Galaxy.

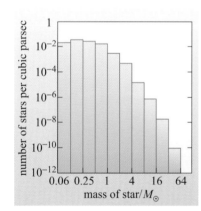

Figure 6.10 The number of stars observed as a function of stellar mass (based on observations of stars in the vicinity of the Sun). Note that both axes are logarithmic. In particular, each mass interval is twice the size of the preceding one. On a linear scale the dominance of the lowest mass stars would be even more pronounced.

If the mass distribution shown in Figure 6.10 continues to even smaller masses then we should have detected some extremely low-mass stars close to the Sun. Let's consider if there is a theoretical basis for defining a lower limit (and an upper limit) to the mass of a star.

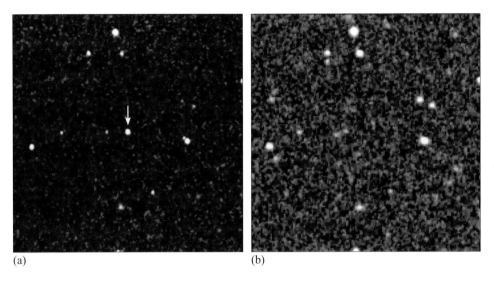

(a)   (b)

Figure 6.11 An L-type star discovered by the 2MASS sky survey. The left-hand image (a) shows the star (arrowed) as seen in the infrared 2MASS survey whereas it is so cool and red that it is not detected in the visible light image (b). (University of Massachusetts and Infrared Processing and Analysis Center/Caltech)

6.4.2 Brown dwarfs

We have already seen that the smaller the mass of a star, the lower the core temperature.

▣ What will be the effect on nuclear processes, such as the pp chains, of reducing the core temperature of a star?

▢ Figure 6.9 shows that the rate of energy generation will be reduced. Eventually, a temperature will be reached at which the nuclear reaction rate is insignificant.

Objects that are more massive than planets but do not have sufficient mass ($< 0.08 M_\odot$) to run nuclear reactions at a rate high enough to match the surface radiation rate are referred to as **brown dwarfs**. This name is often taken to indicate their colour (at the predicted surface temperature of these objects, they would be redder than 'red dwarfs', i.e brown) and their size. However, *brown* was in fact used to imply *no* colour, i.e. beyond the red end of the visible spectrum (the more appropriate *black dwarf* was at the time in use to describe the end state of a more massive star). The likely evolutionary track of a brown dwarf on the H–R diagram is shown in Figure 6.12. It evolves straight past the main sequence because significant nuclear reactions are never triggered.

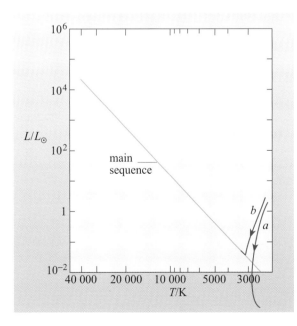

Figure 6.12 The path on the H–R diagram (plotted with the same axes as Figures 5.10 and 5.16), *a*, of a star of very low mass ($0.05 M_\odot$), i.e. a brown dwarf, which evolves straight past the main sequence, and the path, *b*, of a star of mass $\approx 0.3 M_\odot$, as it evolves on to the main sequence.

In addition to their low luminosity, their very low temperatures mean that they emit very little light at *visual* wavelengths. So any brown dwarf would therefore be very difficult to detect with certainty. Several brown dwarf candidates were 'discovered' in the 1990s, but there were difficulties associated with all of them, until 1994, when an object of mass between $0.03 M_\odot$ and $0.05 M_\odot$ was found orbiting the nearby star Gliese 229 (see Figure 6.13). Designated Gliese 229B, its surface temperature is around 1000 K and it is around 2000 times fainter than Gliese 229 which has spectral type M1 V, a temperature of 3800 K and is itself about 50 times less luminous than the Sun.

Figure 6.13 Images of a brown dwarf orbiting at a distance of 44 AU (a similar distance to that of Pluto from our own Sun) from the star Gliese 229. (a) This image was taken in October 1994 from the Palomar Observatory, (b) is from the Hubble Space Telescope in November 1998. The spike running from the over-exposed image of Gliese 229 in the image (b) is an artifact caused by the telescope optics. ((a) T. Nakajima (Caltech)/ S. Durrance (JHU); (b) S. Kulkarni (Caltech), D. Golimowski (JHU)/ NASA)

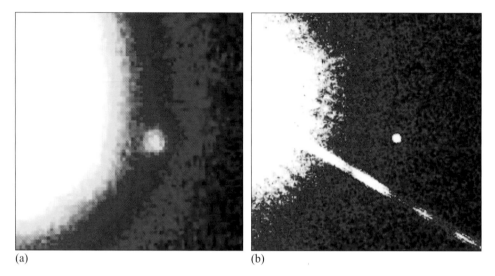

(a)   (b)

What is the difference between brown dwarfs and planets like the gas giant planets Jupiter and Saturn, since in theory they both may have similar masses? One possible answer lies in their formation and composition. Brown dwarfs form directly out of the interstellar medium and will have the same composition as other stars (at the time of formation) whereas planets form by accretion of dust in nebulae surrounding protostars (although distinguishing between a planet and a brown dwarf in a binary system with an ordinary star – see Figure 6.13 – may not be possible). Another definition sometimes used is based on mass alone, with the division being at 0.013 solar masses or 13 Jupiter masses. This is the mass above which deuterium undergoes nuclear fusion, providing a low-level energy source.

Figure 6.14 The central region of the Orion Nebula observed in visual light (a) and infrared (b). The four bright Trapezium stars that illuminate the nebula are clearly visible near the centre of both images. More than 300 stars are visible in the infrared image including about 50 brown dwarfs which are too faint and red to be seen in the visible image. ((a) NASA/K. L. Luhman (HSCA)/ G. Schneider, E. Young, G. Rieke, A. Cotera, H. Chen, M. Rieke, R. Thompson (Steward Observatory); (b) NASA/ C. R. O'Dell, S. K. Wong (Rice University))

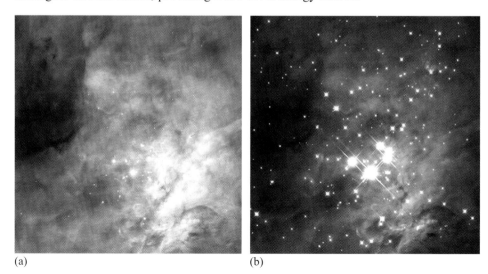

(a)   (b)

Recent surveys using the Hubble Space Telescope have shown that brown dwarfs are indeed common (Figure 6.14) and that, like stars, their numbers increase with decreasing mass. They also appear to be more common in isolation than in orbit around other stars. Many of the L-type stars are likely to have masses below the limit for fusion of hydrogen and are therefore also brown dwarfs.

6.4.3 The most massive stars

What about the situation at the other end of the mass range?

▨ Do you know of any reason why the mass of a star should have an upper limit?

❑ From the discussion so far, the only obvious limitation comes from the mass of the original contracting cloud. As typical clouds can have masses of several thousand solar masses or more, this is not a severe limitation. (However, you will see below that another limit exists.)

There is a process that is not important in most stars but which plays a crucial part in the most massive stars. You have already seen (Section 5.3.4) that electromagnetic radiation exerts a pressure called radiation pressure – the pressure exerted by photons of light, or of any other form of electromagnetic radiation. For photons within a black-body source (as is approximately true for typical stellar material), the radiation pressure is given by

$$P_{rad} = \tfrac{1}{3} \alpha T^4 \qquad\qquad (6.11)$$

where α is a constant, $7.55 \times 10^{-16}\,\mathrm{N\,m^{-2}\,K^{-4}}$, and T is the temperature of the radiation source.

We can use a variant of Equation 6.3 to write the gas pressure as

$$P_{gas} = nkT \qquad\qquad (6.12)$$

where we have used the fact that the number density $n = N/V$.

QUESTION 6.6

Estimate the ratio of the radiation pressure to the gas pressure at the core of the Sun. (*Hint*: in order to calculate n, determine an average value by using the mass and radius of the Sun; also assume that the Sun is composed entirely of ionized hydrogen and that the core temperature is $1.6 \times 10^7\,\mathrm{K}$.) Now use Figure 2.1 for a more realistic estimate of the density at the centre of the Sun. How does this affect the answer? (A qualitative assessment will do.)

We see therefore that radiation pressure is almost negligible in comparison with the gas pressure, at least in a star like the Sun. However, is this likely to be the case for all stars?

▨ What do you think will happen to the effect of radiation pressure in relation to the gas pressure as the temperature increases?

❑ In Question 6.6, it was shown that $P_{rad}/P_{gas} = \alpha T^3/3nk$. Therefore as T increases, the effect of radiation pressure compared with that of gas pressure will increase very quickly.

Whereas the stability of 'normal' main sequence stars results from a balance between the gravitational force and the force due to gas pressure, it appears that for more massive stars stability requires a balance between the gravitational force and the force due to radiation pressure. However, detailed modelling shows that radiation pressure increases so rapidly with temperature that such a star would be easily 'blown apart' by the radiation pressure. Detailed calculations show that the upper limit to the mass of a star is around $100M_\odot$ but its lifetime will be very short due to its rapid exhaustion of nuclear fuel.

6.4.4 Mass loss by stellar winds

Section 2.4.1 introduced us to the idea of the solar wind. It was suggested that it originates from material that has 'boiled off' from the solar corona, perhaps from regions where the magnetic field does not confine the material. If we make the assumption that the Sun is a typical main sequence star, then we might expect that a similar wind is exhibited by most stars – in which case, it should be termed a **stellar wind**.

The result of Question 2.10 shows that the rate of mass loss from a star like the Sun is small. Even if we sum this mass loss over the Sun's main sequence lifetime, it comes to little more than 10^{-4} of the total mass. Observations of the mass loss from other stars are very difficult owing to the low rates of mass loss and the distance to the stars, so data are very sparse. The rate of mass loss is expected to increase with increasing stellar mass such that for a massive star of $50M_\odot$, the mass loss will be of the order of $5 \times 10^{-7}M_\odot$ per year. In view of the very much shorter main sequence lifetime of such a star, the proportion of the total mass lost over the main sequence lifetime is still small.

Despite the relatively small amount of material involved in a typical stellar wind, it is one means by which material is returned from a star into the interstellar medium.

▨ What do you expect the composition of this material to be?

❏ You should recall from Figure 6.5 that whether convection takes place in an envelope (for low-mass stars) or in a convective core (for higher mass stars), the products of the nuclear reactions in the core of a star are not brought to its surface by convection. The material being recycled by main sequence stellar winds is not therefore enriched in heavier elements (predominantly helium) from the nuclear processes occurring in the core, but is representative of the material from which the star first formed. (However, dust, comets and possibly even planets may fall into the parent star's atmosphere and enrich it in heavy elements.)

The main sequence stellar wind isn't the only means by which a star loses material to the interstellar medium. You have already encountered one other episode when significant mass loss takes place (the T Tauri phase, Section 5.3.4) and in Chapter 8 you will encounter others, including red giant stellar winds, planetary nebulae, novae and supernovae, which are many times more effective at removing mass than main sequence stellar winds.

6.5 Summary of Chapter 6

Stellar structure

- Stellar models allow astronomers to predict the variation of such physical properties as temperature, pressure and density within a star during its main sequence lifetime. These properties can be verified only indirectly.

- The internal temperatures of main sequence stars increase with increasing stellar mass.

- The mode of energy transfer within a star depends on its mass. For a star of mass less than about $1.5M_\odot$, convection is confined to an outer envelope. In a more massive star, the temperature gradient in the core is sufficient to allow convection to take place there. In such stars convection is confined to the core and the adjacent region.

- The main sequence on the H–R diagram represents the stable configuration of stars of different mass but similar composition, converting hydrogen into helium through nuclear fusion.

- Nuclear fusion is the source of energy that powers main sequence stars. No other mechanism is known that can provide the observed luminosities over the main sequence lifetime.

- The luminosity of main sequence stars is strongly dependent on their masses. The lifetime of a star on the main sequence decreases rapidly with increasing mass.

Nuclear reactions

- The detailed nuclear reactions that are responsible for converting hydrogen into helium depend on the core temperature, and therefore the mass of a star. For stars of mass less than about $1.5M_\odot$ (lower main sequence stars), the pp chains predominate. For more massive stars (upper main sequence stars), reactions involving carbon, nitrogen and oxygen as catalysts are dominant, and comprise the CNO cycle.

Limits to stellar masses

- Very low mass stars are much more common than those of high mass.

- Stars of mass less than $0.08M_\odot$ never achieve a sufficiently high core temperature to sustain hydrogen fusion and so become brown dwarfs.

- An upper limit of around $100M_\odot$ is found because more massive stars will not be stable against the force of radiation pressure.

- Stellar winds during the main sequence lifetime lead to a small proportion of a star's mass being ejected into the interstellar medium during this phase of evolution.

Questions

QUESTION 6.7

Outline the different means involved in the transport of energy from the centre of the Sun to the skin of a person standing on the surface of the Earth.

QUESTION 6.8

Assuming that the age of the Earth is 4.5×10^9 years, determine the mass of the most massive star now on the main sequence that was also on the main sequence at the time of the Earth's formation.

QUESTION 6.9

Discuss whether the fusion reaction

$$^{12}_{6}C + {}^{12}_{6}C \rightarrow {}^{24}_{12}Mg + \gamma$$

(a) is possible and (b) could be an appreciable source of energy in main sequence stars. (At each of the mass numbers 12 and 24, these are the nuclides with the lowest rest energy per nuclide.)

CHAPTER 7
THE LIFE OF STARS BEYOND THE MAIN SEQUENCE

7.1 Introduction

The distribution of stars on the Hertzsprung–Russell diagram tells us that the majority of them lie on the main sequence. Stellar models (Chapter 6) tell us that their stability is a result of the balance between the inward force of gravity and the outward force of gas pressure (and for higher mass stars, radiation pressure). The outward pressure is sustained by the energy provided from fusion of hydrogen to helium. The rate of energy production, and hence the lifetime of the star on the main sequence is highly dependent on the temperature and hence the mass of the star.

■   What determines the end of a star's lifetime on the main sequence?

❏   The end of the main sequence lifetime is marked by the exhaustion of hydrogen in the core.

What happens to a star when this occurs? The position on the H–R diagram of a star undergoing hydrogen burning is determined by its mass and composition. When the process that maintains the star's structure no longer operates, the structure of the star and hence its temperature and luminosity will change. It will therefore move to a new location on the H–R diagram. Even before such drastic changes occur the star will still move slightly on the H–R diagram since its composition is continually changing as hydrogen is being converted into helium in the core. The main sequence therefore has a finite width on the H–R diagram with stars of the same mass but different ages and hence different compositions being in slightly different locations.

In this chapter we will look at the changes that occur in stars of different masses after the supply of hydrogen in their cores has been exhausted. In particular, we will examine the changes in structure and the changes to energy sources that occur after the main sequence life of stars.

7.2 Post main sequence lifetime of low-mass stars

7.2.1 Hydrogen shell burning

Generally, it is only in the core that it is hot enough for fusion reactions to occur. So the question arises – does the core get replenished with hydrogen through convective mixing between the core and the rest of the star? Convection in the core region occurs only in the more massive stars, but even there, as Figures 6.4 and 6.5 show, it does not extend much beyond the core. The rate at which hydrogen is depleted in the core is thus determined by the rate at which nuclear reactions take place there, and a critical point will be reached when the hydrogen has all gone. Once this stage has been reached, nuclear reactions in the core will stop. The core will then start to contract (slowly) as it is no longer releasing energy at a sufficient rate to generate a pressure gradient sufficient to support the surrounding layers. As a result of this contraction, gravitational energy is converted into thermal energy and

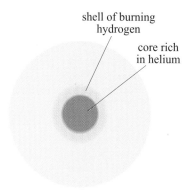

shell of burning
hydrogen

core rich
in helium

Figure 7.1 Schematic drawing
of a star with a shell of burning
hydrogen.

the temperature will rise. This means that a shell of unprocessed material surrounding
the original core will be heated sufficiently for hydrogen burning to start. This is
illustrated in Figure 7.1.

7.2.2 Helium burning

While the hydrogen shell burning occurs the core continues to contract and, in doing
so, continues to heat up as more gravitational energy is converted into thermal energy.
When a temperature of around 10^8 K is reached, a new range of nuclear reactions
becomes possible.

▨ Bearing in mind what you know of the composition of the core at this stage and
what you learned from Box 6.1, what do you think may be the next set of nuclear
reactions?

❑ The core now consists predominantly of helium, and from Box 6.1, we suspect
that **helium fusion** reactions should now occur – these would be exothermic
(Figure 6.6).

We might expect therefore that two helium nuclei would combine in some way to
produce $^8_4$Be . However, $^8_4$Be is very unstable and almost immediately decays back to
two helium nuclei. So is there any way in which helium burning can take place and
nuclear reactions continue? It was not until the 1950s that the full details were worked
out. It was appreciated that, at the temperature and density likely to prevail in the
helium core of a post main sequence star, it would occasionally happen that the short-
lived beryllium nucleus would meet another helium nucleus before it decayed. The
result is the formation of a $^{12}_6$C nucleus. The overall scheme is therefore:

$$^4_2\text{He} + {}^4_2\text{He} \rightleftharpoons {}^8_4\text{Be}$$

$$^8_4\text{Be} + {}^4_2\text{He} \rightarrow {}^{12}_6\text{C} + 2\gamma$$

The ⇌ symbol emphasizes the
fact that this is a two-way reaction.

This chain is known as the **3α (triple alpha) process**, α being the commonly used
symbol for the helium nucleus, because the net effect is the conversion of three helium
nuclei into a $^{12}_6$C nucleus. The onset of the 3α process – helium burning – halts the
contraction of the core and stabilizes the star.

Helium burning is initiated at temperatures of around 10^8 K and furthermore is
remarkably sensitive to temperature: the rate of energy release by this process is
proportional to the 40th power of temperature! This sensitivity arises because the third
α-particle needs to fuse with the beryllium nucleus to form the carbon nucleus before
the beryllium decays back to two helium nuclei. At high temperatures the nuclei are
moving at high speeds and therefore there is a greater probability of interaction.

The 3α process releases 1.17×10^{-12} J of energy per $^{12}_6$C nucleus, or 3.9×10^{-13} J
per $^4_2$He nucleus. This latter figure is only about 10% of the energy released in
forming the helium nucleus from hydrogen. This fact, coupled with the higher
luminosity during this phase, ensures that the time until the helium in the core is
exhausted, which we take to define the end of this phase, will be considerably shorter
than the main sequence lifetime. As a rough rule of thumb, the length of the helium-
burning phase is about 10% of the main sequence lifetime.

7.2.3 The giant phase

What has been happening to the rest of the star during this post main sequence phase? The contraction of the core, which was initially slow, has speeded up under the pressure of the outer regions of the star. Simultaneously, however, the radius of the star as a whole *increases*. Although this surface expansion which accompanies the core contraction is predicted by the equations of stellar structure, the explanation is not straightforward. It is also predicted that this expansion is not accompanied initially by a significant change in luminosity.

QUESTION 7.1

What will happen to the temperature of the outer layers of the star at this phase, bearing in mind the comment above about the star's luminosity in this immediate post main sequence phase?

As hydrogen burning in the shell around the core progresses, the luminosity is expected eventually to increase as convection carries energy to the surface. As an example, for a star of $1M_\odot$, the core will be compressed to about 1/50th of its original size and the core temperature will rise from 15×10^6 K to about 100×10^6 K. At the same time, the diameter of the star will increase by about a factor of 10, with a surface temperature falling to about 3500 K. This causes the star to glow with an orange-red hue, and this, together with its size, gives it its name: red giant. Once the core temperature has risen sufficiently, helium starts to burn in the core and the star now enters the next phase of its red giant life.

What about the paths of these stars on the H–R diagram? These are shown in Figure 7.2 for stars of different masses. Notice that as the red giant phase is approached, the tracks tend to crowd together. This indicates a fairly narrow range of conditions for red giants.

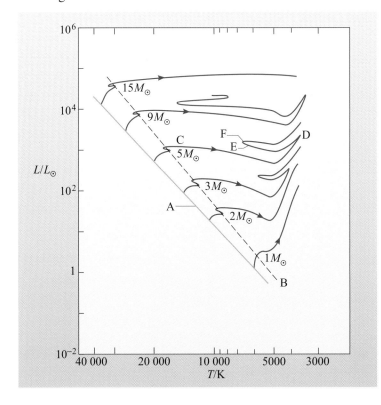

Figure 7.2 The predicted paths of stars on the H–R diagram as they evolve off the main sequence to the red giant (or supergiant) phase. The letters on the $5M_\odot$ track refer to different stages of nuclear reactions in the star. The line marked A denotes the onset of hydrogen core fusion – the start of main sequence life. The dashed line B denotes the cessation of hydrogen core fusion – the end of main sequence life, and the onset of hydrogen shell fusion. Subsequent stages are labelled on the $5M_\odot$ track only: (C) hydrogen shell fusion continues; (D) helium core fusion starts; (E) helium core fusion continues; (F) helium shell fusion starts. The small loops to the left for the $1M_\odot$ and $2M_\odot$ stars have been omitted for clarity.

After helium fusion starts in the core, the tracks retreat from a peak value of luminosity and migrate backwards and forwards on the H–R diagram (the extent of migration depends on the mass) as the red giant readjusts to its new sources of energy. Figure 7.2 also shows that post main sequence stars of mass greater than $\sim2M_\odot$ are not always at temperatures that give them an orange hue – they can be far hotter than this. Such stars, with luminosities comparable with those of red giants, plus the red giants themselves, are the *giant* stars introduced in Section 4.2.2.

Stars lying on the H–R diagram between the main sequence and the giants (i.e. to the right of the line labelled B in Figure 7.2) are, unsurprisingly, called **subgiants**. In evolutionary terms, subgiants consist of stars en route to becoming red giants.

Figure 7.3 illustrates the evolutionary track of a $1M_\odot$ star indicating key regions on the H–R diagram. After leaving the main sequence the star ascends the **red giant branch** (often abbreviated to RGB) with energy production from hydrogen shell burning. (Note from Figure 7.2 that stars of a wide range of masses reach this zone of the H–R diagram at the same stage of their evolution.) After the helium flash (explained in Section 7.2.4 below) the star undergoes helium core burning and moves approximately horizontally (i.e. with relatively constant luminosity) to higher temperatures. This places the star in a region of the H–R diagram called the **horizontal branch** (HB). Once core helium burning ceases, the star once again cools, but expands and increases in luminosity, approaching the red giant branch from the left. This region is called the **asymptotic giant branch** (AGB). Figure 7.3b shows an H–R diagram for a globular cluster (an old cluster which therefore contains many evolved low-mass stars; Sections 3.2.4 and 4.2.5) which illustrates

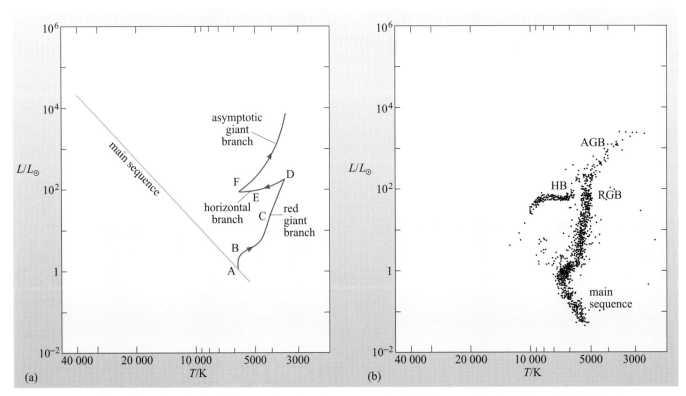

Figure 7.3 (a) The predicted path of a $1M_\odot$ star, plotted on the same scale with the same labels as Figure 7.2, (A) hydrogen core fusion; (B) onset of hydrogen shell fusion; (C) hydrogen shell fusion continues; (D) helium core fusion starts; (E) helium core fusion continues; (F) helium shell fusion starts. (b) The H–R diagram of a globular cluster which illustrates how stars tend to concentrate in these regions.

the tendency for stars to concentrate in these regions. For example stars of different masses all cluster in a horizontal bar, the horizontal branch, during helium core burning. (Note that the main sequence is not in quite the same place as in Figure 7.3a due to the fact that the chemical composition of the interstellar material was somewhat different at the time these stars formed. This will be discussed further in Section 8.4.3.)

We will now look in more detail at what happens within the star during these stages of its evolution.

7.2.4 The helium flash

The manner in which the helium burning starts depends on the *mass* of the star, with an important difference between stars with masses below and above about $2.25M_\odot$.

At the root of this difference is a phenomenon known as **degeneracy**. A detailed description of degeneracy is beyond the scope of this book – it requires a knowledge of quantum mechanics. However, we do need to know something about this phenomenon in order to understand several features of the evolution of stars after they leave the main sequence. Up to now we have been able to assume that the gas inside a star behaves like an ideal gas. For such a gas, simple equations, such as Equation 6.1, can be used to describe the relationships between pressure, temperature and density for example. At the extreme densities that exist deep inside some stars, the matter may be so compressed that a different set of equations must be used to describe the physical properties of the *electrons* in the gas – remember that the atoms are ionized, so we have an electron gas mixed up with a gas of atomic nuclei. This is the so-called degenerate electron gas. It has various properties that differentiate it quite clearly from the more normal gases that we are familiar with.

We shall focus on one property in particular and see how this affects the behaviour of red giants of low mass. Equation 6.1 shows that, if the temperature of an ideal gas is increased, its pressure will increase proportionally if other properties remain unchanged. The increase in pressure leads to expansion, and hence to cooling. For a degenerate gas, however, the situation is different. When the temperature changes, the pressure is unaffected. The pressure (called **electron degeneracy pressure** for a degenerate electron gas) depends on density and composition rather than temperature.

Why is this relevant to the situation we find in red giants? If helium burning starts in an ideal gas, this is basically a stable process. If a small temperature rise occurs in the helium-burning core we can expect an increased rate of release of energy because the nuclear reaction rate depends on a high power of the temperature. If the stellar material is fairly opaque, the energy may not be able to escape. Therefore, the local temperature will rise – in an ideal gas, this will result in the pressure rising, the gas expanding and cooling, and therefore the rate of nuclear reactions falling. There is, in other words, an in-built stability to the whole process.

However, if helium burning starts in a degenerate gas, the situation can be very different. Because the pressure in a degenerate gas is now nearly independent of temperature, the rise in temperature on initiation of nuclear processes does not produce a consequent rise in pressure, expansion and cooling to control the initial rise in temperature. This rise in temperature therefore causes the helium burning to continue even faster. In addition, degenerate material is a very good conductor of heat so once the temperature is high enough for helium fusion to start in one part of the core, the turn-on spreads throughout the core very rapidly. This is an unstable situation – the process can start to 'run away' and produce an explosive release of energy in the degenerate core of these lower mass stars. This is usually termed the (core) **helium flash**. This is, incidentally, believed to be one of the few cases in the history of a star where an event occurs over a timescale perhaps as short as a matter of hours or less. However, although the helium flash happens very quickly in the core, the release of energy takes very much longer to reach the surface.

One result of the helium flash is to raise the core temperature to the point where the degenerate conditions are removed (see Figure 7.4). Once degeneracy is removed in this way, then the core can expand and cool and the situation will be stabilized.

Under exactly what conditions does degeneracy occur? Figure 7.4 shows a plot of density against temperature divided into two regions corresponding to normal (i.e. non-degenerate) and degenerate conditions. The transition between the two is not sharp and so is indicated by a fuzzy boundary. The region occupied by the cores of main sequence (i.e. hydrogen burning) stars is indicated. Stars of lower mass, after their main sequence lifetime, evolve into the region corresponding to degenerate conditions, whereas higher mass stars achieve core temperatures at which helium burning is initiated before degenerate conditions are reached.

To summarize, the evolution of stars after they leave the main sequence depends on whether degeneracy sets in before helium burning (the 3α process) is initiated. If the mass is greater than about $2.25M_\odot$, the helium reactions start before the core can become degenerate. For stars with masses less than about $2.25M_\odot$, helium burning starts in a degenerate core in a violent reaction known as a helium flash. This will probably remove the degenerate state of the core.

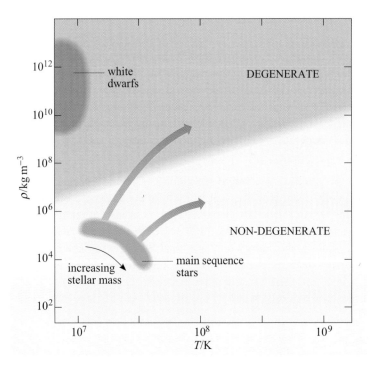

Figure 7.4 Conditions for degeneracy in an electron gas. The conditions in the cores of main sequence stars are shown together with arrows indicating the change in core temperature and pressure as they evolve up to the point of onset of helium burning. White dwarfs are discussed in Section 9.2.

7.2.5 Internal structure

The changes in the structure of a star during the post main sequence phase are quite complicated. Figure 7.5 shows what happens in a star of $5M_\odot$ after it leaves the main sequence. The different processes that occur at the various zones in the cross-section of the star are indicated in the lower panel by the different coloured regions. The horizontal axis indicates the time since the start of the star's life on the main sequence (i.e. since hydrogen burning commenced). The quantity plotted on the vertical axis is the mass fraction, M_R (the fraction of the total mass inside a given radius as we move outwards from the centre of the star). $M_R = 0$ at the centre of the star and $M_R = 1$ at the surface of the star. This quantity is used rather than the radius because the nuclear reactions are taking place in a region that is very small and yet contains an appreciable proportion of the star's total mass. Also, the radius of a star changes enormously as the star evolves but the total mass remains constant (at least until the later stages of evolution – see Chapter 8). The compact nature of the helium-burning inner core is clear from Figure 7.5. It is also clear that (a) hydrogen burning will continue in a thin shell, which will be at the surface of the helium zone, though well away from the helium-burning inner core, and (b) after depletion of helium in the core, helium burning will continue in a shell that moves progressively outwards as further helium is used up. These factors are partly responsible for the tortuous path predicted for some stars across the H–R diagram during the giant phase (Figure 7.2).

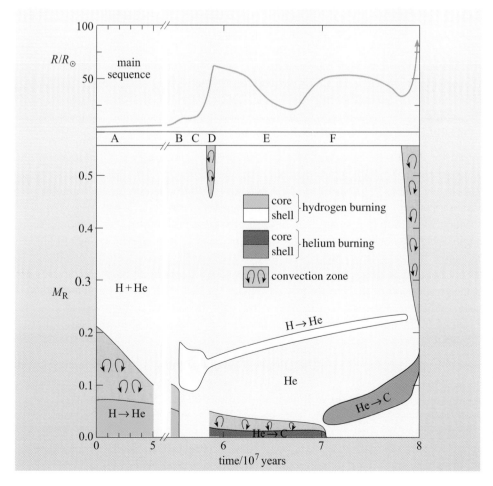

Figure 7.5 Schematic representation of the internal structure of a star of mass $5M_\odot$ during and after its main sequence lifetime. The upper panel shows the change in radius of the star with time. Note there is a change of scale in the time axis between 5 and 6×10^7 years to reflect the faster evolution of the star after it leaves the main sequence. The lower panel shows the change in composition and nuclear reactions in the star as it evolves. The vertical axis is the mass fraction M_R (the fraction of the total mass inside a given radius as we move outwards from the centre of the star), with the centre of the star at the bottom. The coloured regions indicate the locations of nucleosynthesis and the grey zones are convection zones. The labels A to F indicate the times of significant changes in the nuclear reactions as shown on the evolutionary track on the H–R diagram in Figure 7.2.

7.2.6 Pulsating variables

Giants that have evolved to the stage at which core helium burning has started are often intrinsic variables. One class of intrinsically variable star – the Cepheids – was introduced in Section 3.3.5. The variability in Cepheids and related stars results from a large amplitude pulsation of the star – in which the radius of the star periodically varies with an amplitude which may exceed ten per cent. Although this is a much larger effect than the five-minute oscillations of the Sun that you met in Section 2.2, the way in which these pulsations propagate through the star is similar. The principal difference between the two effects arises from the mechanism that drives the pulsations.

■ What is the mechanism that drives small-scale global oscillations of the Sun?

❏ Convective motions are the likely source of disturbance that drives global solar oscillations (Section 2.2.6)

The much larger pulsations that are observed in Cepheids arise from an instability in the envelope of the star. The way in which this instability gives rise to oscillations can be appreciated by considering a thin (spherical) layer of gas within the envelope of the star. Compressing this layer results in an increase in temperature and this gives rise to an increase in opacity (a measure of the absorption of radiation). The rate at which energy is transported by radiation through the layer is thus reduced, and this results in heat being trapped below the layer. Thus the temperature and pressure below the layer increase, and the increase in pressure drives the layer outwards. The layer now expands, its density drops and it becomes more transparent hence allowing heat to pass out through the layer. The pressure below the layer then drops and so the layer falls and becomes compressed. This cycle then repeats itself. The underlying physical cause is the change in opacity with temperature, and this comes about because of the ionization of He^+ to He^{2+}. This ionization increases the density of free electrons, and because electrons interact strongly with radiation, the result is a drop in transparency. This condition arises at a certain temperature (about $40\,000\,K$), and so the layer that exhibits this instability exists at a particular radius within the star.

In order for the star to show large-scale pulsations, it is necessary that the location of the unstable layer is at a position that can drive the oscillation. This behaviour is somewhat similar to the behaviour of the type of swing that is commonly found in children's playgrounds. You may know from experience that in order to achieve a large amplitude oscillation in a swing, you have to provide a push at the correct moment – at the time when the swing has momentarily stopped and is at its maximum height above the ground. If you try applying a push at other parts of the cycle, then the oscillation does not build up to large amplitudes. By analogy, in a Cepheid, the unstable layer is in just the correct part of the star to provide regular 'pushes' that give a large amplitude global oscillation of the star. If the layer is too near the surface or too close to the core, the motion of the unstable layer does not match itself to the global oscillation of the star and pulsation does not occur.

Because the phenomenon of large amplitude pulsation requires a particular combination of conditions in the envelope of the star, we shouldn't be surprised to learn that these stars appear to occupy certain well-defined positions on the H–R diagram. The pulsating classes of stars include Cepheids and RR Lyrae stars, as well as lower luminosity examples such as the so-called delta Scuti stars. The region on the H–R diagram in which the pulsating stars lie is termed the **instability strip**, which extends from the supergiant region to below the main sequence as shown in Figure 7.6.

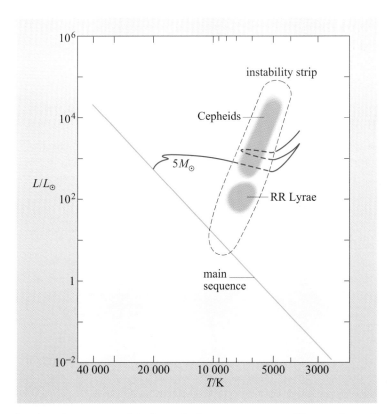

Figure 7.6 The position of the instability strip on the H–R diagram. The evolutionary track of a $5M_\odot$ star is shown dashed where it is exhibiting pulsations. At this time its position on the diagram will oscillate due to its changes in temperature and luminosity.

During the course of their evolution, many stars will pass through this region and thus display this type of instability. Comparison of Figures 7.2 and 7.6 shows that, in the post main sequence phase, the part of the instability strip that will be crossed depends on a star's mass, and thus stars of different mass might be expected to fall into different pulsating classes. The Cepheids correspond to stars of mass greater than about $2M_\odot$; these are giants and supergiants that cross the instability strip as their evolutionary tracks loop across the H–R diagram.

The RR Lyrae stars are those that lie at the intersection of the instability strip with the horizontal branch, and hence correspond to stars with a mass of between $0.7M_\odot$ and $2M_\odot$. RR Lyrae stars all have periods of less than one day and have roughly the same luminosity of around $100L_\odot$. These stars are found in globular clusters (Section 3.2.4) and are used to determine distances to such clusters in much the same way as Cepheids are used to determine distances (Section 3.3.5).

Although there is also an intersection of the instability strip with the main sequence, the numbers of such stars that have been identified is relatively low because the variation in luminosity is small and consequently these stars are hard to detect. Also, as shown in Figure 7.6, a star can pulsate when it crosses the instability strip for the first time during hydrogen shell burning. However, this phase is much shorter than the helium core burning phase normally associated with Cepheid variables. However, across the entire luminosity range of the instability strip, many thousand stars are known to be variables of the pulsating envelope type, so they clearly are relatively common.

The study of variable stars is a very important tool that enables astronomers to probe the structure of stars, and can potentially yield more information than observations of non-varying stars. The time a star spends on the instability strip is thought to be not particularly long, and ends once the conditions that are necessary for pulsation are removed by the continued evolution of the star.

Table 7.1 Data for Question 7.2.

Star	Spectral type	M_V
X	F0	3
Y	K5	−6
Z	A0	1

QUESTION 7.2

Which of the stars listed in Table 7.1 could be pulsating variables on the basis of their position on the H–R diagram shown in Figure 7.6?

7.2.7 Very low mass stars

For stars of mass less than about $0.5M_\odot$, we find a very different story. Theoretical calculations show that the critical mass below which helium burning is unlikely to start is around $0.5M_\odot$. The evolutionary track of a star with a mass somewhere between $0.1M_\odot$ and $0.5M_\odot$ is shown in Figure 7.7. Initially, the star evolves in a similar direction (compare with Figure 7.2) to a star of slightly higher mass. However, helium burning never starts in such a star, so the luminosity soon peaks and then declines rapidly.

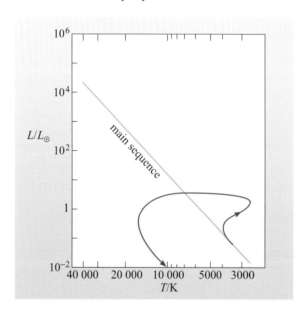

Figure 7.7 The post main sequence evolutionary track of a star of mass between approximately $0.1M_\odot$ and $0.5M_\odot$.

7.2.8 Further reactions in giants

Is there anything more that can happen to a star in the giant phase? In addition to the helium-burning reactions, there is another reaction that occurs in all giants. The $^{12}_{6}C$ nuclei produced by helium burning can capture an additional α-particle in the following reaction

$$^{12}_{6}C + {}^{4}_{2}He \rightarrow {}^{16}_{8}O + \gamma$$

to produce the heavier nucleus, $^{16}_{8}O$. This reaction and the 3α process in red giants are thought to be the main sources of oxygen and carbon in the Universe. (Note that this reaction producing oxygen from carbon should not be confused with the CNO cycle in which carbon, nitrogen and oxygen are involved in the process of conversion of hydrogen into helium in massive stars as described in Section 6.3.)

The addition of another α-particle results in the production of neon:

$$^{16}_{8}O + {}^{4}_{2}He \rightarrow {}^{20}_{10}Ne + \gamma$$

although this is only a minor reaction in stars of this mass.

Giants with a mass less than about $8M_\odot$ do not develop core temperatures high enough to trigger any further nuclear reactions beyond that which produces oxygen. The cores of such stars accumulate carbon and oxygen, the remnants of helium burning. When the helium in the core is exhausted, the core begins to contract again. This heats the helium surrounding the core sufficiently to trigger helium burning, the 3α process, in a shell. This situation is shown schematically in Figures 7.5 and 7.8. The star is now on the asymptotic giant branch (AGB).

Shell helium burning has caused the giant to expand further and move upward on the H–R diagram for the second time. The star is now even larger – its size can be as large as the orbit of Mars! However, this situation isn't stable. The helium-burning shell is rather thin and this causes a thermal runaway to occur. The result is another helium flash, but quite different from the core helium flash that you have already met. In that case, the cause was degeneracy but now, in the helium-burning shell, the material is not compressed enough for degenerate conditions to be reached. These **shell helium flashes** are due to the fact that the shell is too insubstantial to lift the material above it. Thus, as shell helium burning gets underway, the shell cannot expand and so the temperature rise is not moderated. The helium burning rate increases, increasing the temperature further. This leads to a rapid release of energy, sometimes called a **thermal pulse**, which lasts a few hundred years. These flashes are thought to be approximately periodic events but separated by intervals of 10^4 to 10^5 years. During the thermal pulse, conditions in the helium-burning shell are right for another type of nuclear reaction to take place in parallel; this is called the **s-process reaction** (where s is for slow; you will meet the r or rapid process in Section 8.3.1).

In the s-process a neutron is added to an existing nucleus to make a heavier one.

- Why should neutrons be able to react easily with a nucleus?

- They are neutral particles, so there will be no electrostatic repulsion as would be the case if they possessed a positive charge. Thus reactions can take place at low temperatures.

In the s-process nuclei capture neutrons slowly in the sense that for any nucleus, the typical time between neutron capture reactions is very long. The reason for this is that the density of free neutrons is relatively low. If this addition results in an unstable nucleus there is time for some radioactive decay (usually by the emission of an electron) before the next neutron capture takes place. By the slow, steady addition of neutrons, interleaved with some radioactive decay, nuclei as heavy as

$^{209}_{83}\text{Bi}$ are created. Some isotopes such as $^{80}_{36}\text{Kr}$ and $^{82}_{36}\text{Kr}$ can only be created through the s-process.

The thermal pulse has other effects as well, which help us see what is happening in the inner parts of the star. The release of energy in the helium-burning shell alters the pattern of convection in the star, so that a deep convective envelope is formed. Convection reaches right down from the surface to the helium-burning shell, and circulates surface material down to that level while dredging some of the helium burning and s-process products up to the surface (you can see that a deep convective envelope appears at the extreme right-hand edge of Figure 7.5). Spectral lines from these elements can then be seen in the star's spectrum. Due to the s-process plus the 'dredge-up' by convection, the radioactive element technetium

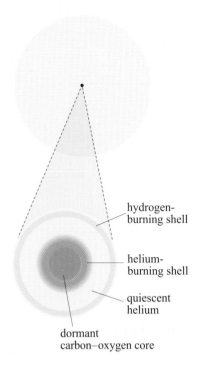

Figure 7.8 The structure of an old red giant of low mass ($<4M_\odot$). The star has a size comparable with the orbit of Mars.

(labels in figure)
hydrogen-burning shell
helium-burning shell
quiescent helium
dormant carbon–oxygen core

has been seen in the spectra of some stars. Technetium is an unstable element which decays typically in a few million years. If the surface of the star was not refreshed from below then radioactive elements like technetium would have decayed long ago and would no longer be seen in the star's spectrum. If the s-process were not currently taking place in the interior of the star there would be no technetium to bring to the surface. Short lifetime (on an astronomical timescale) elements on the surface of a star, far from the regions where nuclear reactions take place, testify to the large-scale circulation of material in the star. The actual elements found tell us which nuclear processes are taking place inside the star.

The thermal pulses may play a part in the ejection of large amounts of material, a phenomenon observed to take place towards the end of the lives of some stars (as you will see in Chapter 8).

The various processes that have been discussed in the giant phase explain why the evolutionary track on the H–R diagram at this time is predicted to be quite complicated (Figure 7.2). Every time a new energy source dominates the star's evolution, the direction on the H–R diagram is likely to change. We should remember, however, that this part of a star's lifetime is short compared with the main sequence lifetime – for a $1M_\odot$ star, 10^9 years as opposed to 10^{10} years. This is why we would expect to observe far fewer stars in the giant stages of their evolution compared with those on the main sequence.

In the final stages of their lives, low-mass stars become variable stars and undergo mass loss culminating in the loss of their outer envelopes and formation of dense stellar remnants called white dwarfs. This final stage of their evolution will be discussed in Chapters 8 and 9.

7.2.9 Winds from red giants

Before we finish discussing the main phase of the life of a giant, we should consider one important effect of its increased radius. The star's gravity at its surface will be very much less than what it was during its main sequence lifetime since the mass has not changed significantly but the radius has increased considerably. This means that atoms in the red giant's atmosphere can more easily escape, resulting in a copious stellar wind, though at a lower velocity than that of the main sequence stellar wind. Mass loss rates of up to $10^{-6}M_\odot$ per year are possible for the most luminous red giants. Winds from red giants can therefore be responsible for a more significant loss of the star's total mass, in contrast to the very small loss by this mechanism during the main sequence lifetime. Mass loss in highly evolved stars will be discussed in more detail in Chapter 8.

7.3 Post main sequence lifetime of high-mass stars

We have seen that stars like the Sun, and those of up to a few times the mass of the Sun, when they run short of hydrogen in their cores, swell to become red giants and undergo helium fusion. What is the fate of stars much more massive than the Sun?

You saw in Section 6.2.5 that massive stars have a much shorter lifetime on the main sequence than lighter stars. Because they are more massive they have a higher central temperature, and so the hydrogen-burning nuclear reactions proceed faster. Because their nuclear reactions proceed faster, energy is released more quickly and so these stars are also brighter. For a short while they are hot and bright, and are found in the top left-hand part of the H–R diagram (spectral types O and B).

Some stars (those that formed relatively recently) are still to be found in that part of the H–R diagram. But where are those that formed earlier? Stars that were once on the main sequence in the top left-hand corner of the H–R diagram are today's **supergiants**. (See Figure 4.5 to remind yourself where the supergiants are located in the H–R diagram.)

You have already met two of the prominent stars in the constellation of Orion (see Figure 3.1) that are supergiants. Betelgeuse, which to a northern hemisphere observer is the bright star in the top left-hand corner of the constellation, is termed a red supergiant. The colour is perceptible, but is not very dramatic; as one's eyes become dark-adapted after 15 minutes or so, an orange tint becomes noticeable. A colour photograph taken with an ordinary camera will show it clearly. Rigel, at the opposite corner of the constellation, is a blue supergiant, shining with a blue-white light.

QUESTION 7.3

If a star moves from the upper main sequence in the top left-hand part of the H–R diagram to become a supergiant, what changes take place in its luminosity and surface temperature?

QUESTION 7.4

A main sequence star initially has a surface temperature of $25\,000\,K$ and radius $10R_\odot$. The temperature drops (without change of luminosity) to $5000\,K$ as it becomes a (yellowish-white) supergiant. Use Equation 3.9 to determine its new radius. Express your answer in units of solar radius and AU.

As the result of Question 7.4 shows, supergiants may have very large radii. In fact, Betelgeuse has such a large diameter and is sufficiently close that it has been possible to resolve the stellar disc using the Hubble Space Telescope (Figure 3.17).

7.3.1 Massive stars on the main sequence

As with the less massive stars, the reason for a massive star moving off the main sequence is that a significant fraction of the hydrogen in the core has been consumed and converted into helium by nuclear fusion reactions.

▓ What is the name of the set of nuclear fusion reactions that dominate in the heavier main sequence stars?

❏ The CNO cycle (Section 6.3.2).

QUESTION 7.5

Can you suggest a reason why there are so few stars in that part of the H–R diagram between the upper main sequence and the area where the supergiants lie?

The evolution of a massive star to a supergiant parallels the evolution of a less massive star to a red giant. In lower mass stars the pressure of the hot gas supports the star, whereas in more massive stars (over about $5M_\odot$) radiation pressure provides the dominant supporting force (Section 6.4.3).

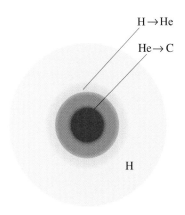

H → He

He → C

H

Figure 7.9 The structure of a helium-burning supergiant (not to scale).

7.3.2 After the main sequence

As you saw in Section 7.2.1, the nuclear burning slows as the fuel in the centre of the star is used up. Let's consider this process in a general sense. There is still burning where fresh material is available in a shell around the outer edge of the core, but the core is becoming choked with 'ash', i.e. nuclei that are the product of the nuclear reaction. As the nuclear reactions diminish, there is no longer the pressure gradient from the escaping radiation to balance the gravitational force and so the core contracts under gravity. A consequence of this contraction is that the density and the temperature of the core of the star rise, and there comes a point where what has been inert 'ash' can itself start to burn. Another nuclear fusion reaction can start, converting the material in the core into yet more massive nuclei, and once again producing energy. So the star keeps shining and again is in balance.

The first time this happens, after the main sequence, it is the fusion of helium to carbon (the 3α process, described in Section 7.2.2) that starts. While this phase lasts the star has two sources of energy from nuclear reactions (Figure 7.9): the fusion of helium to carbon in the core, and the fusion of hydrogen to helium in a shell outside the core.

By this stage the star has become a supergiant. Its surface temperature (and hence whether it is called blue, yellow or red) depends on the star's mass and the rate at which it loses mass through its stellar wind (see Section 7.3.5). Future changes in luminosity and temperature also depend on these properties. Some supergiants are always blue, some track from blue to red and stop there, and some will track back again from red to blue.

7.3.3 Astronomical alchemy

What happens next in the star, when the amount of helium available in the core for conversion into carbon diminishes noticeably? As the nuclear reaction wanes and the pressure gradient due to the escaping energy diminishes, the core of the star once again contracts under gravity. The core temperature goes up yet again and, at temperatures of about $3 \times 10^8 \, \text{K}$, the reaction

$$^{12}_{6}\text{C} + {}^{4}_{2}\text{He} \rightarrow {}^{16}_{8}\text{O} + \gamma$$

can commence. The star then has three sources of energy from nuclear reactions.

■ What are these three sources of energy?

❏ $^{12}_{6}\text{C} + {}^{4}_{2}\text{He} \rightarrow {}^{16}_{8}\text{O} + \gamma$ in the core; the 3α process in a shell around the core; and hydrogen burning in a shell outside that again.

QUESTION 7.6

Sketch the structure of the star at this stage. (You might find it helpful to have Figure 7.9 in front of you as you do this.)

The supergiant star goes through this pattern of steps a number of times as successively heavier elements become scarce in the core. In contrast to lower mass stars, in stars initially over about $8M_\odot$, each new burning starts in a non-degenerate core. However, these later stages of evolution of a supergiant star are not fully understood, and what follows should be treated with caution; as our understanding continues to grow the details may well change. The following account describes the main phases of nuclear burning that a high-mass star is believed to experience, and highlights some of the more important nuclear reactions that are thought to take place in the core. In reality, many more nuclear reactions take place than can be described here.

When helium burning stops, the carbon and oxygen core of the star contracts and at a temperature of about 5×10^8 K **carbon burning** commences. The fusion of two nuclei can have different outcomes. The simplest outcome, which actually occurs rather infrequently, is the formation of magnesium:

$$^{12}_{6}\text{C} + ^{12}_{6}\text{C} \rightarrow ^{24}_{12}\text{Mg} + \gamma$$

The two most likely reactions are those that produce sodium and neon:

$$^{12}_{6}\text{C} + ^{12}_{6}\text{C} \rightarrow ^{23}_{11}\text{Na} + ^{1}_{1}\text{H}$$

$$^{12}_{6}\text{C} + ^{12}_{6}\text{C} \rightarrow ^{20}_{10}\text{Ne} + ^{4}_{2}\text{He}$$

The reaction that produces sodium also results in the emission of a proton ($^{1}_{1}\text{H}$), while that which forms neon also forms an α-particle. Thus, very light nuclei tend to be formed as by-products.

QUESTION 7.7

What would happen to a proton or an α-particle in this environment? Suggest a reaction that has already been described that might occur. Is it likely that a significant amount of hydrogen or helium would build up in the core as a result of carbon burning?

The effect of carbon burning is to create elements with mass numbers (A) of around 20 and to produce protons and helium nuclei which undergo fusion reactions with any of the heavier elements (from carbon to magnesium) that are present in the core.

As carbon becomes depleted, the core contracts and heats up, and at a temperature of about 10^9 K a new type of process starts to become important. As an introduction to this process it is useful to consider the following question.

QUESTION 7.8

Calculate the wavelength of the peak of the black-body spectrum corresponding to a temperature of 10^9 K. In what part of the electromagnetic spectrum does this peak lie?

As Question 7.8 revealed, the core is so hot that the black-body spectrum extends into the γ-ray part of the spectrum. This is significant because many fusion reactions that release energy in the form of a γ-ray can be reversed. For example, a reaction that occurs at temperatures of about 3×10^8 K (i.e. before the onset of carbon burning) is the formation of neon by the reaction

$$^{16}_{8}O + {}^{4}_{2}He \rightarrow {}^{20}_{10}Ne + \gamma$$

However, this reaction can be reversed if nuclei of $^{20}_{10}Ne$ are in an environment where there are γ-rays:

$$^{20}_{10}Ne + \gamma \rightarrow {}^{16}_{8}O + {}^{4}_{2}He$$

Since the effect is that of a nucleus being split up by a γ-ray photon, this process is called **photodisintegration**. At the core temperatures that occur after carbon burning, this photodisintegration reaction plays an important role in the next stage of nuclear burning.

■ What does the core of the star consist of at the end of carbon burning?

❏ From the carbon burning reactions, the core will consist of neon, sodium and (some) magnesium.

At temperatures greater than 1.5×10^9 K some of the neon will undergo photodisintegration to oxygen according to the reaction described above. The other product of this reaction is α-particles, which can react with nuclei of $^{20}_{10}Ne$ (that have not undergone photodisintegration) to form magnesium:

$$^{20}_{10}Ne + {}^{4}_{2}He \rightarrow {}^{24}_{12}Mg + \gamma$$

These processes have the effect of changing the composition of the core to being a mixture of oxygen and magnesium. This stage is sometimes referred to as **neon burning** although it should be noted that the reaction is not a simple fusion reaction between two nuclei of neon.

After neon burning, the core yet again contracts and the temperature rises. At about 2×10^9 K the oxygen nuclei start to react to form silicon by a process called **oxygen burning**:

$$^{16}_{8}O + {}^{16}_{8}O \rightarrow {}^{28}_{14}Si + {}^{4}_{2}He$$

As in the carbon and neon burning phases, the α-particles formed in this reaction quickly disappear as they take part in fusion reactions with heavier nuclei.

Once the oxygen in the core is exhausted, the core contracts yet further and the temperature reaches about 3×10^9 K. At this temperature, the photodisintegration of silicon starts:

$$^{28}_{14}Si + \gamma \rightarrow {}^{24}_{12}Mg + {}^{4}_{2}He$$

As was earlier the case with the photodisintegration of neon, this provides a source of α-particles. These α-particles rapidly undergo fusion reactions with silicon and with the subsequent products of fusion, leading to sequences of reactions such as:

$$^{28}_{14}Si + {}^{4}_{2}He \rightarrow {}^{32}_{16}S + \gamma$$

$$\ce{^{32}_{16}S} + \ce{^{4}_{2}He} \rightarrow \ce{^{36}_{18}Ar} + \gamma$$

$$\ce{^{36}_{18}Ar} + \ce{^{4}_{2}He} \rightarrow \ce{^{40}_{20}Ca} + \gamma$$

This type of reaction will proceed as far as producing elements with atomic masses up to $A \sim 56$, such as iron, chromium, manganese, cobalt and nickel (these are termed the **iron group** elements). As we will see below, this is the limit of nuclear burning. This phase of the star's life is often called **silicon burning**.

While the photodisintegration that occurs during silicon burning absorbs energy, the nuclear reactions that form nuclei up to the iron group will produce somewhat more energy. The overall effect is a net release of energy and the conversion of most of the silicon core into iron. When this process is complete, the core temperature is about 7×10^9 K. Surrounding this core, like the layers in an onion, are shells consisting mainly of Si and S, O and C, He, and H, as shown in Figure 7.10.

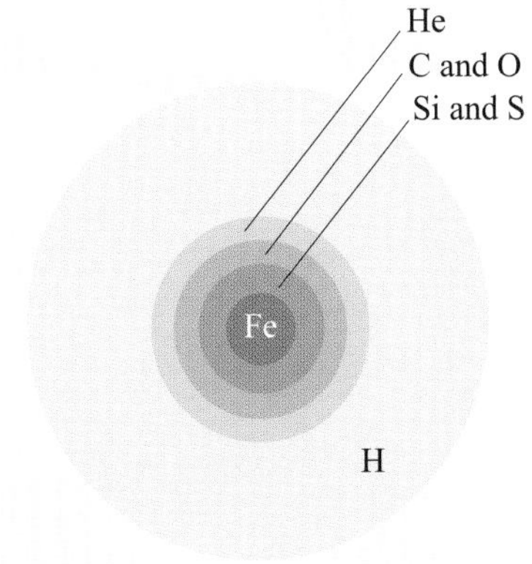

Figure 7.10 The structure of a highly evolved supergiant (not to scale) on the last day of its life.

QUESTION 7.9

In each of the stages of a supergiant's life cycle the conversion of hydrogen into helium has been taking place somewhere in the star. Describe how the site of this reaction moves as the star evolves.

7.3.4 Diminishing returns

We shall next see that each new reaction is less efficient than the previous at releasing energy, so reactions have to go faster (i.e. the number of nuclei that undergo reactions every second has to increase) to produce the necessary radiation to balance gravity. The faster reaction rates in the later stages of the star's life also mean there are more neutrinos produced, and these carry away a growing proportion of the energy generated. Nearly all the neutrinos escape without interacting with the outer layers of the star (Section 2.2.4) so do not contribute to its pressure balance. So the star goes through its life cycle at an ever-increasing pace, squandering its reserves faster and faster.

This is reflected in the fact that the time spent in each core nuclear burning phase gets dramatically shorter as the burning progresses to heavier elements. We saw in Section 7.2.2 that the time that a star spends in the core helium burning phase of its life is about 10% of the time that the star spends on the main sequence. For example, in a $25M_\odot$ star the hydrogen and helium burning phases are approximately 7×10^6 years and 5×10^5 years respectively. For this star the carbon burning phase is likely to last about 600 years, the neon burning and oxygen burning phases last for about a year and for about six months respectively. The final stage, that of silicon burning is estimated to last only about a day!

Broadly speaking, as the star progresses to the fusion of more massive elements, the energy released per kilogram of material undergoing the reaction diminishes. The full proof of this statement is beyond the scope of this book, but one can sense the correctness of it by looking at the left-hand side of Figure 6.6, plotted here as Figure 7.11.

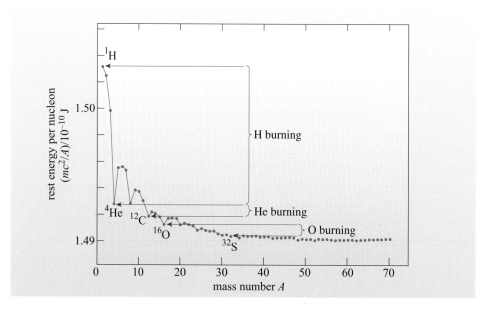

Figure 7.11 The variation of rest energy per nucleon with mass number, up to $A = 70$. As the mass of the nucleons undergoing fusion in a star increases, the available energy decreases.

You learnt in Chapter 6 that the fusion of light elements is exothermic, i.e. energy is released. Energy is released by a reaction if the product is at a lower value of rest energy per nucleon than the reactant(s). However, the curve in Figure 7.11 is gradually flattening – its gradient is becoming shallower – so the energy produced by a fusion reaction decreases with mass, as indicated by the examples shown.

■ What do you suppose will happen when the fusion reactions have built up to the element iron ($A = 56$), which is where the curve in Figure 6.6 has its minimum?

❑ In some sense, yet to be spelled out, the star has reached the end of the road. Any change in nuclear composition would consume rather than produce energy, whether fusion to an element of higher A or fission to an element of lower A.

We will investigate what happens next in Chapter 8.

7.3.5 The most massive stars

The evolution of stars of very high mass (i.e. over $50M_\odot$) occurs extremely rapidly. Their evolution is also dramatically affected by mass loss (remember that radiation pressure rather than gas pressure is the dominant force opposing gravity for such stars). Whereas main sequence stars like the Sun lose mass through their stellar winds at rates of $10^{-14}M_\odot$ year^{-1}, massive O-type stars lose mass through their stellar winds at a rate of up to $10^{-6}M_\odot$ year^{-1}.

QUESTION 7.10

Calculate and compare the fraction of the mass lost by a star of solar mass and one of $50M_\odot$ during their main sequence lifetimes (10^{10} years and 10^6 years respectively).

Your answer to Question 7.10 shows that the total mass loss during the main sequence phase, as a fraction of the mass of the star, increases as the star's mass increases, despite the very much shorter lifetime.

The most massive stars of around $100M_\odot$ may lose up to half their mass before the end of their main sequence lifetime. The outer layers of these stars are lost and the core is exposed so that the products of nucleosynthesis (in this case the CNO cycle), are revealed. Their spectra therefore imply a much greater abundance of helium near the surface than most stars. In addition, their evolution is drastically changed since they will not be able to undergo shell burning (there is no longer any region above the core for it to occur) and never become red supergiants. These objects are known as **Wolf–Rayet stars** after their discoverers (C. S. E. Wolf and G. Rayet). The expanding shells of gas around the stars as a result of their mass loss may be observable directly, as in Figure 7.12, or be inferred from the structure of the emission lines seen in their spectra (see Figure 7.13).

The final stages of the evolution of high-mass stars, which involves variability, significant mass loss and a catastrophic finale in a supernova, are described in Chapter 8.

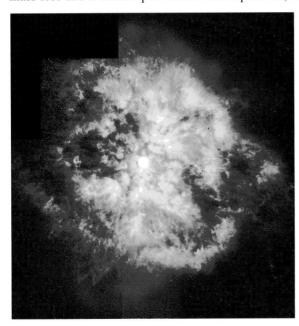

Figure 7.12 The Wolf–Rayet WR124 surrounded by gas ejected during the last 10 000 years. (Y. Grosdidier (University of Montreal and Observatoire de Strasbourg), A. Moffat (University of Montreal), G. Joncas (Université Laval), A. Acker (Observatoire de Strasbourg)/NASA)

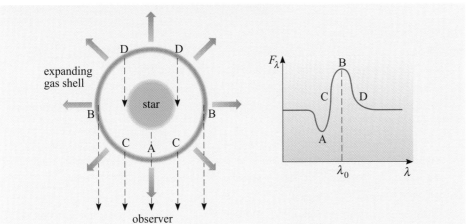

Figure 7.13 Schematic illustration of a star with an expanding shell of gas (such as a Wolf–Rayet star) showing how the characteristic spectral lines are produced. Material at position C has a component of motion towards the observer so emission lines from this glowing gas are blue-shifted relative to those emitted by gas in region B, which is moving perpendicular to the line of sight (and therefore has no Doppler shift relative to the star, which defines the wavelength λ_0). Similarly, light from region D is moving away from the observer and is red-shifted relative to B. In region A, the maximum blue-shift occurs but an absorption line is seen because some of the radiation from the hotter star behind is absorbed and re-emitted in other directions.

7.4 Summary of Chapter 7

Low-mass stars

- The main sequence lifetime of a star ends when the hydrogen in the core is exhausted.

- The core contracts and the temperature rises sufficiently for hydrogen shell burning to start in a region surrounding the helium core.

- At a temperature of around 10^8 K, helium burning is initiated. This is the 3α process. In stars of mass less than $2.25M_\odot$, the electrons in the core first become degenerate. This leads to a core helium flash.

- This transition phase is accompanied by a contraction and heating of the core but a swelling of the diameter by typically a factor of 10, and a cooling of the surface. The star becomes a red giant.

- For stars of mass less than $0.5M_\odot$, the temperature in the core does not rise high enough for helium burning to start.

- After depletion of helium in the core, helium burning continues in a shell surrounding the core, accompanied by periodic shell helium flashes. This shell helium burning causes a further swelling of the red giant.

- These changes in energy source and structure cause the star to move on the H–R diagram from the main sequence to the red giant branch (RGB), horizontal branch (HB) and asymptotic giant branch (AGB).

- During their helium core burning stage the conditions in the envelopes of stars lead to an instability and they undergo pulsations. This occurs in a region of the H–R diagram called the instability strip. Stars of mass above $2M_\odot$ become Cepheid variables.

- The giant phase lasts for approximately 10% of the main sequence lifetime.

- In giants, fusion reactions can also produce oxygen and neon. The capture of neutrons (s-process) produces heavier elements.

- Red giants suffer more significant mass loss due to stellar winds than they did when they were main sequence stars.

High-mass stars

- Massive stars spend less time on the main sequence than less massive stars, and then evolve across the H–R diagram more quickly.

- When a significant fraction of the hydrogen in the core of a massive star has been converted into helium, the star moves off the main sequence to become a supergiant.

- In stars of initial mass greater than about $8M_\odot$, carbon burning (to produce magnesium, sodium and neon) is followed by a range of reactions dependent on mass which can result in the formation of nuclei as massive as iron.

- These reactions may be fusion of identical nuclei (e.g. in oxygen burning to produce silicon), photodisintegration of a nucleus by a γ-ray photon to produce a lighter nucleus (e.g. the production of magnesium from silicon), or fusion with a helium nucleus (e.g. production of sulfur from silicon).

- Each new reaction produces energy less efficiently than the previous one; to compensate, the reaction rate is greater.

- The most massive stars lose mass at a prodigious rate through stellar winds, which drastically changes their evolution. Wolf–Rayet stars lose their entire outer layers to expose the core.

QUESTION 7.11

Sketch an H–R diagram and on it mark the area occupied by the supergiants, identifying the positions of red, yellow and blue supergiants. Show the evolutionary tracks of supergiants that evolve from blue to red, and of supergiants that change from blue to red and back again. Show from what part of the main sequence the supergiants have come.

QUESTION 7.12

For a $25M_\odot$ star; (a) arrange the following stages of core nuclear burning in chronological order of occurrence: silicon burning, neon burning, helium burning, carbon burning, hydrogen burning, oxygen burning. (b) State which of these stages involve fusion of two (or more) identical nuclei, and those in which photodisintegration plays an important role. Briefly explain in words what the effect of photodisintegration is.

QUESTION 7.13

Explain in your own words why a massive star goes through its life cycle at an ever-increasing pace.

CHAPTER 8
THE DEATH OF STARS

8.1 Introduction

In this chapter we consider what happens to stars in the very last stages of their lives. We have already seen that stars are powered by thermonuclear reactions in their cores, and that over time, the fuel supplies for this nuclear burning become depleted. Eventually all stars reach a stage where energy can no longer be generated by nuclear reactions – either because the core temperatures never become high enough to trigger the fusion of heavier elements, or because the core has been completely converted into elements of the iron group.

- ■ What physical property of a star determines which of these outcomes occurs?

- ❏ Its initial mass. In stars with initial masses greater than about $11 M_\odot$, core nuclear burning progresses to form iron group elements; in lower mass stars, core temperatures never reach the values required for nuclear burning to progress to this stage.

As you will see later, the formation of a core that comprises iron group elements gives rise to a spectacular instability in the star. This leads to the star being totally disrupted in a rapid and dramatic supernova explosion. In stars that do not form such a core, the end of stellar life is much more sedate. However these stars are subject to instabilities such as large-scale pulsations and intense outflows of stellar material.

Given the distinct differences between the ends of the lives of low- and high-mass stars, they are considered separately in the next two sections of this chapter. However, as you will see, there is a common theme to stellar death in both mass regimes: the enrichment of the interstellar medium with elements heavier than hydrogen and helium. Section 8.4 considers the effect of stellar processes on the composition of the gas and dust that constitute the material from which subsequent generations of stars are formed. The issue as to what remains of the core of the star – the stellar remnant – is an exciting topic in its own right and is the subject of Chapter 9.

8.2 The death of low-mass stars

In this section we look at the processes that mark the final stages in the life of stars with initial mass less than about $11 M_\odot$. In Chapter 7 you saw how such stars evolve to form giants with helium burning cores. Following the exhaustion of helium in the core, the track on the H–R diagram moves back towards the region that was occupied when the star was a hydrogen-shell burning red giant (i.e. before core helium ignition) but now at a somewhat higher luminosity. This region is the asymptotic giant branch (stars at this stage of evolution are commonly referred to as AGB stars). As a star evolves up the asymptotic giant branch it starts to undergo thermal pulses due to shell

Figure 8.1 Evolutionary tracks on the Hertzsprung–Russell diagram for stars of masses of $1M_\odot$ (red line) and $5M_\odot$ (blue line) showing the progression from the main sequence to beyond the end of the AGB phase. The dashed lines at the end of the AGB phase indicate tracks that are not directly observable because of the presence of obscuring circumstellar material. Note that the main sequence is shown here as a thick band (rather than a line) to reflect the evolution that occurs during core hydrogen burning. (Adapted from Iben, 1991)

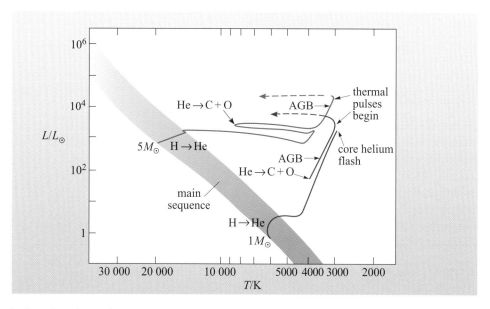

helium burning. These stages are indicated on an H–R diagram in Figure 8.1, which shows the typical evolutionary tracks of stars of masses of $1M_\odot$ and $5M_\odot$.

In this section we will look at processes that occur when a giant becomes a thermal pulsing AGB star and the immediate aftermath of this stage – the planetary nebula phase.

8.2.1 Pulsations and mass loss in evolved giants

Stars that have evolved to the stage at which they are undergoing shell helium flashes may exhibit global pulsations which are similar to those exhibited by Cepheid variables. These stars are called **Mira variables** after the prototype Mira (omicron Ceti) which varies in brightness by about 6 magnitudes with a period of about 331 days. Note that although Mira variables undergo global pulsations, their position on the H–R diagram is well away from the instability strip that is populated by Cepheids and RR Lyrae variables. In many Mira variables, the pulsation seems to be erratic and the relationship between period and luminosity is not as well defined as it is in the case of Cepheids. Furthermore the pulsation of Mira variables seems to cause considerable disturbance to the stellar envelope. Indeed, it is believed that the envelopes can become highly distorted from the spherical shape that the star had up until this time. Some indication of this is shown in Figure 8.2 which shows an image of Mira itself – its outline is clearly not circular. Although Mira is part of a binary system that may be partly responsible for distortion of its envelope, it is likely that such irregular shapes are common for stars at this stage of their evolution.

Figure 8.2 The star Mira as imaged at visible wavelengths using the Hubble Space Telescope. The envelope is clearly not spherical. In this image Mira is about 7 AU in extent. (M. Karovska (Harvard-Smithsonian Center for Astrophysics)/NASA)

A second characteristic of thermally pulsing AGB stars (whether or not they exhibit Mira-like pulsations) is that they show very high rates of mass loss. As has been mentioned in Section 7.2.9, the mass loss rates during the very late stages of giant evolution may exceed $10^{-6}M_\odot\,\mathrm{yr}^{-1}$. Such stars are typically very strong sources of infrared emission, which arises from dust grains that condense in the out-flowing material. These grains are formed primarily from graphite or silicates. The presence of this dust results in the visible light from the star being very strongly absorbed or scattered. The out-flowing material is also a site for the formation of molecules such as hydrogen cyanide (HCN) and silicon carbide (SiC). In cases where a large amount of ejected material surrounds a highly evolved

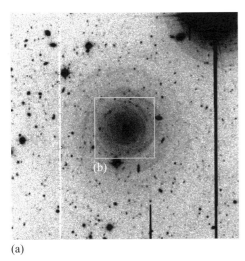
(a)

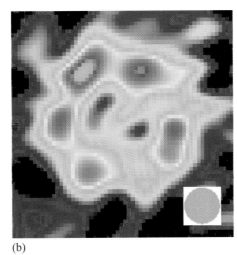
(b)

Figure 8.3 Maps of circumstellar material around the evolved star IRC+10216. (a) An image at visible wavelengths that primarily shows light that is scattered by circumstellar dust. This image shows a region around the star that is about 3×10^4 AU across. The dust cloud appears as a series of faint concentric shells which suggests that the star has undergone erratic episodes of mass loss. The square shows the area shown in (b). (b) A map made at microwave wavelengths that shows the emission from rotational transitions of a carbon-rich molecule called butadiynyl (C_4H). This map shows a region, again centred on the star, that is about 9000 AU across. The grey circle in the lower right-hand corner of the image shows the angular resolution of the millimetre-wave telescope that was used to generate this map. ((a) Mauron and Huggins, 2000; (b) Dayal and Bieging, 1993)

giant star, it may form a **circumstellar shell**. Figure 8.3 shows an example of an evolved red giant called IRC+10216 which is surrounded by a shell of material that contains copious amounts of dust and carbon-rich molecules.

Thus a key characteristic of the very late stages of the evolution of a giant is that it undergoes a very high rate of mass loss. This episode is the precursor to the next stage of stellar evolution – the appearance of a planetary nebula.

8.2.2 Planetary nebulae

Extended gaseous envelopes of the type shown in Figure 8.4 (overleaf), when observed in the 18th century, were given the name **planetary nebulae** because in small telescopes they appeared disc-shaped, like planets. However, the name, although universally accepted, is entirely inappropriate. These objects have nothing to do with planets! They are now known to be large, somewhat tenuous gas shells, also containing some dust, expanding with typical speeds of a few tens of kilometres per second. Their mass is typically between $0.1M_\odot$ and $0.2M_\odot$, and they are often associated with hot stars that are contracting while the envelopes themselves are expanding. Observations also seem to indicate that planetary nebulae are generally 'disconnected' from the star. They are also relatively common, with about 1500 having been detected in our Galaxy.

Although the first planetary nebulae to be discovered were those which appear circular, it is now recognized that many have a more complex structure, such as the example of IC 418 (Figure 8.4b). Many appear to be bipolar as shown in Figure 8.4e and 8.4f. It is a current matter of debate as to whether the bipolar nature is a result of the mass-losing star being in a binary system.

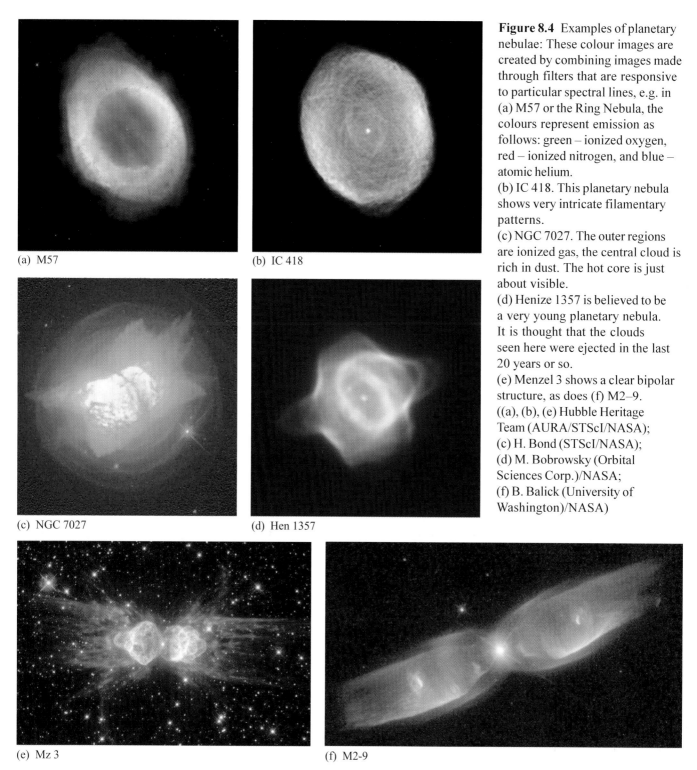

Figure 8.4 Examples of planetary nebulae: These colour images are created by combining images made through filters that are responsive to particular spectral lines, e.g. in (a) M57 or the Ring Nebula, the colours represent emission as follows: green – ionized oxygen, red – ionized nitrogen, and blue – atomic helium.
(b) IC 418. This planetary nebula shows very intricate filamentary patterns.
(c) NGC 7027. The outer regions are ionized gas, the central cloud is rich in dust. The hot core is just about visible.
(d) Henize 1357 is believed to be a very young planetary nebula. It is thought that the clouds seen here were ejected in the last 20 years or so.
(e) Menzel 3 shows a clear bipolar structure, as does (f) M2–9.
((a), (b), (e) Hubble Heritage Team (AURA/STScI/NASA); (c) H. Bond (STScI/NASA); (d) M. Bobrowsky (Orbital Sciences Corp.)/NASA; (f) B. Balick (University of Washington)/NASA)

(a) M57

(b) IC 418

(c) NGC 7027

(d) Hen 1357

(e) Mz 3

(f) M2-9

How do planetary nebulae fit into the picture of stellar evolution? All the evidence points to their being generated at the end of the giant phase or afterwards. The exact method of their expulsion is not completely clear. It is possible that they result from the pulsations of the outer layers of a star during a phase as a Mira-like variable, maybe being the result of pulsations that have grown very large. Alternatively, perhaps their release is triggered by the thermal pulses which arise because of shell helium burning. Another possibility is that, rather than a sudden expulsion of matter, they may simply represent the continuing expansion and ultimate mass loss from a giant during its advanced stages of evolution. Whatever the generating mechanism, it seems that planetary nebulae dissipate into the interstellar medium about 20 000 years after ejection, leaving only the hot dense cores of their parent stars. The nature of these cores – stellar remnants – will be taken up in Chapter 9.

8.3 The death of high-mass stars

In Chapter 7, you saw how supergiant stars evolve rapidly to form an 'onion-skin' structure – concentric zones of nuclear burning around the core. On the final day of the life of such a star, the conditions in the core are such that silicon burning takes place and forms iron group elements (for brevity, this will be referred to as the 'iron core' in the discussion that follows). Since the iron group elements are the most energetically stable nuclei, the star runs out of fuel in a very abrupt fashion. It is at this point that we take up the account of the events leading to the dramatic end of the lives of high-mass stars.

8.3.1 Supernova explosions

When the reactions that produce iron diminish, the iron core must contract under gravity. The temperature and density increase still more, but now with no hope of further nuclear reactions that can release energy to provide a force to balance gravity. The collapse brings the central regions to a sufficiently high density that the electrons become degenerate. However, even though the electron degeneracy pressure is large, in these high-mass stars it is unable to halt the collapse; there is a maximum mass that can be supported by electron degeneracy pressure. Once the iron core has grown to more than about $1.4 M_\odot$ it has exceeded this limit. Most stars whose main-sequence mass was greater than about $11 M_\odot$ form, in the last stages of evolution, iron cores that exceed this limit.

So the electron degeneracy pressure will only temporarily delay the collapse, and then the core continues shrinking inexorably. The core temperature continues to rise, until at about 10^{10} K, the iron nuclei begin to photodisintegrate, producing α-particles, protons and neutrons. (These neutrons may be important for a process that will shortly be described: the r-process.) Unlike the earlier processes in which photodisintegration played a part in reactions that produced energy, the net effect now is one in which energy is absorbed. The core collapse gets faster and faster, reaching supersonic speeds. As the collapse continues, the density rises, as does the energy of the degenerate electrons. There comes a point where these electrons (e^-) have enough energy to make possible the reaction

$$e^- + p \rightarrow n + \nu_e$$

where n stands for neutron. This reaction removes electrons, so the electron degeneracy pressure drops and the core collapse proceeds in earnest. It stops, finally, when the core density becomes comparable with the density of the nucleus of an atom! The core temperature has risen to 10^{12} K and the core density has become approximately 3×10^{17} kg m^{-3}!

At these densities a new form of degeneracy pressure, **neutron degeneracy pressure**, comes into play. This pressure, due to the neutrons, builds up quite quickly, causing the collapse of the inner part of the core to come to a sudden halt and rebound slightly. The overlying layers of the star are still falling inwards at speeds that may be as high as 70 000 km s^{-1}, so there must be a region where the gas flow undergoes a dramatic decrease in speed – a shock front (Section 5.3.1). This shock front moves radially out from the collapsed core and forms a shock wave.

The Earth has a mass of 6×10^{24} kg. What would be its radius if it had a density of 3×10^{17} kg m^{-3}, like the collapsed core of the star?

Exactly what happens next is still not completely clear, but it seems that some, or all, of the following things happen:

- The shock wave itself may blow apart the outer layers of the star, which consist mainly of lighter elements;
- The shock wave may heat the outer layers to a temperature of about 10^{10} K initiating explosive nuclear fusion reactions, which release enormous amounts of energy and throw off the outer layers of the star;
- Enormous numbers of neutrinos are produced and, although most of them escape without interacting with the outer layers of the star, sufficient may interact to lift off the material.

Whatever the detailed mechanisms, the net result is that the star suffers catastrophic self-destruction. The inner core of the star collapses to an enormous density, forming a body called a **neutron star** that is supported by neutron degeneracy pressure. The nature of neutron stars will be explored in more detail in Chapter 9; here we note that they have masses in the range of about $1.4M_\odot$ to $3M_\odot$ and a radius of only about 10 km or so. The collapse that forms a neutron star releases a huge amount of gravitational potential energy. Consequently, the outer layers of the star are blown off in a gigantic explosion. The star spent millions of years evolving to the point where it had a massive iron core, and then went through these last stages in seconds. The energy released by the core collapse is about 10^{46} J. At least 99% of this is carried away by neutrinos; the remainder goes into the kinetic energy of expansion (10^{44} J) and into the sudden brightening of the star (10^{42} J), with a little bit probably also going into the production of high-energy particles called cosmic rays. Typically, the star's luminosity brightens by a factor of 10^8, and it may for a while outshine the entire galaxy in which it is situated. An example of a supernova, supernova 2001CM, is shown in Figure 8.5. If the star lies within our Galaxy then it may be bright enough to be seen in daylight for a few weeks. It is called a supernova; nova means new

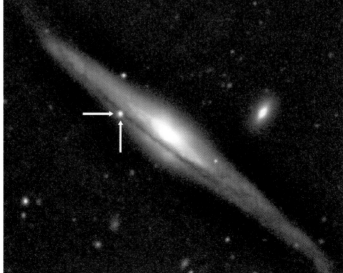

Figure 8.5 Supernova 2001CM in the galaxy NGC 5921. The supernova is the blue object indicated in the disc of the galaxy. Note that all of the other points of light in this image are foreground stars that lie within our Galaxy. This image was obtained about two months after maximum brightness. This supernova was classified as Type II on the basis of its spectrum. (H. Dahle (NORDITA))

(star) in Latin and a supernova is an extra-bright new star. This is a misnomer because it is the death throes of an *evolved* star that we are witnessing, but the name does remind us that what was previously an inconspicuous star suddenly brightens up and becomes noticeable.

Type I and Type II supernovae

Supernovae are classified into two broad groups depending on whether or not their spectra show hydrogen lines: Type I supernovae are those that do not show hydrogen lines, whereas Type II supernovae do. The supernova resulting from the explosion of a supergiant that has a significant amount of hydrogen in its envelope will contain hydrogen spectral lines and is thus likely to be of Type II.

For stars more massive than $30–40M_\odot$, it is suspected that radiation pressure and the stellar wind may cause the star to shed its hydrogen envelope (see Section 7.3.5). If such a star subsequently undergoes a supernova explosion it will lack hydrogen lines in its spectrum and will be classed as Type I. In fact, there is a sub-division of the Type I class based on the presence of silicon absorption lines in the spectrum. Those with silicon lines are classed as Type Ia, whereas those that lack silicon lines are classed as Type Ib or Type Ic. The supernovae that result from supergiants that have lost their envelopes produce spectra of Type Ib and Ic.

The Type Ia supernovae are found in those areas of galaxies where there are old, slowly evolving stars, whereas Type II, Ib and Ic supernovae are associated with massive, young stars that have evolved rapidly to the supernova stage.

▓ What does this suggest about the mass of most objects that become Type Ia supernovae?

▢ Because slowly evolving stars have low mass it suggests that most Type Ia supernovae are the explosions of low-mass stars.

At first sight this may seem a rather puzzling result, but the key to understanding this observation is that stellar evolution can be influenced dramatically if a star has a close binary companion. The influence of binary interaction on stellar evolution and the origin of Type Ia supernovae will be considered in more detail in Chapter 9.

8.3.2 Creation of heavy elements in supernova explosions

We have seen that fusion of light elements in stars can build nuclei as far as the iron group ($A \sim 56$), but cannot build further. However in Section 7.2.8 you saw that some elements can be built by another process, the s-process (s for slow), in which neutrons are added to nuclei. Some, but not all, of the isotopes beyond the iron group may be built this way.

To appreciate why the s-process cannot create certain nuclei, consider the following example. If we were to analyse a sample of the metal tin (symbol Sn) that had been obtained from tin-ore deposits on Earth, we would find that 4.6% of the mass of the sample would be in the form of the stable isotope $^{122}_{50}\text{Sn}$. As we will see, it is not possible for this isotope to be produced by the s-process. The stable isotope of tin of lowest mass number (A) that is formed by the s-process is $^{116}_{50}\text{Sn}$. If a nucleus of $^{116}_{50}\text{Sn}$ were to capture a neutron, then a nucleus of the stable isotope $^{117}_{50}\text{Sn}$ would be formed. To predict the effect of further slow neutron capture events, we need to know about the stability of isotopes to radioactive decay. Table 8.1 shows this information for isotopes of tin from mass number 116 to 122.

β^--decay is a type of radioactive decay in which a neutron is converted into a proton with the emission of an electron (e^-) and a particle called an electron antineutrino ($\overline{\nu}_e$).

Table 8.1 Stability data for some isotopes of tin (Sn). (Note that there are more isotopes of tin than are shown here.)

	Mass number A						
	116	117	118	119	120	121	122
type of radioactive decay	none	none	none	none	none	β^--decay	none
product of radioactive decay	–	–	–	–	–	$^{121}_{51}\text{Sb}$	–
half-life	stable	stable	stable	stable	stable	27 hours	stable

▓ Using the information in Table 8.1, which isotopes of tin can be formed by successively adding neutrons by the s-process to a nucleus of $^{117}_{50}\text{Sn}$? (Remember to assume that in the s-process, a neutron is captured by a nucleus about once every 10^4 years.)

❑ The isotopes of tin with mass numbers 118, 119, 120 and 121 will be formed by the s-process. However, on forming $^{121}_{50}\text{Sn}$ the nucleus will undergo β^--decay to $^{121}_{51}\text{Sb}$ (antimony) with a half-life of 27 hours.

Thus a nucleus of $^{121}_{50}\text{Sn}$ will almost certainly decay before another neutron can be captured. So, there is no route to forming the isotope $^{122}_{50}\text{Sn}$. Since this isotope is present on Earth, there must be another mechanism by which it can be synthesized.

▓ How could a nucleus of $^{122}_{50}\text{Sn}$ be formed in a process that involves the addition of neutrons to a nucleus of $^{121}_{50}\text{Sn}$?

❑ If neutrons could be supplied at such a rate that a nucleus of $^{121}_{50}\text{Sn}$ was more likely to capture a neutron than to undergo β^--decay, then this would result in the formation of a nucleus of $^{122}_{50}\text{Sn}$.

In creating some isotopes nature does just this. It is possible to rapidly add another neutron to an unstable nucleus before it has time to decay. This technique works where there is an abundant supply of neutrons, and involves a mechanism that allows the rapid absorption of neutrons by nuclei. Reactions embodying this mechanism are called **r-process reactions** (r for rapid), and are believed to occur in supernova explosions. The supply of neutrons comes from the break-up of the iron nuclei in the core, resulting in a flood of neutrons (10^{36} neutrons per square metre per second!). The r-process occurs for a few seconds only during the explosive expansion of the outer layers of the star, which have been heated to over $10^{10}\,\text{K}$, and builds elements beyond iron. Moreover, the outer layers of the star, which are rich in hydrogen and helium, will very probably undergo a rapid sequence of nuclear fusion reactions, building elements predominantly of the iron group. Heavier elements ($A > 56$) are created in the explosion by the r-process, and certain of these, such as gold and plutonium, are created predominantly by this means.

However, it is important to appreciate that elements with mass number up to $A = 209$ ($^{209}_{83}\text{Bi}$) can also be created by the s-process in lower mass giant stars, although as we saw in the example of tin, there are often particular isotopes of an element that cannot be formed in this way. Terrestrial samples of a given element are typically mixtures of nuclei that were formed in s- and r-process reactions long before the formation of the Solar System. So, for example, the origin of the nuclei in a terrestrial sample of barium (Ba) is estimated to be 89% from the s-process and 11% from the r-process. In contrast, most nuclei in a sample of europium (Eu), were formed mainly by the r-process (97%) with only a small contribution (3%) from the s-process.

A Type II supernova explosion provides not only an environment in which heavier elements ($A > 56$) are created by the r-process, but also the mechanism for distributing them through a large volume of space. Elements, such as silicon, sulfur and magnesium which were formed earlier by nuclear fusion in the core of the star are also ejected as the explosion shatters most of the star. Supernova explosions are the most important way in which the chemical composition of the interstellar medium is enriched with elements heavier than neon.

8.3.3 The supernova light curve

Type II supernovae have hydrogen emission lines in their spectra; Doppler shifts (Box 3.1) of these lines observed in recently exploded supernovae show material expanding outwards at speeds of up to $10\,000\,\text{km s}^{-1}$. For about a month the visible surface of the star expands steadily at several thousand kilometres per second; as the surface area increases so the amount of light radiated increases. Then the brightness begins to fade as shown in the light curve in Figure 8.6. When the visible surface has expanded to a radius of about $2 \times 10^{10}\,\text{km}$ it becomes transparent, and the amount of light produced drops markedly.

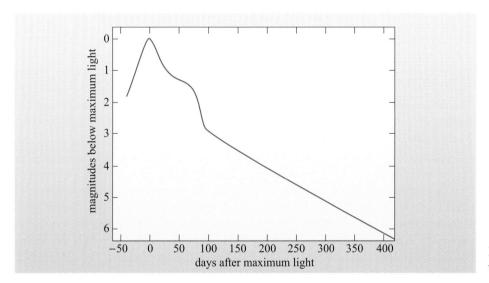

Figure 8.6 The light curve of a typical Type II supernova.

QUESTION 8.2

How does this radius of $2 \times 10^{10}\,\text{km}$ compare with the radius of a supergiant, the radius of the Solar System, and the distance to the nearest star?

We discussed earlier how, when building heavier elements step by step from lighter ones, unstable (that is, radioactive) nuclei can bring the building process to a halt. Radioactive nuclei have other effects too. In particular, the type of radioactive decay in which a γ-ray is emitted can be important for the brightness of a new supernova. It has been suggested that such γ-rays are the energy source governing the brightness of a supernova for the period starting six or eight weeks after the explosion. The most likely decay sequence is $^{56}_{28}\text{Ni}$ rapidly decaying to $^{56}_{27}\text{Co}$, which in turn decays more slowly to $^{56}_{26}\text{Fe}$ with the emission of γ-rays. We shall see later, when we discuss supernova 1987A, that this suggestion has been confirmed.

Note the usage of the words visible and optical. Optical wavelengths are the visible wavelengths plus the near infrared and the near ultraviolet wavelengths. This corresponds to a range of wavelengths from 300 to 900 nm that can be observed through the Earth's atmosphere.

Note that the brightness or luminosity plotted, in light curves such as Figure 8.6, is that measured in the optical part of the spectrum. Why should it be the optical data that are used? Supernovae are rare occurrences and our understanding of them has been achieved by putting together all available data. Because optical telescopes have been and are more numerous than, for example, neutrino detectors or far-ultraviolet telescopes, most of the available data are from the optical wavelength band. Also, although most energy is lost through neutrino emission, neutrinos are difficult to detect. Thus for studying the evolution of the supernova over the subsequent weeks and months the optical radiation emitted is a more useful diagnostic.

Often we miss the initial rise in optical brightness as the photosphere of the star explodes, noticing the supernova only when it is close to maximum brightness, some weeks after the explosion. The shape of the light curve after maximum brightness depends on both the radius and the temperature of the visible surface (remember Equation 3.9). For the first 25 days or so after maximum brightness the visible surface is still expanding but, nevertheless, the effects of falling temperature dominate and the luminosity drops. The majority of Type II supernovae have a curious shoulder, or plateau, in their light curves between days 25 and 75 after the maximum. The outer layers of the star are thinning so we are seeing further into it; but meanwhile the star is expanding in a way that just balances our ability to see deeper into it. Like walking down the up-escalator, the net result is that the surface we see does not appear to move much – the radius is roughly constant. So too is the temperature, and hence the brightness does not change much, giving the shoulder. After about day 75, we start to see further into the star, so while the outer layers of the supernova are actually expanding outwards, the *visible* surface appears to retreat rapidly. Although the temperature remains roughly constant, the shrinking radius causes the luminosity to fall abruptly. From 100 days after the maximum, radioactive heating by γ-rays controls the brightness, and the shape of the light curve is governed by the half-life of the radioactive decay.

8.3.4 How common are supernova explosions?

The last supernova seen in our Galaxy was in AD 1604 – before the telescope was invented! And before that, we know of ones in AD 1572, AD 1054 and AD 1006. So, at first glance, a rate of roughly 4 in 1000 years would seem appropriate for our Galaxy. However, supernovae at maximum brightness can be as bright as a whole galaxy, so we can find them also in other galaxies. For these, the rate seems to be one every 25 to 50 years per galaxy – a very different rate.

Why the difference? Are there factors that bias the numbers observed? It is believed that the discrepancy is caused by absorbing material in the interstellar medium that is concentrated in the plane of our Galaxy. The supernovae that have been observed are all quite close to us in the Galaxy; we suspect that ones further away, if they are near the plane of the Galaxy, are not seen because of obscuration by the interstellar medium. Correcting this bias as well as possible suggests that the real rate in our Galaxy is much the same as in other galaxies. There are approximately equal numbers of Type I and Type II supernovae.

8.3.5 Supernova 1987A

For almost 400 years no bright, nearby supernova had given astronomers the chance to check out their conjectures about the death of massive stars. During these centuries the available instrumentation grew: first the optical telescope was invented, then the other wavelengths of the electromagnetic spectrum were recognized and, since the middle of the 20th century, telescopes have been developed to detect these too. Neutrino detectors, similar to those described in Section 2.2.6, were in place and waiting. What was really needed was a good supernova explosion to allow astrophysicists to check out their theories on how supernovae occurred, whether nucleosynthesis actually took place, and what happened to the core of the star.

Then on 24 February 1987 it happened. A Canadian astronomer, Ian Shelton, using a mountain-top telescope in Chile, found an unexpected bright star on a photographic plate he had just exposed, something that hadn't been there the previous night when he had also photographed that part of the sky. After 20 minutes trying to explain away the spot (OK, nobody's too bright in the cold at 3000 metres altitude at 3 o'clock in the morning) he looked outside and saw it was real. In a nearby galaxy in the southern sky, called the Large Magellanic Cloud (Figure 8.7), a star had exploded. Figure 8.8 shows this part of the sky before and after the explosion.

Figure 8.7 The Large Magellanic Cloud. This irregular galaxy is probably the nearest galactic neighbour to our own Galaxy. It is believed to be at a distance of about 52 kpc from the Sun. (D. Malin/AAO)

(a) (b)

Figure 8.8 Supernova 1987A. The left-hand image (a) shows a region of the Large Magellanic Cloud (Figure 8.7) which contains the supergiant star (indicated) that gave rise to Supernova 1987A. The panel on the right (b) shows the same region as it appeared after the eruption of the supernova. (D. Malin/AAO)

For every discovery there are, usually, several near misses. Some 19 hours earlier in Australia, Robert McNaught had photographed that part of the Large Magellanic Cloud at the Anglo-Australian Observatory. Although he developed the plate he did not examine it that night. Subsequently he found that he had the first photograph of the supernova – it was beginning to happen and at that stage was just becoming bright enough to be visible to the naked eye. Shelton is therefore credited with the discovery of **SN 1987A** (SN being the abbreviation for supernova, and 'A' indicating that it was the first one found in 1987.) However, McNaught's observation, combined with Shelton's negative observation the previous night, is scientifically important in pinning down the start of the event.

This supernova had already, unknowingly, been observed by neutrino detectors, and would be observed in all the major wavelength bands over the next few years. During the previous year, two underground neutrino detectors, not primarily designed for the detection of neutrinos from supernovae, had been sufficiently improved in sensitivity that they were able to play an important part in the study of this explosion. (Note that the Homestake mine experiment that was described in Chapter 2 was not sufficiently sensitive to detect neutrinos from SN 1987A, and that the Sudbury Neutrino Observatory, or SNO, had yet to be built in 1987.) One of the detectors was in a zinc mine in Kamioka in Japan (Figure 8.9), the other in a salt mine near Lake Erie in Ohio, USA. Simultaneously they detected a short burst of a few neutrinos lasting about 10 seconds, and well above the normal background rates. About 3 hours before McNaught's observation 12 neutrinos were counted in the Japanese detector and 8 in the American one. The number and energy of the particles in the neutrino burst fitted well with the prediction of what would be produced if the central core of a supergiant collapsed to nuclear densities. It appears that about 3×10^{46} J of energy were carried away by 4×10^{58} neutrinos. About 170 000 years after the explosion, as the burst of neutrinos swept though the Earth, 20 of them interacted in the two large underground detectors.

Figure 8.9 The Kamiokande II neutrino detector that detected 12 neutrinos from supernova 1987A. The heart of the experiment was this tank of ultra-pure water. Occasionally neutrinos would interact with electrons in the tank of water, resulting in a flash of light that could be detected by an array of photocells that are arranged around the tank. This experiment has now been superseded by a larger and more sensitive neutrino observatory called *Super-Kamiokande*.

About 1% of the energy of the collapsing star went into the shock wave, which slowly travelled out through the star and caused the increase in the luminosity. By that time the neutrinos had gone, at the speed of light, and so the neutrinos were detected a few hours before the star was seen to brighten.

QUESTION 8.3

Is it necessary for a cosmic source of neutrinos to be above the horizon for the detection of neutrinos here on Earth? Was this source above the horizon at the time of the detection?

It had been assumed that massive stars exploded as *red* supergiants, so it was something of a surprise when it became clear that the progenitor, the star that exploded producing SN 1987A, was the *blue* supergiant catalogued as Sanduleak $-69°202$, a $20 M_\odot$ star of spectral class B. Ultraviolet observations made when the supernova had faded somewhat established that Sk $-69°202$ no longer existed, confirming the identification. This stimulated new work on the evolution of massive stars, and the circumstances that led up to the supernova are still a matter of research effort.

The spectrum of the supernova contains hydrogen lines, and it must therefore be classified as a Type II supernova, but SN 1987A was fainter, by a factor of ten, than the Type IIs that, up until then, were thought to be typical. This relatively low luminosity is thought to be connected to the unusual nature of the progenitor. The light curve in Figure 8.10 is a plot of the optical luminosity of the supernova against time. It strikingly confirms the theory that material is heated by γ-rays that are produced by the radioactive decay of $^{56}_{27}\text{Co}$. From about 100 days to about 700 days after the outburst, the fading of the supernova follows the rate of radioactive decay of cobalt nuclei. Further evidence that the nuclear reactions are reasonably well understood came from infrared and γ-ray observations. Infrared spectrometers found spectral lines due to iron and cobalt, while γ-ray telescopes detected γ-rays with the energies expected from the decay of $^{56}_{27}\text{Co}$. This was the first time that a direct check of the theory of element formation in supernovae had been possible.

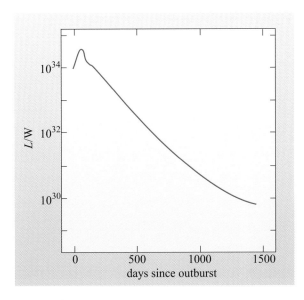

Figure 8.10 The light curve of SN 1987A.

QUESTION 8.4

If the decay of $^{56}_{27}$Co is solely responsible for a supernova's luminosity, and each decay produces two γ-rays, one of energy 1.3×10^{-13} J and one of energy 1.9×10^{-13} J, what mass of cobalt nuclei must decay per second to produce a supernova luminosity of 10^{33} W?

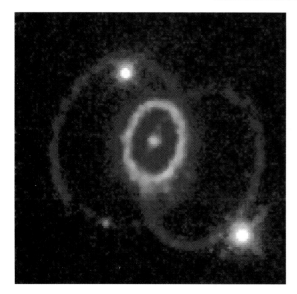

Figure 8.11 Rings of material around supernova 1987A that were ejected prior to the supernova explosion. The emission of light arises from the illumination of this material by electromagnetic radiation from the supernova, and was first seen in an observation made in February 1994 using the Hubble Space Telescope. The structure of these rings, which comprise an inner ring and two symmetrical outer rings, was an unexpected feature of SN 1987A. (STScI/NASA)

An intriguing aspect of supernova 1987A arises from the interaction between the supernova and shells of material that are believed to have been expelled by the pre-supernova star over the last 30 000 years or so. The first effect to be seen was the illumination of this material by the electromagnetic radiation from the supernova. The structure that was revealed comprised an inner ring and, surprisingly, two outer rings (Figure 8.11). This complex structure probably represents different episodes of mass loss from the pre-supernova star. The second interaction between the supernova and its environment arises from the impact of the expanding shock wave. The shock has now reached the material that appeared as the inner illuminated ring, and localized hot spots of shock-heated gas have been seen to emerge over the last few years. Figure 8.12 shows the appearance of the ring in February 2000 and the shock-heated features that appeared between 1997 and 2000.

How will SN 1987A appear thousands of years hence? Most probably it will develop a circular structure, such as that shown by the Cassiopeia A (Cas A) supernova remnant. The appearance of that supernova remnant at radio, X-ray and optical wavelengths is shown in Figure 8.13. Alternatively it could evolve into a rather irregular shape such as happened in the case of the Crab Nebula (Figure 8.14). In the Crab, the collapsed core has formed a rapidly rotating *neutron star* which is detectable from its pulsed emission of radio waves. Neutron stars that exhibit such pulses of electromagnetic radiation, typically at radio or X-ray wavelengths, are termed *pulsars*. The properties of pulsars will be explored in more detail in Chapter 9.

Figure 8.12 The inner ring of material around SN 1987A that was ejected prior to the supernova explosion is now being heated by the shock wave from the supernova. (a) shows the appearance of the inner ring in February 2000. (b) shows the features in the ring that had appeared between observations made in 1997 and in 2000, showing the emergence of several bright hot spots around the ring. These hot spots are attributed to the shock heating of clumps of material in the ring. (P. Challis, R. Kirshner, P. Garnavich/NASA)

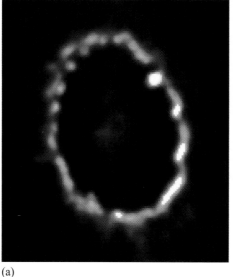 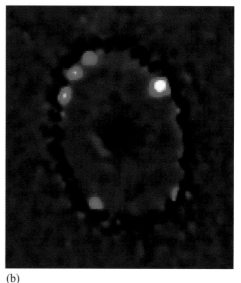

(a)   (b)

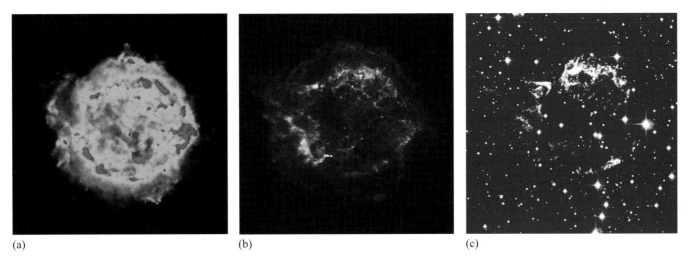

(a)   (b)   (c)

Figure 8.13 The Cassiopeia A (Cas A) supernova remnant. The supernova is believed to have erupted about 300 years ago, although it is not recorded as having been seen at the time. These images show the remnant at (a) radio, (b) X-ray and (c) visible wavelengths. While the remnant is very clearly seen at radio and X-ray wavelengths, the optical image is somewhat less distinct. ((a) VLA; (b) NASA/CXC/SAO; (c) MDM Observatory)

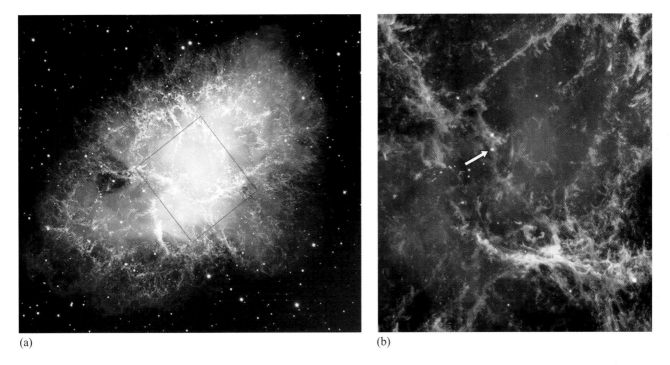

(a)   (b)

Figure 8.14 The Crab Nebula. The supernova that gave rise to the Crab Nebula was recorded by Chinese astronomers in AD1054. (a) This image at visible wavelengths was taken from a ground-based telescope and shows the supernova remnant as a rather irregular nebula that is now about 2 pc across. The square field shows the area that is shown in (b) observed using the Hubble Space Telescope. The pulsar, which is the remnant of the core of the star that gave rise to the supernova, is indicated. Note that both colour images were created by combining images taken though filters that are sensitive to particular spectral lines. ((a) ESO; (b) W. P. Blair (John Hopkins University))

8.4 Feeding the interstellar medium

In this section we look at how stars of any initial mass return material to the interstellar medium (ISM). Much of this material has been subject to nuclear processing in stars, and this has led over time to a change in the chemical composition of the ISM and of stars.

8.4.1 Dispersal of stellar material

▪ Name three processes that you have already met that return stellar material to the interstellar medium.

❑ Stellar winds, the shedding of planetary nebulae, and supernovae.

These three mechanisms will be considered in turn, before we consider how the interstellar medium changes with time and the effect that this has on the compositions of stars.

Stellar winds and circumstellar shells

Stellar winds are a relatively gentle form of dispersal. All stars lose some matter in this way, but in our Galaxy the overall amount of matter returned by stellar winds to the ISM is dominated by the copious winds from hot (class O and B) stars, and by strong winds from cool red giants. These latter winds give rise to circumstellar shells similar to that shown in Figure 8.3. In their inner regions, circumstellar shells are kept moderately warm by the star's luminosity, with temperatures around 1000 K, not much less than that of the star's photosphere.

▪ What is the range of photospheric temperatures of cool red giants?

❑ From Section 4.2.2, about 2000 to 3000 K.

Figure 8.15 A region of interstellar cirrus as seen using IRAS. This image shows the emission from interstellar cirrus at a wavelength of 100 μm. The area of sky covered by this image is 12° square. (Data provided by IPAC, Caltech)

Circumstellar shells are, as you might expect, less dense than the photospheres of red giants, but are enormously dense by ISM standards, with $n \sim 10^{17}\,\mathrm{m}^{-3}$ in their inner regions. As has already been noted, these conditions result in the formation of dust grains and molecules. However, as the matter in such shells moves outwards it becomes less dense, and also cools, and the next we see of it might be the widespread but thin **interstellar cirrus**, so called because of its resemblance to the terrestrial cloud type called cirrus. An example of the interstellar variety is shown in Figure 8.15 – what we are seeing is thermal radiation from dust at about 20 K. Interstellar cirrus was discovered in 1983 from observations using the space-borne Infrared Astronomical Satellite (IRAS).

Ultimately, all matter from stellar winds becomes thinly dispersed.

Shedding of planetary nebulae

Slightly more violent than mass loss via stellar winds is the shedding of a planetary nebula. In Section 8.2 you saw that this is the fate of a giant, which becomes unstable and throws off a significant fraction of its mass. All but the lowest and highest mass stars become giants, and so planetary nebulae are an important source of matter for the ISM. The remnant star is very hot, and also emits a powerful wind. As a result, planetary nebulae are hotter than circumstellar shells, with temperatures of order 10 000 K. They are also fairly dense, with $n \sim 10^9\,\mathrm{m}^{-3}$. As in the case of stellar winds, the matter in planetary nebulae moves away from the star, at speeds of the order of several kilometres per second, cooling and thinning as it proceeds. It takes about 20 000 years for a planetary nebula to disperse – not long on the cosmic timescale.

Supernovae and their effects

Yet more violent ejection of stellar material occurs as a result of supernovae. We have already seen that Type II supernovae are cataclysmic stellar explosions, involving massive stars, in which most of a star's mass is flung into space. Supernovae of Type Ia, despite arising from a different explosion mechanism, also result in the dispersal of a large amount of stellar material. The explosive nature of supernovae might lead you to think that they are major sources of interstellar matter, but this is not really the case.

▓ Why not?

❑ Supernovae are rare.

However, though supernovae are not major sources of *mass* for the ISM, they are important sources of heavy elements, a point to which we return later.

Supernovae also release huge amounts of *energy* into the ISM. In all supernovae it is believed that about 10^{44} J is carried off as kinetic energy in a radially expanding gas shell and about 10^{42} J in the form of electromagnetic radiation. The kinetic energy of the shell means that the expelled material does not simply cool and vanish into the general ISM in the manner of stellar winds and planetary nebulae, but wreaks far-ranging and profound changes. For example, the Vela supernova remnant (Figure 8.16), which is about 11 000 years old, has grown to a size of roughly 50 pc in diameter, and continues to violently disturb the interstellar medium that it impinges upon.

To explore the effects of Type II supernovae further, consider the speed at which the shell is initially expelled, given that

- of the ~5 to ~10 solar masses of material returned to the ISM, only about $0.25M_\odot$ carries most of this kinetic energy,
- the kinetic energy is almost entirely in the radial motion of the shell, rather than in the random thermal motions within it.

Figure 8.16 The Vela supernova remnant. This is the remnant of a supernova that occurred about 11 000 years ago. (a) An optical image that is sensitive to Hα emission. The remnant is visible as the filamentary structure in the central region of the field of view. Note that there are other Hα emitting regions also visible in this image which are not associated with the supernova remnant. (b) An X-ray image of the same field. The Vela supernova remnant is the large roughly circular feature that fills the field of view. The bright region at the upper right-hand edge of the remnant is the Puppis supernova remnant. This lies well beyond the Vela supernova remnant, and is not visible in the Hα image. The two supernova remnants are not physically associated with one another. ((a) Royal Observatory Edinburgh; (b) Max Planck Institute for Astrophysics)

(a)

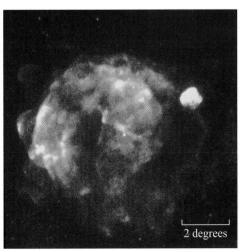

2 degrees

(b)

Calculate the initial radial speed of such a shell.

Thus, a shell can have a radial speed that is a few per cent of the speed of light.

These high speeds mean that the kinetic energy of an atom in the shell is *far* higher than the thermal energy of a typical atom or molecule in the surrounding ISM. Because of this we get the following effects, which are summarized in Figure 8.17.

- The shell sweeps up much of the gas that it encounters, the swept-up gas being greatly compressed, greatly heated, and trapped within the shell, of which it becomes a part. The shell, which started hot and highly ionized, thus remains in this condition. It also remains thin compared with its radius.

- A small amount of gas finds its way into the volume of space enclosed by the shell. This gas has also been greatly heated, and expands to fill the volume, giving rise to a roughly spherical cavity pervaded by hot, low-density gas. The shell constitutes the supernova remnant. The temperatures within it range from 10^5 K to 10^7 K.

- The gas ahead of the shell gets no warning of the shell's approach, and thus its transition from a peaceful life to the tortured conditions in the shell is abrupt. Hence a shock front forms at the thin transition zone between these two regimes, and the swept-up gas is said to have suffered a **strong shock**.

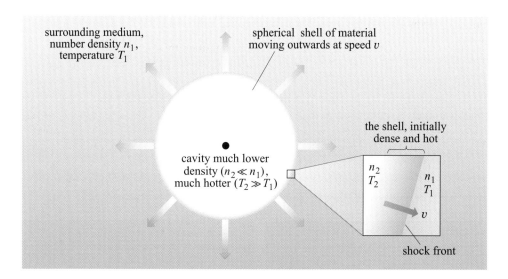

Figure 8.17 Conditions produced by the shell expelled by a supernova.

The interior of the supernova remnant is at a high temperature and so is a source of thermal radiation. However, the density of gas within the shell is very low, and so the spectrum of emission does not have a black-body spectrum. For a supernova remnant with a temperature of 10^6 K, estimate (a) the photon energy (in electronvolts), and (b) the wavelength at which you would expect the thermal emission from the supernova remnant to be strong. Which part of the spectrum does such emission correspond to?

The answer to Question 8.6 shows that the thermal emission from supernova remnants would be expected to give rise to X-rays. This is indeed the case as is illustrated in the X-ray images of the Cas A and Vela supernova remnants shown in Figures 8.13b and 8.16b, respectively.

Supernova remnants also emit copious amounts of radio waves. These radio waves originate from electrons that interact with the magnetic fields within the shell. Any charged particle that moves within a magnetic field is forced to move in a spiral around magnetic field lines (Figure 8.18). The spiralling motion of electrons in a magnetic field gives rise to the emission of electromagnetic radiation. If, as is the case in supernova remnants, the electrons are moving at speeds that are a substantial fraction of the speed of light, the emission is called **synchrotron radiation**. The spectrum of such emission is usually a rather featureless continuous spectrum. Such emission is common in very energetic environments that contain magnetic fields. An example of another astrophysical situation that you have already come across that generates synchrotron radiation is the radio emission that accompanies solar flares (Section 2.3.2).

Supernova remnants are strong sources of synchrotron emission at radio wavelengths, as Figure 8.13a illustrates. Radio waves are not appreciably absorbed by the ISM, and so most remnants have been discovered through the radio waves they emit.

As the remnant expands its temperature falls and its X-ray emission also diminishes. However the synchrotron radio emission persists and is prominent for much of the 10^5 or so years before the shock front grinds to a halt, whereupon the supernova remnant dissolves into the general ISM.

Let's now briefly consider the growing cavity contained by the shell. The high temperatures (10^5 K to 10^6 K) ensure a high degree of ionization. Furthermore the density within the cavity is relatively low with number densities typically in the range $10^2 \, m^{-3}$ to $10^4 \, m^{-3}$. Under such conditions, the rate at which the gas can lose energy by emission of electromagnetic radiation is relatively low (for instance, it is far lower than the rate at which energy would be radiated by a black-body source of a similar size at the same temperature). Thus the material in the cavity remains very hot as the shell expands. The low radiative efficiency and the low densities result in such a slight amount of radiation from the cavity that there is little to see. A single cavity can grow to hundreds of parsecs across, and cavities can overlap and merge. Moreover, because the material in the cavity cannot cool very readily, it can outlast the supernova remnant. A consequence of this is that a large volume of the interstellar medium comprises hot, low-density material, not necessarily bounded by supernova remnants. This component of the ISM is the hot intercloud medium that was introduced in Section 5.2.1 and summarized in Table 5.1.

8.4.2 Evolution of the interstellar medium

Since stars return material that is, in many cases, enriched in heavy elements that are formed over the lifetimes of those stars, the ISM has evolved in its chemical composition. To summarize briefly the effects of stellar nucleosynthesis that have been discussed so far: stars with masses up to around $8M_\odot$ produce elements with atomic numbers up to about that of oxygen; stars with masses up to $11M_\odot$ produce elements up to and including magnesium, plus small amounts of the heavier elements; supernovae produce, during the explosion, a wide range of elements including those of the iron group and heavier elements.

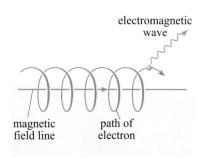

Figure 8.18 The spiralling motion of an electron in a magnetic field results in the emission of electromagnetic waves. (Note that the electron need not travel in the same direction as the field line – it could travel from right to left in this figure, but it would still spiral around the field line and still emit electromagnetic radiation.)

Before considering what effect stellar nucleosynthesis might have had on the composition of material in our Galaxy, it is necessary to ask what the primordial composition of galactic material was. It is believed by the vast majority of astronomers that the origin of the Universe can best be described by a theory called the hot **big bang model**. It would be rather a lengthy diversion to explore this model in detail here, but a key feature is that early in the history of the Universe, all matter was in a very hot and dense plasma, and this plasma cooled with time. At a time when the Universe was about 1 second old, the temperature was about 10^{10} K and matter was in the form of protons, neutrons, and electrons. One of the theoretical predictions of the hot Big Bang model is that these conditions led to a process of **primordial nucleosynthesis** throughout the Universe that formed helium and tiny amounts of lithium but no heavier elements. The outcome of this is that the composition of matter just after the Big Bang would have been a mixture of hydrogen ($X \approx 0.78$) and helium ($Y \approx 0.22$) and extremely small quantities of lithium ($Z < 10^{-9}$). The uncertainty over the exact value of hydrogen mass fraction (X) and the helium mass fraction (Y) depends on the details of the model, but the limit that the metallicity (Z) is less than 10^{-9} is a firm prediction. It is a reasonable approximation to say that the primordial composition of matter was simply a mixture of hydrogen and helium.

■ What is the value of Z for a mixture that contains only hydrogen and helium?

❏ The definition of metallicity for a sample of material is

$$Z = \frac{\text{mass of heavy elements in sample}}{\text{mass of sample}}$$

Since the sample contains no heavy elements, $Z = 0$.

One way of characterizing the effect of stellar nucleosynthesis on the composition of the ISM and the stars that form from this material is to see how the metallicity has changed from its primordial value of $Z = 0$. The easiest place to measure this is in our local neighbourhood.

■ What is the metallicity of the material that the Sun formed from?

❏ In Section 2.2.3 it was stated that the value of Z throughout the Sun is about 2%, i.e. that $Z = 0.02$. Since the Sun has not (yet) synthesized any heavy elements, this must have been the metallicity of the material that the Sun formed from.

In fact, this value for the metallicity is typical for stars and the ISM in the part of the Galaxy where the Sun resides. A value of $Z = 0.02$ may not seem particularly large (although your perspective on this might change once you consider the prospects for your own existence in a Universe where $Z = 0$!), as it shows that only a small fraction of the mass that was originally in hydrogen and helium has been processed to heavy elements.

This might seem surprising in the light of the nuclear processing that has occurred in generations of stars, but is less so when one considers that much of the material returned to the ISM is from the envelopes of stars of low and medium mass, in which little enrichment of the heavy elements has occurred; most of the enriched material remains locked in the cores of these stars. Supernovae do lose more of their mass, and are massive to start with, but they are rare, and so return a relatively small amount of mass to the ISM.

The metallicity of material is only a crude measure of its composition; a fuller description is given by quoting the relative number of nuclei of different elements. This information on abundances of elements in interstellar material in our part of the Galaxy is shown in Figure 8.19. This graph shows the number of nuclei of different elements that would be present if we had a hypothetical sample of material from the interstellar medium that contained 10^6 nuclei of Si. This is essentially the same information as is given in Appendix A5, but displayed in this way it shows some interesting properties that reflect the processes of stellar nucleosynthesis. The most abundant elements are, of course, hydrogen and helium. There is a general trend that the abundance of elements seems to decrease with increasing mass number, which fits in well with some of the ideas of element formation that have been developed in the last two chapters:

- The end point of nuclear burning depends on stellar mass; the higher the stellar mass, the higher mass number of the main product of nuclear burning (up to a limit of $A \sim 56$).

- There are far more low-mass stars than high-mass stars, and this results in a greater enrichment in low mass-number elements than those of higher mass number.

There are clear discrepancies from this general trend and to some extent these can be attributed to the properties of the nuclei of individual elements. So, for instance, the high nuclear stability of the iron group elements is reflected by the high peak of the iron abundance.

While the general picture of the synthesis of elements is now well established, much research effort continues to be directed towards explaining the abundance distribution, as shown in Figure 8.19, in terms of the outcomes of particular stellar processes.

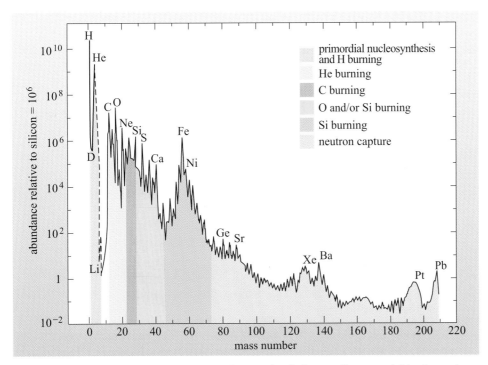

Figure 8.19 The distribution of elements by number in interstellar material in the region of the Galaxy that is local to the Sun. The stages of nuclear burning that give rise to elements at different mass numbers are indicated. (Adapted from Pagel, 1997)

8.4.3 Metallicities and stellar populations

The effect of stellar nucleosynthesis is to increase the metallicity of the ISM as time passes. The observed metallicity of a star provides an indication of the level of enrichment in heavy elements that had occurred in the ISM prior to the formation of that star.

■ Would you expect the original metallicity of a very old star to be higher or lower than that of the Sun?

❑ It is likely to be lower. The metallicity of the ISM by stellar nucleosynthesis increases with time. Hence an old star would have formed at a time when the metallicity of the ISM was lower than it was at the time that the Sun formed.

This general trend is indeed observed for stars in the Galaxy. Young stars tend to have high metallicities whereas older stars tend to have low metallicities. Astronomers categorize stars into classes called 'Populations' which correspond approximately with age. Roughly speaking, Population I stars are those that are young or of moderate age, and these stars are found to have metallicities of up to 2–3%. Population II stars appear to be intermediate age and old stars, and typically have metallicities of less than 0.8%.

The picture that emerges from the study of stellar populations is that there are some regions of the Galaxy, such as the halo (see the Introduction to the book), where star formation stopped a long time ago. The metallicity of stars (Population II) that we see in those regions corresponds to the level of enrichment of the ISM at that time. There are also other regions, such as the disc of the Galaxy, where star formation is recent or even on-going and the metallicity of stars in these regions (Population I) has been enhanced by the greater degree of cosmic recycling that has taken place.

■ Do the lowest metallicity stars correspond to material that formed from the primordial ISM?

❑ No, the lowest metallicity stars have a metallicity of about 0.1% of that of the Sun, hence $Z \sim 10^{-5}$ or so, whereas the primordial ISM would have had a metallicity of $Z < 10^{-9}$.

So even the oldest stars that have been observed in our Galaxy were formed from material that had already been enriched by a previous generation of stars. No stars have ever been observed that correspond to a primordial composition.

8.5 Summary of Chapter 8

Low-mass stars

- After core helium burning commences in a giant, its position on the H–R diagram will evolve towards a low temperature, high luminosity state called the asymptotic giant branch (AGB).

- AGB stars undergo a very high rate of mass loss. The exact reasons for this are unclear, but it is most likely linked either to large-scale global pulsations of the star (as exhibited by Mira variables) or to the onset of thermal pulses.

- After this stage, some stars eject a visible shell of material, of mass $0.1M_{\odot}$ to $0.2M_{\odot}$, called a planetary nebula.

High-mass stars

- On the last day of the life of a supergiant, the conditions in the core are such that nuclear fusion results in the formation of iron group elements.

- In the last few seconds of the life of a supergiant, the iron core collapses to nuclear densities with the copious emission of neutrinos.

- A shock wave is launched through the outer layers of the star; the effect of this, with assistance from the neutrinos, causes the outer layers to be explosively expelled. This is called a Type II (or possibly a Type Ib or Ic) supernova explosion.

- In the explosion, nuclear processes (fusion and the r-process) take place producing, in the main, iron group and heavier elements.

- Subsequent radioactive decay of some of these elements affects the post-explosion light curve of the supernova.

Supernova remnants

- Supernovae are a major source of energy for the interstellar medium. The kinetic energy of the shell of the supernova remnant causes a strong shock that heats material to a temperature of about 10^6 K.

- Supernova remnants grow to sizes of several hundred parsecs in extent. Merging of remnants occurs and forms an interconnected volume that is filled with hot gas from the supernova remnant cavities.

Cosmic recycling and enrichment in heavy elements

- Dispersal of stellar material may occur by stellar winds, planetary nebulae or by supernovae explosions.

- Stellar winds and planetary nebulae result in the return of a significant mass of material to the interstellar medium. This material is only modestly enriched in elements that are heavier than iron.

- Supernovae result in the return of a small mass of material to the interstellar medium, but are a major source of elements that are heavier than iron.

- As time has progressed, the metallicity of the interstellar medium has increased from the very low levels ($Z \sim 10^{-9}$) that resulted from primordial nucleosynthesis, to a value of about $Z \sim 3\%$ in star-forming regions at present.

- The stars in our Galaxy show different levels of enrichment in heavy elements that represent different epochs of star formation. Stars that have been formed recently tend to have higher metallicity than very old stars. However, no star has been observed that has a primordial composition.

Synchrotron radiation

- In the presence of a magnetic field, electrons will spiral around magnetic field lines. As they do so, they will emit electromagnetic radiation.

- If electrons are moving at speeds more than a few per cent of the speed of light, the radiation that is emitted is called synchrotron radiation. Synchrotron emission usually produces a featureless continuous spectrum.

Questions

QUESTION 8.7

A planetary nebula has a diameter of 0.4 pc and the speed of out-flowing material is $20 \, km \, s^{-1}$. Estimate the age of the planetary nebula and state any assumptions that you have to make in calculating your answer.

QUESTION 8.8

At its brightest, a hypothetical supernova has a luminosity of $5 \times 10^9 L_\odot$. It is sufficiently bright that it turns night into day – that is, it shines in the night sky as brightly as the Sun shines in the daytime sky, and delivers the same flux density to the Earth as does the Sun. How far away is it? Give your answer in astronomical units and parsecs. How many supergiant stars are there within that distance from the Earth?

QUESTION 8.9

Identify the following phases on the supernova light curve shown in Figure 8.6: (i) shock heating and expansion of the photosphere; (ii) temperature dropping but surface expanding; (iii) temperature and radius of visible surface approximately constant; (iv) visible surface shrinking at constant temperature; (v) γ-ray heating.

QUESTION 8.10

Use the data in Table 8.2 to decide whether the following isotopes of palladium (Pd) could be formed by the s-process starting from a sample of pure $^{104}_{46}Pd$: (a) $^{108}_{46}Pd$ and (b) $^{110}_{46}Pd$.

Table 8.2 Stability data for some isotopes of palladium (Pd, $Z = 46$). (Note that there are more isotopes of palladium than are shown here.)

	Mass number A						
	104	105	106	107	108	109	110
type of radioactive decay	none	none	none	β^--decay	none	β^--decay	none
product of radioactive decay	–	–	–	$^{107}_{47}Ag$	–	$^{109}_{47}Ag$	–
half-life	stable	stable	stable	6.5×10^6 years	stable	13.7 hours	stable

QUESTION 8.11

If there were no supernova explosions, and massive stars quietly collapsed in on their iron cores, how would the chemical composition of the interstellar medium be different?

CHAPTER 9
THE REMNANTS OF STARS

9.1 Introduction

In this chapter we look at what remains of a star after nuclear burning processes cease. We shall use the term 'stellar remnants' to describe these objects, which essentially form from the dead cores of stars. (This is distinct from the term 'supernova remnant' that refers to the envelope that is ejected following a supernova explosion.) These stellar remnants are some of the most bizarre objects in the Universe: white dwarfs, neutron stars and black holes. Their common features are small radii, extremely high densities and intense gravitational fields. We shall see that the type of remnant that is left at the end of a star's life depends on its initial mass; that white dwarfs are the end point of evolution of low-mass stars, while neutron stars and black holes arise from the catastrophic events that mark the end of the life of a high-mass star.

Because of their small radii, isolated stellar remnants tend to have very low luminosities and hence are difficult to detect. This is exemplified by the case of black holes; they are effectively black, and it has been a major challenge to astronomers to prove their existence. As we will see, the key to solving this problem lies in the observation of interacting binaries, which can be considered to be laboratories for studying not just black holes, but all kinds of stellar remnant.

9.2 White dwarfs

We saw in Chapter 8 that stars with an initial mass of less than about $11M_\odot$ will shed a planetary nebula as they reach the final stages of their lives. It is as this phase of stellar evolution comes to an end that white dwarfs are formed. The class of stars that are called white dwarfs were introduced in Section 4.2.2 on the basis of their observational properties.

- ◼ What properties of white dwarfs can you recall?
- ❏ They are very small (Earth-sized), have low luminosity and have surface temperatures typically in the range 5000 to 25 000 K.

Why should we connect these objects with the late stages of stellar evolution? One clue comes from looking at the central stars of planetary nebulae on the H–R diagram. These are shown in Figure 9.1, together with the positions of typical white dwarfs. If we assume that the star at the centre of a planetary nebula is the hot core of a giant that is running out of nuclear fuel, and that the only course left open to it is to cool and contract under gravity, then we can predict that it would evolve along the track shown, ending in the region of the white dwarfs. This is the type of evolutionary picture that is generally accepted today. Let's assume that this is broadly speaking correct, and take a closer look at some properties of white dwarfs.

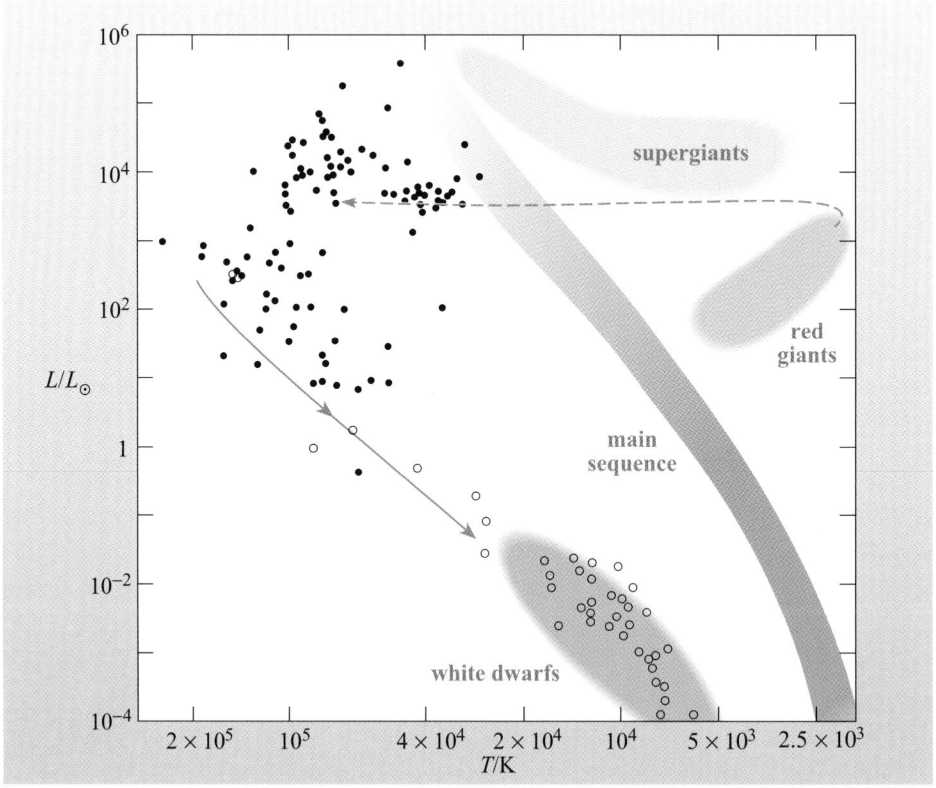

Figure 9.1 The positions of central stars associated with planetary nebulae (dots) and of white dwarfs (open circles) on the H–R diagram. Also shown (solid line) is the evolutionary track that would be followed by a star of constant radius as it cools and (dashed line) a schematic evolutionary track between the regions occupied by AGB stars (Section 8.2.1) and by the central stars of planetary nebulae.

The core of a white dwarf consists of electron degenerate material. With no possibility of initiating any further nuclear reactions, the remnant of the giant, possibly devoid of its outer layers through a combination of the copious stellar wind in the AGB phase and the ejection of a planetary nebula, contracts under the force of gravity. The density in the core shoots up until electron degenerate conditions are created. This assertion is supported by reference to Figure 7.4. The density and temperature conditions believed to pertain in the cores of white dwarfs are shown – they are clearly in the area where material is electron degenerate.

Such material gives rise to a degeneracy pressure (in this case electron degeneracy pressure – see Section 7.2.4) and this plays a vital role in counteracting the collapse of the core of the dying red giant. The pressure gradient due to degeneracy pressure within the core halts any further contraction and forms a stable star – a white dwarf. However, this halted contraction doesn't prevent a typical white dwarf from having some fairly exotic properties.

QUESTION 9.1

Assuming that a typical white dwarf has a radius of 7×10^3 km and a mass of $0.6M_\odot$, calculate its average density. How does this compare with the value for the Sun?

In principle, the equations of stellar structure can be solved for a white dwarf just as for a main sequence star.

■ What are the main differences in the data used for solving the equations of stellar structure for a white dwarf, as opposed to a main sequence star?

❏ The major differences are in the size and the composition.

What emerges from solving these equations for a typical white dwarf? First, it is found that for a surface temperature of about 10^4 K, the core temperature will be in the region of 10^7 K. At first sight this might appear anomalous, as we have already seen that a temperature of this order is able to trigger nuclear fusion reactions in a star like the Sun.

■ Given its likely composition, why shouldn't such fusion reactions also occur in a white dwarf?

❏ A white dwarf, being essentially the collapsed core of a giant, is likely to be almost devoid of hydrogen, much of it having been used during the main sequence lifetime or thrown off into the interstellar medium in the form of a planetary nebula. The white dwarf, depending on the mass of its 'progenitor' or parent, is likely to consist largely of helium, or carbon and oxygen, or heavier nuclei. Such nuclei require higher temperatures before fusion reactions can occur.

Solution of the relevant equations does show, not surprisingly, that the outer layers of a white dwarf are not degenerate and must therefore be treated with different equations for certain properties. Perhaps more surprising is the prediction that as the mass increases, the radius decreases. Following from this is the prediction of an upper limit to the mass of a white dwarf, above which electron degeneracy pressure can't halt the contraction. This value is about $1.4M_{\odot}$ and it is known as the **Chandrasekhar limit** after the Indian astrophysicist who predicted it. How does this rather surprising prediction compare with observations? Actually quite well – no white dwarfs have been found above this mass limit.

SUBRAHMANYAN CHANDRASEKHAR (1910–1995)

Subrahmanyan Chandrasekhar (Figure 9.2) was brought up in India. While he was travelling by boat from India to England on his way to Cambridge in 1930 he worked out that there would be a maximum mass that a white dwarf could have. In 1936 he moved to the University of Chicago, where he spent the rest of his working life, and was notable for the mathematical detail with which he explored astrophysical topics. In 1983 he was awarded the Nobel prize, and in 1999 a major NASA X-ray observatory was named Chandra in his honour.

Figure 9.2 Subrahmanyan Chandrasekhar. (University of Chicago)

Figure 9.3 The results of a search for white dwarf stars in the globular cluster M4. The globular cluster is at a distance of about 2 kpc from the Sun and contains more than 10^5 stars. The left-hand image shows the entire cluster as viewed from a ground-based telescope; the area of the detailed search made using the Hubble Space Telescope is indicated. The right-hand image shows the seven isolated white dwarf stars (circled) that were found. The other stars in the image are main sequence stars and red giants. ((left) M. Bolte (University of California, Santa Cruz); (right) H. Richer (University of British Columbia)/NASA)

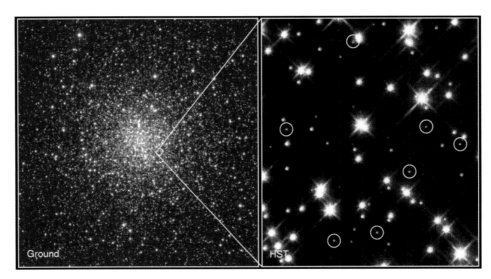

What about the frequency of occurrence of these objects? Here we run into the problem of selection effects. Because white dwarfs are rather faint, we probably are not able to observe many of them. When we take account of this fact, it appears that white dwarfs are really rather common, perhaps constituting 10% of all stars in our Galaxy. As an illustration of the ubiquity of white dwarfs, Figure 9.3 shows the results of a sensitive survey of stars in the globular cluster Messier 4 (M4) made with the Hubble Space Telescope: the right-hand panel shows seven isolated white dwarfs (circled) that were discovered within the survey area.

What about stars with intermediate mass, above the Chandrasekhar limit of $1.4M_\odot$ and below about $11M_\odot$? Remember the prediction that white dwarfs cannot have a mass above $1.4M_\odot$. Therefore, if these intermediate mass stars are to become white dwarfs, as seems to be the case, they must lose mass. The shedding of mass, as we have seen, plays a significant part in these later stages of stellar evolution, through stellar winds and planetary nebulae. However, from what we know of them, it doesn't seem that these mechanisms are able to remove enough mass to bring the more massive stars below the Chandrasekhar limit. So some difficulties still remain to be explained in this picture of the final stages of evolution of intermediate mass stars.

QUESTION 9.2

If the luminosity of a white dwarf drops by a factor of 10^4, and its radius stays constant (a feature of degenerate matter), what has happened to its surface temperature? (Answer this question from theory, and check your answer by referring to Figure 9.1.)

Is there any future for a star after the white dwarf phase? With the exception of cases where a white dwarf exists in a close binary system (Section 9.5) the answer is almost certainly 'no'. The fate of the isolated white dwarf is to cool and fade. The white dwarf has a small surface area, and hence its luminosity is relatively low. This means that the thermal energy of the white dwarf is lost slowly: it cools on a timescale of 10^9 years or more. As the white dwarf cools, it follows a track on the H–R diagram similar to that shown in Figure 9.1 and moves to the lower right. This is generally regarded as a final resting point in stellar evolution, and its material is now essentially lost from taking any further part in the cosmic cycle. This is to be the final fate of our Sun – a dark, cold, dense sphere of degenerate material, rich in carbon and oxygen and only about the size of the Earth!

9.3 Neutron stars

What is left after a Type II supernova explosion? We saw in Chapter 8 that the explosion ejects the outer layers of the star into space, and leads to the formation of a supernova remnant. Here we will consider what is left of the core of the star that underwent the explosion. (Note that the term *supernova remnant* is used by some authors to include any central object; here the term refers only to the material that is ejected.)

In earlier sections we followed the evolution of the core of the star as it collapsed rapidly. The implosion had been triggered by the photodisintegration of the iron group nuclei in the core, and the reaction

$$e^- + p \rightarrow n + \nu_e$$

which absorbed electrons (removing their contribution to the pressure supporting the star) and produced neutrinos (which carried away energy). You saw that, even if the core mass exceeds the Chandrasekhar limit of $1.4M_\odot$, the collapse could be brought to a sudden halt when the density became comparable with the density of an atomic nucleus, and a new form of degeneracy pressure due to the neutrons came into play.

Neutron degeneracy pressure is the same phenomenon as the electron degeneracy pressure that you met in connection with the cores of red giants (Section 7.2.4) and with white dwarfs (Section 9.2), except that the particles providing the pressure in this case are neutrons, and the density is much higher. The existence of this pressure allows the presence of another stable form of matter, more dense than that of the white dwarf star, in which the strong forces of gravitational contraction are balanced by the pressure from the neutrons.

We saw earlier that the Chandrasekhar limit is the maximum mass that can be supported by the electron degeneracy pressure. Similarly for neutrons there is a limit, but because the composition of the star is different, the limit is different. There is some uncertainty over exactly what the limit is since the extreme environment of the inside of a neutron star is not fully understood. The limit is estimated to be between $3M_\odot$ and $5M_\odot$ and most astrophysicists suspect that the true value lies close to $3M_\odot$. Despite the uncertainty over the physics of neutron stars, it is well established that no star could be supported against gravity by neutron degeneracy pressure if its mass exceeded $5M_\odot$.

▪ Suppose the mass of the collapsing core is bigger than this limit, what happens?

❑ The collapse of the imploding core of the supergiant is not halted and it continues to shrink under gravity to even greater densities.

What it becomes, we shall see in the next section. In this section we shall concentrate on cores where the mass is less than $3M_\odot$. So, after the collapse, what do we have? Something small, dense and very rich in neutrons! This is a neutron star, a body of up to a few solar masses, packed into a sphere about 10 km in radius.

QUESTION 9.3

Calculate the average density of a $1.5M_\odot$ neutron star of radius 10 km. How many tonnes would a thimble-full contain? (Take the volume of a thimble to be 1 cm³. Also note that 1 tonne = 1000 kg.)

These densities are hard to envisage, and hard to believe! Something comparable would be achieved if Mont Blanc were compressed to thimble size, or if the whole Earth shrunk to a few hundred metres across.

The gravitational field at the surface of the neutron star is correspondingly enormous.

QUESTION 9.4

The acceleration due to gravity (g) at the surface of a spherical body of mass M_* and radius R_* is given by,

$$g = \frac{GM_*}{R_*^2}$$

Calculate the acceleration due to gravity at the surface of:

(a) The Earth (mass $= 5.98 \times 10^{24}$ kg, radius $= 6378$ km).

(b) A neutron star of mass $1.5M_\odot$ and radius 10 km.

QUESTION 9.5

Use your answers from Question 9.4 to calculate the speed of an object at a time of 1.0×10^{-5} s after it is dropped from rest from a position just above the surface of:

(a) The Earth.

(b) A neutron star of mass $1.5M_\odot$ and radius 10 km.

(Assume that the object is dropped from such a height that it does not hit the surface during that time, and that the acceleration due to gravity can be assumed to be constant throughout the motion.)

So neutron stars are small objects, rich in neutrons, with very high densities and huge surface gravitational fields. You may, justifiably, feel that this is already plenty of unusual properties for any type of astronomical object, but the bizarre nature of these objects does not end here. The next property we shall consider is the rotation of the star, but first it is necessary to introduce the concept of angular momentum (see Box 9.1).

BOX 9.1 CONSERVATION OF ANGULAR MOMENTUM

If you have ever watched a high diver you may have noticed that, when tight-tucked, the diver somersaults rapidly during the dive, but when stretched out tumbles more slowly and makes fewer, graceful turns before entering the water. This is an illustration of the conservation of angular momentum.

A pirouetting skater is another example. With the arms outstretched the skater spins slowly; pull them in and the rotation speeds up.

Angular momentum relates both to how fast a body is rotating and how compact or how spread out is the mass of the body around the rotation axis. **Conservation of angular momentum** means that, unless the body is interfered with, its angular momentum will not change. For an isolated body, left on its own, the angular momentum at the end of an event is the same as at the beginning.

It is important to define clearly the way in which the rate of rotation is measured. The rate of rotation of a body is called its **angular speed** and is measured in the number of radians turned through per second. The angular speed is usually denoted by the symbol ω (the lower-case Greek letter 'omega'), and has the SI unit of rad s^{-1}.

◼ If a body rotates through one complete revolution, how many radians has it turned through?

❏ 2π (or 6.283) radians corresponds to one complete revolution of a body.

So, for instance, a body that undergoes one complete revolution in 1 s, will have an angular speed of 2π (or 6.283) rad s^{-1}. As this example suggests, it is often useful to be able to convert between the time taken for a body to undergo a full rotation (called the rotation period T) and the angular speed , and the two quantities are related by

$$\omega = \frac{2\pi}{T}$$
(9.1)

It is also useful to be able to convert between the frequency f with which a body undergoes complete revolutions and the angular speed

$$\omega = 2\pi f$$
(9.2)

QUESTION 9.6

Calculate the angular speeds of the following:

(a) A body that rotates through a half turn in one second.

(b) A body that undergoes a full rotation in 0.25 s.

(c) A body that rotates with a period of 4.0 s.

(d) A body that rotates with a frequency of 4.0 Hz.

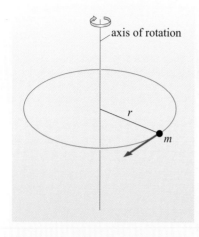

Figure 9.4 A simple system consisting of a mass m that is concentrated at a point that lies a distance r from the axis of rotation.

The second factor that determines the angular momentum of a body is a quantity that describes how compact or how spread out is the mass of the body around the rotation axis. This is called the **moment of inertia** around that axis. A more compact body has a smaller moment of inertia than a more extended body of the same mass. We will not be concerned too much here with the derivation of moments of inertia, but we will consider the following simple situation that illustrates how moments of inertia change if the distribution of mass changes. Figure 9.4 illustrates a system that consists of a mass m that is concentrated at a point that is a distance r from the axis of rotation. In this case the moment of inertia is given by $I = mr^2$. If, for example, the system shown comprises a 1 kg mass that rotates at a distance of 2 m from the axis, then its moment of inertia would be 4 kg m². If the system is made 'more compact', by reducing the distance from the axis to 1 m, the moment of inertia would drop to a value of 1 kg m². Note that the equation given here relating the moment of inertia to mass and distance applies only to this simple case: for other distributions of mass, such as that found within a star, the moment of inertia will be given by a different equation. Even so, the general principle still holds that if a distribution of mass is made more compact with relation to the axis of rotation, the moment of inertia will decrease.

The magnitude of the angular momentum L is related to the angular speed ω and the magnitude of the moment of inertia I by

$$L = I \times \omega \tag{9.3}$$

The diver has smaller moment of inertia when tight-tucked and so must have a larger angular speed (must turn faster) to keep the angular momentum constant. With arms outstretched the skater has larger moment of inertia and so must pirouette more slowly than with arms drawn in. Because the angular momentum is conserved, the angular speed can change only if the moment of inertia changes.

- Returning to the system shown in Figure 9.4: as before, the system initially comprises a 1 kg mass at a distance of 2 m from the axis of rotation, and changes such that the distance of the mass from the axis of rotation becomes 1 m. If the system is initially rotating, would you expect its angular speed to increase or decrease as a result of this change?

- As a result of the change of distribution of mass, the moment of inertia has decreased. By the conservation of angular momentum, the angular speed must increase. (In fact, because the moment of inertia has decreased by a factor of 4, the angular speed must *increase* by a factor of 4 for angular momentum to be conserved.)

When the core of a supergiant collapses, its moment of inertia drops dramatically and hence the star spins up. To give a feel for the size of the effect, suppose that the Sun collapsed to the size of a neutron star; how rapidly would it be rotating then? If angular momentum is conserved

$$I_\odot \omega_\odot = I_n \omega_n \qquad (9.4)$$

where the subscript n denotes the quantities for the neutron star. The moments of inertia are:

$$I_\odot = 6 \times 10^{46} \, \text{kg m}^2 \quad \text{and} \quad I_n = 10^{38} \, \text{kg m}^2$$

The Sun rotates on its axis once every 25.5 days, hence its period of rotation is

$$T_\odot = 25.5 \times 24 \times 3600 \, \text{s} = 2.20 \times 10^6 \, \text{s}$$

Therefore $\quad \omega_\odot = \dfrac{2\pi}{T_\odot} = \dfrac{2\pi}{2.20 \times 10^6} \, \text{rad s}^{-1}$

So $\quad \omega_n = \dfrac{(I_\odot \, \omega_\odot)}{I_n} = \dfrac{2 \times \pi \times 6 \times 10^{46}}{2.20 \times 10^6 \times 10^{38}} \, \text{rad s}^{-1}$

$$= 545\pi \, \text{rad s}^{-1} \, (= 1.71 \times 10^3 \, \text{rad s}^{-1})$$

The period of rotation of the neutron star is

$$T_n = \dfrac{2\pi}{\omega_n} = \dfrac{2\pi}{545\pi} \, \text{s} = 3.67 \times 10^{-3} \, \text{s}$$

Therefore the neutron star rotates on its axis in about 4 milliseconds.

In fact it is not the Sun, but between 1.4 and about 3 solar masses in the centre of the supergiant that collapses to become the neutron star. While the values we have used in the calculation may not be strictly relevant, they are estimates that are reasonable to within an order of magnitude (i.e. within a factor of ten). So this calculation indicates that we would expect a newly formed neutron star to be spinning rapidly.

So now we see that neutron stars are small objects, predominantly made of neutrons, with high density and large gravitational fields, which rotate extremely rapidly. Do they have yet more unusual features?

We have seen how the rotational properties of the star become concentrated and produce rapid spin as the star collapses. In an analogous manner, any magnetic field that there was in the core of the supergiant would become intensely concentrated in the collapse. Once again, let's suppose that it is the Sun that shrinks to form a neutron star, and ask what would happen to the Sun's magnetic field strength in that process.

We saw in Chapter 2 (Box 2.2) that we can represent magnetic fields by lines – the direction of the lines gives the direction of the field, and the strength of the field is shown by how closely packed they are (that is, by the number of lines per unit area). Figure 9.5 shows how the magnetic field of a star changes as the star shrinks. The magnetic field strength, in SI units, is measured in tesla (T). At the Earth's surface the magnetic field strength of the Earth is about 4×10^{-3} T. The Sun has a general magnetic field strength at the solar surface of about 2×10^{-4} T, which we can think of as being through a cross-sectional area of $\pi R_\odot^2$. If the

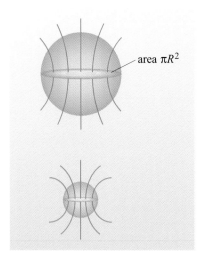

Figure 9.5 The magnetic field of a star through a cross-sectional area of πR^2, and how it becomes concentrated as the area shrinks.

radius of the Sun were to decrease, the magnetic field would become compressed (or concentrated), and its strength would increase as the cross-sectional area decreased. So the magnetic field strength of the neutron star would be larger by the factor $R_{\odot}^2 / R_n^2$, where R_n stands for the radius of the neutron star. This factor is equal to $(7 \times 10^5)^2 / 10^2$, or about 5×10^9. So the expected field of the neutron star is $(2 \times 10^{-4}\,\text{T}) \times (5 \times 10^9)$, or about $10^6\,\text{T}$. This is an enormous magnetic field! In fact it is an underestimate – for reasons that are beyond the scope of this book we believe that the magnetic field of a neutron star can be $10^8\,\text{T}$ or even $10^9\,\text{T}$.

So neutron stars are small dense objects, rich in neutrons, with huge gravitational and magnetic fields, and they carry all this round with them as they spin many times per second. But that does not complete the list of bizarre properties that neutron stars possess. Neutron stars stretch physicists' understanding of material at high densities, test Einstein's General Theory of Relativity, and produce copious radiation in ways that are not yet understood.

9.3.1 Pulsars

Not surprisingly, most astrophysicists had never dreamt that such unlikely objects as neutron stars could exist. There were a few notable exceptions to this statement, such as the famous physicist Robert Oppenheimer – but they were not taken too seriously! The unexpected discovery of neutron stars as pulsing radio stars in the late 1960s therefore produced some excitement and amazement. (A first-hand account of this discovery is given in Box 9.2, at the end of this section.)

ROBERT OPPENHEIMER (1904–1967)

Robert Oppenheimer (Figure 9.6) was one of the leading theoretical physicists of his time. He was born and brought up in the United States, but studied in Europe as quantum mechanics was being developed. He returned to the States in 1929 and throughout the 1930s he worked on various topics in theoretical physics, including a study, with George Volkoff, of the nature of neutron stars. In 1942 he was appointed as director of the Manhattan project, with the aim of developing an atomic bomb. His management skills were no less impressive than his talent for physics: he successfully steered a disparate group of scientists, technicians and military personnel to achieving this goal.

Figure 9.6 Robert Oppenheimer. (Caltech)

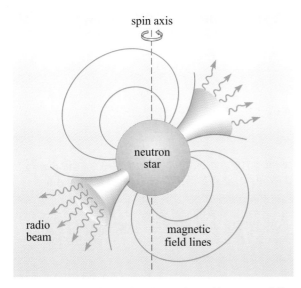

Figure 9.7 A schematic illustration of how a rapidly spinning neutron star produces a beam of radio waves.

The neutron star, swinging its immense magnetic field around each time it rotates, behaves like a huge lighthouse. Some distance away from the surface of the neutron star, electrons that are trapped in the magnetic field generate radio waves by the synchrotron process (Section 8.4.1). These radio waves are emitted in a beam whose direction is determined by the magnetic field, and this beam is swept around by the neutron star as it spins. The essential ingredients in producing the radio signal are the intense magnetic field and the rapid rotation; it is probably also important that the magnetic axis is at an angle to the axis of rotation as shown in Figure 9.7.

At this point it may help you envisage what is happening if you carry out a demonstration with a pair of scissors (preferably straight-bladed ones) as shown in Figure 9.8. Fix the blades open with sticky tape and twirl them about one shaft, which is held upright. With a bit of experimenting, you should be able to find a position for the scissors and an appropriate opening angle for the blades so that as you twirl the scissors there will be one point in each revolution where the slanted blade is point-on to you and you see right along its length (Figure 9.8).

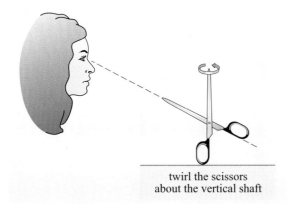

Figure 9.8 A demonstration using a pair of scissors, of the visibility of the beam of radio waves from the neutron star.

Similarly with the neutron star; if the relative orientation of the Earth and the neutron star is correct then, once in every revolution of the star, a line of sight from the Earth will look along the beam of radio waves.

■   What happens if the orientation of the Earth and the neutron star is not right?

❏   The radio beam misses the Earth.

■   If a neutron star rotating at four times a second is orientated so that its beam sweeps across the Earth, how many flashes of radio waves per second are received at Earth from this beaming star?

❏   The beam sweeps across the Earth every time the star rotates. So if the star rotates four times a second then the observer on Earth with a suitable receiver will detect four pulses per second of radio emission.

An observer on Earth with the appropriate radio-receiving equipment can detect the signal produced by a suitably aligned neutron star. The signal received is a string of regular pulses, a set of equally spaced bursts of radio emission as illustrated in Figure 9.9. These neutron stars are called **pulsars** – the name is an abbreviated form of *pulsating radio source*, although it should be noted that the neutron stars themselves do not pulsate but, as we have seen, they do produce pulses of emission.

QUESTION 9.7

It is believed that the time-averaged radio luminosity of a pulsar is about 10^{20} W. Compare the signal picked up from a pulsar at a distance of 10 kpc with that detected from a 100 kW radio transmitter 100 km away.

The accidental discovery of radio pulses from one of these objects caused considerable surprise. First, the repetition rate of the pulses (typically several per second) was much faster than anything previously known in astronomy. An object cannot change its brightness in a time less than the time it takes light to travel across the object. If something is producing (say) four pulses per second then it must be less than a quarter of a light-second across.

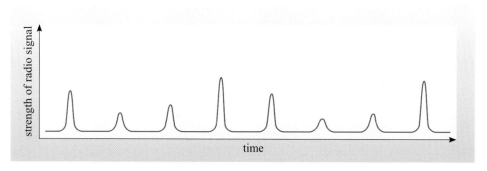

Figure 9.9 The radio signal received from a pulsar.

■ How far is a quarter of a light-second? How does that distance compare with the size of the Sun?

❏ A quarter of a light-second = 1/4 (3×10^8) m = 75 000 km. This is about a tenth of the solar radius.

The only objects of that sort of size or smaller that were known at the time were not normally radio emitters, and certainly not rapidly pulsed radio emitters.

The second cause for surprise was that the pulse rate was accurately maintained. This meant that, whatever the source was, it had large reserves of energy so that it could radiate pulse after pulse after pulse without showing any sign of slowing. If an object has large energy reserves, that usually means it is big.

So the object producing these radio pulses was big, and yet it was small!

This is a good demonstration of the need for precision in scientific language. Our use here has been a little loose and resulted in a conundrum. Let's sharpen up the terminology a little.

■ In what respect is the source of the radio pulses small? (Small in mass, in diameter?) And in what respect is it big?

❏ It is small in radius, in overall dimension. Our deductions about it being big came from noting that it must have considerable energy reserves. Most likely therefore it is big in the sense of being massive.

With hindsight (our best developed faculty!), we can see that neutron stars fit the bill exactly – they are very dense so they are both small in radius and massive.

There are now well over 1000 pulsars known; they lie far outside the Solar System but within the bounds of our Galaxy (the Milky Way). If their positions in the Galaxy are plotted they tend to lie in the nearer half. This distribution is apparent, not real; there are pulsars throughout the Galaxy, we believe, but they are sufficiently weak that we can only detect the nearer ones.

All the pulsars *emit* a beamed signal *continuously* at radio wavelengths which we *receive* as a stream of regular *pulses*. The pulse period is unique to each pulsar. The fastest produce pulses at a rate of hundreds of times per second, the slowest have periods of several seconds. All are believed to have been formed in Type II supernova explosions, although only a handful have an unambiguous association with a supernova remnant, so their origins are still under investigation.

Some pulsars can be very good time-keepers. As you will see below, the pulse periods of all (radio-emitting) pulsars are gradually getting longer. However once this effect is accounted for, the regularity of pulses from some pulsars can match the time-keeping from atomic clocks. International standards of time are defined by networks of atomic clocks and such systems are able to keep time to an accuracy of about 1 part in 10^{14} – this corresponds to uncertainty in time measurement of 1 μs over a period of about 3 years. The pulsars that have very short periods (typically of a few milliseconds) appear to be very smooth running, and some examples of this type of pulsar have been found to keep time as well as atomic time standards. Indeed, it has even been suggested that a combination of such pulsars could be used to define a new time standard that would be more accurate than the atomic clock networks that are currently in use.

QUESTION 9.8

An accuracy of 1 part in 10^{14} corresponds to how many seconds per century?

When we remember that the pulse period is (presumably) the rotation period of the neutron star, then these accuracies are not so surprising, for it takes a lot of energy to change the rotation rate of a star. However, the energy radiated comes ultimately from the rotational kinetic energy of the neutron star (although we do not yet understand exactly how). We can, through studying changes in the pulse period of the neutron star, monitor its loss of rotational energy as the pulsar ages.

Putting together all the available information, the picture we have is of a neutron star formed in a supernova explosion, initially rotating tens of times per second, but slowing as the star loses energy. After about a million years it will be a middle-aged, typical pulsar, with a pulse period of around 0.5 second. It continues to slow as it ages, until the mechanism for generating radio waves ceases to be effective when the star is rotating only once every few seconds, and the radio emission stops. The pulsar then becomes invisible, some 10^7 or 10^8 years after the supernova exploded.

■ What sort of object is left when the pulsar has stopped pulsing?

❏ We are left with a slowly rotating neutron star which is too faint to be detectable.

Considering that the supernova explosion was almost totally devastating, it is amazing that there should have been this lively object left. What is even more amazing is that, for some pulsars, the run-down just described is not the end – for some there is a rejuvenating mechanism that stirs them into life yet again!

Later in this chapter we shall discuss briefly interacting binary systems – examples of stars that are paired with a companion, and so closely paired that they affect each other's evolution. If a pulsar is paired in this manner then it is possible for its companion to transfer matter on to the pulsar, and transfer it in such a way that the pulsar is made to rotate faster. This way, we believe, the rapidly rotating pulsars that have periods of milliseconds or tens of milliseconds have been produced.

While the pulsar's gravitational field is drawing material off the companion and spinning up, several other things are happening. First, any radio radiation that might be produced at this stage is blanketed by the material streaming between the companions and so no radio pulses are seen. Secondly, the transferred material is compressed as it approaches the compact neutron star and is heated. Often it is heated so much that there is copious emission of X-rays. Most of the strongest X-ray sources in our Galaxy are of this kind and such is their strength that examples in other galaxies, external to our own, can also be seen by X-ray telescopes. These X-ray sources will be considered in more detail in Section 9.5.

BOX 9.2 THE DISCOVERY OF PULSARS

Figure 9.10 Jocelyn Bell Burnell, who together with Antony Hewish discovered pulsars in 1967. She later went on to become Professor of Physics at the Open University, a post that she held until 2001.

The following account of the discovery of pulsars was written by Jocelyn Bell Burnell (Figure 9.10), who, as a research student supervised by Antony Hewish, was involved with the discovery of pulsars in 1967.

In the mid-1960s Tony Hewish, a radio astronomer at the University of Cambridge (UK), was awarded a grant to build a special radio telescope to map the quasars in the sky visible from Cambridge and to determine their angular diameters. A full discussion of quasars is beyond the scope of this book, but briefly they are powerful radio sources that are believed to be in the distant reaches of the Universe; studying them gives valuable information on the Universe at an earlier age.

Quasars apparently have very small angular diameters, but Tony Hewish was exploiting a newly discovered technique which allowed these diameters to be determined. It had been noticed that the signal from some radio sources tended to fluctuate rapidly – they 'twinkled', or scintillated – and these were all quasars. The larger angular diameter radio galaxies did not show this fluctuation. The scintillation is produced by turbulence in the solar wind; detailed study of the scintillation gives the angular diameter of the quasar.

The first purpose of the experiment was to scan the sky for objects that scintillated – they were presumed to be quasars. The scintillation is a rapid flickering and so the telescope had to be able to follow rapid variations in the radio signal. If the signal were to be detectable the radio telescope had to have a large collecting area. The telescope Tony Hewish designed covered 4.5 acres (which is an area that could accommodate 57 tennis courts). I joined Tony Hewish as a research student just as construction of this telescope was about to start. We put up over a thousand posts and strung more than 2000 antennae like TV aerials between them. The whole thing was connected by 120 miles of wire and cable. It took five of us two years to build; when finished it looked like a hop field, but it worked beautifully (Figure 9.11).

The construction was finished in mid-1967 and the construction crew melted away, leaving me to operate the telescope as it surveyed the sky for scintillating quasars. Computing power was very limited so the telescope output was on pen chart, 96 feet of it every day. As a mere research student the job of analysing these charts fell to me. As the telescope repeatedly scanned the sky it detected the quasars, but also, inevitably, it picked up interference from local sources, and one of the skills quickly acquired was the ability to distinguish between them on the charts.

However, after a few weeks' operation I realized that there was, very occasionally, a third type of signal. When present it occupied about a quarter-inch in the four hundred feet it took for a complete sky scan, and it wasn't always present. After a few sightings it clicked that this curious signal (nicknamed a piece of 'scruff') had been seen before *from the same part of the sky*. Curiosity raised, we decided to explore further, but at that point the 'scruff' faded and for a month could not be detected! Finally, perseverance paid off and a signal like that in Figure 9.9 with a pulse period of 1.33 seconds was traced out by the chart recorder pen.

Figure 9.11 The radio telescope used in the discovery of pulsars (the upright posts are about 2.5 m high). (G. Pooley, University of Cambridge)

Discoveries are rarely straightforward, and this is where our problems began. The pulses were too fast and too accurately maintained to be any known type of star, so it seemed logical to search for their origin within our equipment. No fault could be found, and when a colleague and his research student using their own telescope and receiver also picked it up this suggested its origin was beyond the observatory. It was suspiciously like a man-made signal, but when we found that it kept a fixed place among the stars that seemed to rule that out. We dubbed it LGM, for Little Green Men, and argued that if it were another civilization signalling to us they would probably be on a planet orbiting their star. Through studying accurately the pulse arrival times, we should be able to detect the Doppler effect as their planet went round their star. This experiment did indeed find a Doppler effect, but it was that due to the Earth orbiting the Sun. (Remember the Doppler effect works for movement of the observer as well as for movement of the source.) Using a technique called radio dispersion, we estimated the distance of the source as 65 parsecs – well beyond the Solar System, but well inside the Milky Way.

Several months had elapsed by this time (and several thousand feet of survey chart paper accumulated) and we had reached the point where we didn't really believe it was a signal from little green men, but we didn't have a sound physical explanation to put forward instead. We were wondering what to do next, when routine scanning of the charts surveying a totally different part of the sky suggested that there might be a second source of scruff-like signals. Difficult observations at the dead of night just before Christmas confirmed the pulsing signal – this time with a period of 1.25 seconds and of course from a different direction in space.

This discovery was much more exciting because it looked as if we really had found a new kind of star; it was highly unlikely that two lots of little green men would both choose to signal at the same time to an inconspicuous planet, both using a non-ideal method of communicating. When the third and fourth examples were found just after Christmas 1967 it became clear that these had to be stars, but it was probably another six months before the astronomical community agreed that these objects had to be neutron stars.

9.3.2 The detection of non-pulsing neutron stars

In the previous section we saw how some neutron stars, those that emit steady pulses at radio wavelengths, can be detected as pulsars. Here we briefly consider why it is that neutron stars are so hard to detect by any other means, and examine a remarkable case in which a neutron star that is *not* a pulsar has been detected.

The most direct method of observation of a neutron star would be to measure the thermal emission from its surface. Neutron stars are formed in a very hot state; it is believed that their initial surface temperatures are over 10^8 K. They cool to about 10^6 K over a few thousand years and reach a surface temperature of about 10^5 K about a million years after formation. The emission from the surface of a neutron star is expected to follow a black-body spectrum, and hence its luminosity will be given by Equation 3.9.

QUESTION 9.9

A neutron star has a radius of 10 km and a surface temperature of 5×10^5 K.

(a) In which part of the electromagnetic spectrum will the peak of the black-body emission from the neutron star occur?

(b) Calculate the luminosity of the neutron star. Express your answer in terms of the solar luminosity.

As the answer to Question 9.9 shows, isolated neutron stars may be expected to be low-luminosity sources that have a peak in emission in the soft X-ray or extreme-UV parts of the electromagnetic spectrum.

Unfortunately, the number of such neutron stars that could be expected to be found from surveys of the sky in the X-ray band is very low. Despite this, an isolated neutron star was found accidentally in 1996. Earlier, in 1992, the field of view of an X-ray observation of an interstellar cloud was found to include an unidentified source that was given the rather anonymous name of RX J185635–3754 (Figure 9.12). Initial follow-up observations revealed that the source has a black-body spectrum with a temperature of 6.6×10^5 K and that emission comes from a star with a radius of about 14 km. The only plausible interpretation of those observations is that the star was a neutron star. However, there have been observations that suggest that its true radius is actually only about 6 km. If this were the case, then it might be the first evidence for the existence of an object that is yet more extreme than a neutron star – a **quark star**. Quarks are the fundamental particles that make up neutrons and protons: each neutron and each proton comprises three quarks. In a quark star, it is hypothesized that neutrons lose their individual identities, and that matter exists as a sea of quarks. The theoretical properties of such objects are very poorly understood, and it is not even clear whether such objects should actually exist. Thus, this object could be highly significant. However, the evidence that RX J185635–3754 is a quark star is far from conclusive, and most astronomers require much more convincing proof that it is anything other than a 'normal' neutron star.

Figure 9.12 Images of the field-of-view that contains the isolated neutron star RX J185635–3754. (a) In the soft X-ray band, the neutron star is the bright source in the lower right-hand part of the image. (b) An image made by combining two visible wavelengths observed with the Hubble Space Telescope. The colour of a star in the image corresponds to its temperature. An extremely hot source, that is interpreted as a neutron star is visible as the white object in this image. (The blue spots are due to cosmic rays hitting the detectors of the telescope.) The neutron star is estimated to be about 61 pc from the Sun. (F. Walter (University of New York at Stony Brook)/NASA)

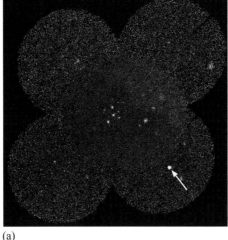

(a)

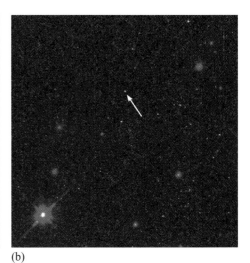

(b)

9.4 Black holes

Case study 1

Imagine a neutron star, a massive one as neutron stars go, close to the maximum mass that can be supported by neutron degeneracy pressure. The star's gravity pulls down onto it some of the gas that is in the vicinity. The extra material takes the mass of the star over the limit, so that the gravitational contraction now overwhelms the neutron degeneracy pressure. The neutron star collapses.

Case study 2

Imagine now the core of SN 1987A. A burst of neutrinos gave a clear indication that a neutron star had been formed, but there has been no subsequent sign of such a star's existence. Suppose the neutron star did exist, briefly, and that it was a massive neutron star, close to the maximum mass that could be supported by neutron degeneracy pressure. The supernova explosion has expelled a lot of the surrounding material, but it has been estimated that the neutron star's gravity could cause about $0.1M_\odot$ of the material to fall back onto the star in the first few hours after the explosion. If this added material caused the gravitational force to exceed the neutron degeneracy pressure then the neutron star would collapse under its own gravity.

Case study 3

Picture now a particularly massive supergiant with a large iron core near the end of its evolution. As described earlier (Section 8.3) when the core contraction starts the iron nuclei break up, electrons merge with protons to form neutrons and in a matter of seconds the inner core has collapsed to neutron star densities. Suppose, however, that the collapsing inner core is more massive than in the case considered in Section 8.3, so massive that the neutron degeneracy pressure cannot withstand the gravitational force. The collapse will not halt at nuclear densities but will continue.

These three case studies all point to the same question: what happens when the gravitational force is greater than the force provided by the neutron degeneracy pressure? Reviewing the story so far, we see that in stars too massive to be white dwarfs, where the electron degeneracy pressure was insufficient to support the star, collapse to another form of matter (the neutron-rich material) ensued. The existence of this other stable form of matter at higher densities allowed the collapse to halt and produced an unusual kind of star.

Will something similar happen in this case, in stars too massive to be neutron stars? Is there another kind of particle and another kind of pressure that will come into play and halt the contraction? We have already mentioned the possibility of quark stars, and the fact that their existence is not yet proven. Even if they do exist, they would also have a limiting mass, above which they would collapse.

There appears to be no mechanism that would halt the collapse of a very massive star. It seems that there is no force that can resist the gravitational contraction. There is nothing to stop the star shrinking under gravity.

The strength of the gravitational force on an object of mass m at the surface of a star of mass M_* and radius R_* is

$$F_g = \frac{GM_*m}{R_*^2}$$

(9.5)

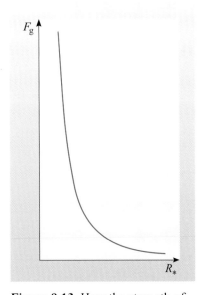

Figure 9.13 How the strength of the gravitational force at the surface of a star depends on the radius of the star.

■ What happens to the gravitational force at the surface of the star if the radius of the star decreases?

❑ As the radius of the star (R_*) diminishes, the gravitational force at its surface, F_g, increases, as shown in Figure 9.13.

The increase in F_g makes it less likely that there can be a force that can effectively resist this gravitational force. More importantly, the increase in F_g produces yet more contraction…which of course increases F_g, which produces more contraction! (Strictly speaking, as the force increases there comes a point where this formula ceases to apply. However, it still serves as an indication of what will happen.)

So the collapse continues relentlessly, apparently until the star has been squashed into an infinitely small space, that is, until it has become a point mass.

■ If the volume of the star has become infinitely small, what has happened to its density?

❑ The density has become infinitely high; the star has zero volume but finite mass!

This is called a singularity; it is a concept that is mathematically and physically difficult to handle (and perhaps you feel that in other ways too it is difficult to handle!) so there is discussion about whether quantum effects or some other effects come into play when the density is extremely high, but not quite infinite, so as to avoid the formation of a singularity. The details need not concern us here – the star collapses down to something not far short of a point, if not an actual point.

During the collapse the gravitational effects increase enormously and we now turn to consider these effects. We shall work with a quantity called the **escape speed**, which is given by:

$$v_{esc} = \sqrt{\frac{2GM}{R}} \tag{9.6}$$

where M is the mass of the (spherical) body from which escape is desired, and R is its radius. Note that the speed of escape does not depend on the mass of the escaping object – it is the same for a ball or a rocket. For escape from the surface of the Earth it is $11\ \text{km s}^{-1}$.

■ As a star of mass M shrinks, how does its escape speed change?

❑ The escape speed increases (apparently without limit as the star's radius shrinks to approximately zero).

A full treatment of black holes requires an understanding of Einstein's Theory of General Relativity, which is beyond the scope of this book. However, you may be aware of one of the conclusions of relativity theory, which is that nothing can travel faster than c, the speed of light in a vacuum.

■ Can you guess what will happen when the star shrinks sufficiently that the escape speed reaches the speed of light?

❏ It looks as if for radii smaller than this, things will not be able to escape from the collapsing star, because to do so would require speeds greater than c, which are not possible.

The radius where the escape speed equals the speed of light is a critical radius in the collapse of a star, called the **Schwarzschild radius** (pronounced Sh-<u>vartz</u>-child). At radii smaller than this, no material can escape from the collapsing star. It also represents a point-of-no-return for the collapsing star itself – once this radius has been passed the collapse of the star cannot be halted. Furthermore, no light waves (or radio, or X-ray, or infrared, or any other electromagnetic radiation) can directly escape either. The collapsing star as it crosses the Schwarzschild radius becomes a **black hole**; black because no radiation gets out, and a hole because matter can fall in but cannot get out of it!

QUESTION 9.10

Derive an expression for the Schwarzschild radius (R_S) in terms of G, M and c.

QUESTION 9.11

Calculate the size of the Schwarzschild radius for a star of $1.0M_\odot$.

Although black holes are not luminous, their gravitational fields still exist. Anything that comes too close to a black hole is pulled towards it by gravity. Anything that is pulled closer than the Schwarzschild radius cannot escape; it is sucked in and squashed to (near) infinite density.

When something falls down a black hole do we see any change in the black hole? If we can ascertain the mass of the black hole (perhaps through measuring the strength of its gravitational field) then we would see that its mass has increased. The only other physical properties possessed by a black hole are its electric charge and its angular momentum. All other information about what went down a black hole is lost. One cannot tell whether it was a one kilogram bag of feathers or a one kilogram bag of lead that has just been swallowed by the hole.

Disembodied gravity is difficult to detect, so is there any observational evidence for the existence of black holes or are they a theoretician's dream (or nightmare)? Isolated black holes would be hard to see, but many stars are in binary systems; if one of the stars in a close binary had evolved into a black hole, then it would be possible to infer its presence from observations of the binary. It is this type of system that we shall consider in the next section.

9.5 Stellar remnants in binary systems

Most stars exist in binary systems or bigger groupings (Section 3.2.3). In many multiple star systems the stars are sufficiently far apart (more than 100 AU) that there is little interaction between the stars apart from their mutual gravitational influence on one another. However, there are systems in which the two stars orbit each other sufficiently closely that matter can be transferred from one star to another. The process of mass transfer has direct observational consequences and, over time, alters the evolution of both stars. Such binary systems are the subject of this section.

9.5.1 Interacting binaries – laboratories for studying stellar remnants

Before considering binary systems, let us first imagine what would happen if we could drop material directly onto an isolated star. Any process by which material is added to a star is termed **accretion**. If a mass m of material falls onto a star of radius R_* and mass M_*, then provided that the material starts from a position that is a long way from the star, the gravitational potential energy released will be $E_g = GM_*m/R_*$. Initially this would be released as the kinetic energy of the in-falling material, but on hitting the surface of the star, this would be converted into thermal energy and cause heating.

■ Imagine that we could drop 1 kg of material onto a $1.4M_\odot$ white dwarf and a $1.4M_\odot$ neutron star. For which type of star would the gravitational potential energy release be greatest?

❑ The energy released is given by $E_g = GM_*m/R_*$, but G, M_* and m are the same in both cases, so the larger amount of energy will be released in the case where R_* is smaller. Since the radius of a neutron star is much smaller than that of a white dwarf, it is accretion onto a neutron star that liberates most energy.

Hence for a given mass, the smaller the radius of star that material can fall onto, the greater the energy that can be released. So we might expect that accretion onto stellar remnants which have relatively small radii might produce energetic phenomena. At this point, you might be wondering how gravitational potential energy might be released in the case of accretion onto a black hole, since there is no surface for the in-falling material to strike. As you will see later, there is a mechanism by which accretion onto black holes can liberate copious amounts of energy.

There are two main processes by which material can be transferred from one member of a binary system to the other. As an introduction to the first type of mass transfer process, consider the situation described in the following question:

QUESTION 9.12

In a certain binary system, one of the stars is a white dwarf with mass $1.20M_\odot$ and radius $0.006R_\odot$ while the other is a main sequence star with mass $0.70M_\odot$ and radius $0.96R_\odot$. The centres of the two stars are a distance of $2.18R_\odot$ apart. Consider a blob of gas that lies on the line that connects the centres of the two stars and at the outermost part of the envelope of the main sequence star (i.e. it is at a distance of $0.96 R_\odot$ from the centre of the main sequence star). This blob is influenced by the gravitational attraction of both stars.

(a) Draw a diagram of this system that shows the direction of the gravitational forces acting on the blob.

(b) Compare the gravitational forces due to the two stars on the blob in this position. Which is the larger?

Question 9.12 indicates that at the surface of the main sequence star the white dwarf may exert a stronger force, and so it is quite possible to have material transferred from one star onto the other. There will be a number of points in the vicinity of the two stars where the blob of gas experiences *equal* gravitational pull towards the two stars; we did the calculation for a point on the line joining the stars, but there are, for example, points either side of this line where this is also true. In practice, the stars are rotating about each other and this complicates the analysis of forces. However, when the system is considered in a frame of reference that is rotating with the orbital period, it is still true that there are positions where a blob of gas experiences no net force. These are called **Lagrangian points** (the term Lagrangian, pronounced 'Lah-grahn-jee-an', refers to the French mathematician Joseph Louis Lagrange, 1736–1813). As far as transfer of material between the two stars is concerned, it is the Lagrangian point on the line joining the centres of the two stars that is of primary importance; this is called the **inner Lagrangian point**. Material that is positioned at the inner Lagrangian point is poised such that with a tiny push, it could fall either towards one star or the other. If one of the stars in the binary is large enough that its surface touches the inner Lagrangian point, then material can overflow this point and start to fall towards the other star.

The shape of the surface that a star occupies when it touches this point is not spherical: the influence of the other star and the rotation of the entire system result in a surface that is somewhat pear-shaped. This surface is called a **Roche lobe** and it represents the maximum volume that a star can occupy before it begins to transfer material through the inner Lagrangian point, and the process of mass transfer is called **Roche lobe overflow**. Roche lobes can be defined around any star in a binary system regardless of whether that star is actually filling the lobe. In diagrammatic representations of close binary stars such as that shown in Figure 9.14, it is usual to show the Roche lobes around both stars.

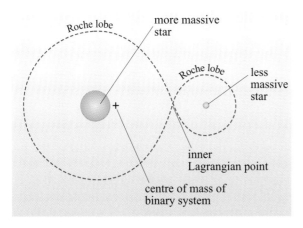

Figure 9.14 A cross-section through a close binary system; the dashed line shows the boundary of the Roche lobes.

Figure 9.15 A simulated view of a binary system in which mass transfer occurs from a star with a relatively large radius by Roche lobe overflow on to a white dwarf. The material that is transferred forms a luminous accretion disc which can, as in this case, dominate the optical emission from the system. (The white dwarf, which is located at the centre of the accretion disc, cannot be discerned against the bright background of the disc.) (R. Hynes, University of Texas at Austin)

When Roche lobe overflow occurs, material is transferred through the inner Lagrangian point and enters the Roche lobe of the other star. Because of the conservation of angular momentum, such material does not fall directly towards this star, but preferentially heads to one side of it. If this material were in the form of a small solid test particle it would go into a highly elliptical orbit around the star. In reality, the material is in the form of a continuous stream of gas. This stream interacts with itself and forms a disc of material around the star, called an **accretion disc**, as shown in Figure 9.15. Interactions within the disc redistribute the angular momentum of material. Most matter in the disc loses angular momentum and consequently spirals down through the disc and onto the star. However, we saw earlier that angular momentum is a conserved quantity, so the angular momentum that is lost by the accreting matter must be transferred elsewhere. In fact, angular momentum is transferred to the relatively small amount of material that forms the outer parts of the accretion disc, and is then fed back into the orbital system of the binary through tidal interactions. So the accretion disc allows most of the matter that is transferred to move inwards towards the central star. In doing so, the gravitational potential energy of this material is converted into thermal energy and heats the disc.

Thus the accretion disc is a source of electromagnetic radiation. Most of the disc emits strongly in the optical and the ultraviolet parts of the spectrum and those parts of the disc that are closest to the stellar remnant tend to be a luminous source of X-rays. In addition, in the case of white dwarfs and neutron stars, there may also be strong X-ray emission from the surface of the stellar remnant as accreting material is dramatically brought to a halt. Thus an interacting binary that contains a white dwarf, neutron star or black hole will tend to be an X-ray source, and consequently they are generically referred to as **X-ray binaries**. These sources dominate the sky at X-ray wavelengths as Figure 9.16 shows.

A second mass-transfer process can occur where one star has a very strong out-flowing stellar wind; as it moves in its orbit, the other star captures some of the circumstellar material as indicated in Figure 9.17. Such a process is called **stellar wind accretion**. Strong stellar winds tend to be associated with high-mass stars, so binary systems in which this process is taking place often include a luminous star of spectral type O or B. Unlike the case of Roche lobe overflow, it is rather difficult to predict whether an accretion disc will be formed as a result of stellar wind accretion. Furthermore, observations of stellar wind accreting systems do not unambiguously show the presence of accretion discs and so there is a lack of detailed knowledge about these systems.

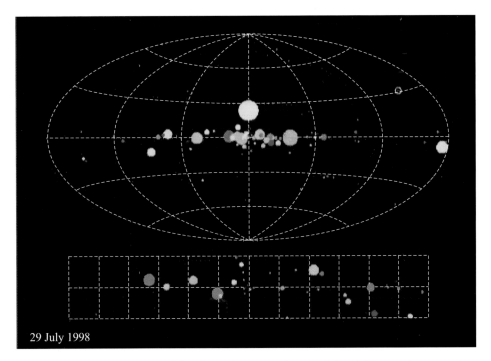

29 July 1998

Figure 9.16 An X-ray map of the sky as it appeared on 29 July 1998 to an instrument on the Rossi X-ray Timing Explorer satellite. The upper map shows the whole sky: the centre of the map corresponds to the direction towards the centre of our Galaxy and the horizontal line across the centre of the map corresponds to the plane of the Galaxy. The lower map shows a central region ($60° \times 10°$ in extent) of the all sky map. The size of a circle indicates the intensity of the X-ray flux while its colour provides information about the spectrum of the X-ray source. Note that this map only shows the brightest X-ray sources – there are many more fainter sources which were not detected by this particular instrument. The most numerous sources in the map are X-ray binaries in which matter is transferred by Roche lobe overflow onto a neutron star. The very bright source that is positioned somewhat above the centre point of the map is one such source: an X-ray binary called Sco X-1 that is the brightest X-ray source in the sky. (M. Muno, Massachusetts Institute of Technology)

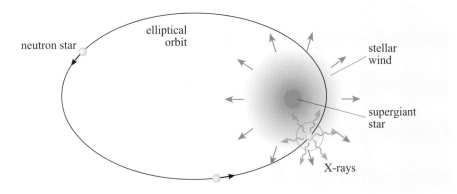

Figure 9.17 The process of mass transfer in which a stellar remnant 'captures' some of the strong stellar wind that emanates from a supergiant companion star. In this case the binary orbit is highly elliptical. The highest rate of accretion and hence X-ray emission occurs when the stellar remnant and the companion star are closest to each other.

9.5.2 Accretion onto white dwarfs – cataclysmic variables

Given the ubiquitous nature of white dwarfs, it is perhaps not surprising that the most common forms of interacting binary are those in which material is transferred to this sort of stellar remnant. Such systems are called **cataclysmic variables**, a name that reflects the often dramatic changes in luminosity that they exhibit. These changes are most prominent in the optical, ultraviolet and X-ray parts of the spectrum. In cataclysmic variables the mass transfer process is exclusively by Roche lobe overflow; there are no known cases of stellar wind accretion onto a white dwarf. There are several subclasses of cataclysmic variable, but two of the more important ones are dwarf novae and novae.

The **dwarf novae** are stars that show erratic outbursts in their optical emission. A typical outburst may involve a brightening of 2–5 magnitudes in a few days, followed by a slower decline to quiescent levels. Figure 9.18 shows the light curve for the dwarf nova SS Aurigae. The outbursts are not periodic, but reoccur over a timescale of weeks to months. The optical luminosity of these systems is dominated by emission from the accretion disc, and it is likely that the outburst arises from an instability in the disc. When the dwarf nova is in a quiescent state, matter is only able to pass through the accretion disc at a low rate and hence builds up in the disc. The outburst represents a change in the disc that suddenly allows the accumulated material to pass quickly through the disc hence giving a sharp increase in the luminosity.

A **nova** (sometimes called a classical nova) is also characterized by a sudden outburst in luminosity, but in this case the star brightens by more than 10 magnitudes (Figure 9.19). The outburst lasts for a matter of days or weeks before slowly returning to its pre-outburst level. The mechanism that drives nova outbursts is thermonuclear burning on the surface of the white dwarf. As a result of the process of mass transfer, hydrogen-rich material accumulates on the surface of the white dwarf. The density and temperature at the bottom of this layer increase over time until conditions are such that nuclear fusion of hydrogen can begin. This proceeds in a runaway fashion, leading to the observed outburst and the ejection of some of the material from the surface of the white dwarf.

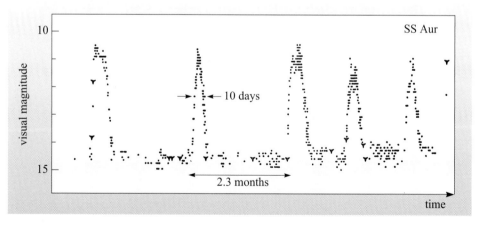

Figure 9.18 The visual light curve for the dwarf nova SS Aurigae. The data for this light curve were collected by amateur astronomers – this is an invaluable service for researchers who study cataclysmic variables. (F. A. Córdova and H. Papathanassiou from data provided by AAVSO)

Figure 9.19 Nova Cygni 1975 erupted on 29 August 1975. (Left) Before and (right) after the nova outburst. (Lick Observatory)

9.5.3 Neutron stars in binaries

Neutron stars are found in binaries where mass transfer occurs both by stellar wind accretion and by Roche lobe overflow. It seems that the mass transfer process disrupts the radio emission that is observed in isolated pulsars, and radio pulses are not observed from these systems. However, accretion in these systems results in material being heated to over 10^6 K and consequently they tend to be luminous sources in the X-ray band. Since neutron stars typically have high magnetic field strengths and are rotating rapidly, it is not surprising to find that some neutron stars in accreting binaries show regular pulses in their X-ray emission, and are described as **X-ray pulsars**. The range of pulse periods in X-ray pulsars is wider than for radio pulsars – from a minimum of about 69 ms up to a maximum of over 800 s. The origin of the X-ray pulsations is linked to the interaction between the accreting material and the magnetic field of the neutron star.

■ The accreting material is in the form of a plasma. What would you expect this material to do as it enters the very high magnetic fields that are close to the neutron star?

❏ A plasma in a very strong magnetic field is forced to move along the magnetic field lines (Section 2.3.1).

This effect may disrupt the inner parts of an accretion disc (if one is present), and material will funnel down onto the magnetic polar regions of the neutron star. These regions will be heated by the impact of accreting material, and so they will emit more strongly than other parts of the neutron star. As in the case with radio pulsars, it is likely that the magnetic axis and the rotation axis of the neutron star are not aligned (as illustrated in Figure 9.20, overleaf), and so rotation of the neutron star gives rise to a modulated pattern of X-ray emission. However, not all binaries that contain neutron stars emit regular pulses; some emit rather weak pulsations that are not strictly periodic, and others simply show rapid flickering that has no periodic variability.

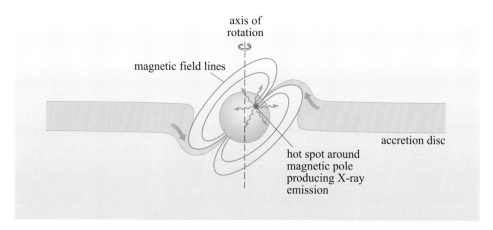

Figure 9.20 A schematic diagram of the inner part of an accretion disc (shown in cross section) and a neutron star. Note that the neutron star and accretion disc are not drawn to scale and that the inner edge of the accretion disc is hundreds of neutron star radii in diameter. The magnetic field of the neutron star affects the in-fall of accreting material – the inner edge of the accretion disc is disrupted by the magnetic field and consequently plasma funnels down the field lines. Hot spots form near the magnetic poles of the neutron star. Since the axis of rotation and the magnetic axis are not co-aligned, the rotation of the neutron star gives rise to modulation in X-rays.

The existence of neutron stars in any binary system, including those binaries that are too widely separated to be interacting systems, offers the possibility of measuring the masses of neutron stars. Thus, the rather poorly determined theoretical limit to the mass of a neutron star can, in principle, be tested against observation.

> ■ Given that no neutron star binaries are visual binaries, what physical property of the stars needs to be measured, and what is the other necessary condition that would allow the masses of the stars in the binary to be determined?
>
> ▢ It must be possible to measure the orbital speeds of each star and the system must be an eclipsing system (see Section 3.3.7).

In neutron star binaries it is often possible to measure the orbital speed of the companion star by spectroscopic means. The orbital speed of the neutron star cannot be determined by spectroscopic methods, since its radio or X-ray emission does not show sharp spectral lines. However, if the neutron star is a radio or X-ray pulsar, then its orbital speed can be determined by measuring the change in frequency of the *pulses* over an orbital cycle. This is because the Doppler effect (Box 3.1) applies to the pulses of emission from the pulsar, and the radial velocity can be found in a similar way to Equation 3.3,

$$v_r = c \times (f - f')/f' \tag{9.7}$$

where f and f' are the rest frequency and the observed frequency of the pulses. The pulses are pulses of electromagnetic radiation, so their speed is c – the speed of light.

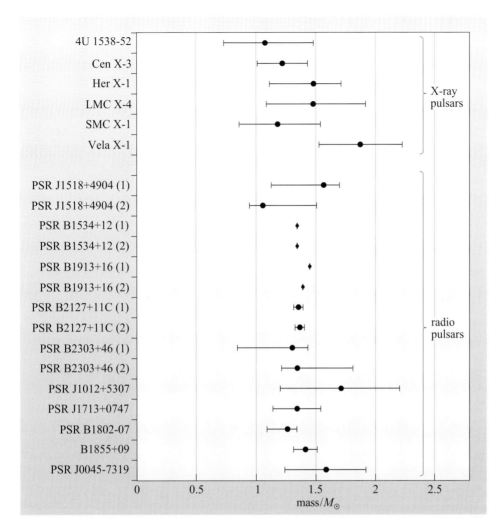

Figure 9.21 The measured masses of neutron stars. The upper six cases are X-ray pulsars that are in interacting binaries, the lower ten are radio pulsars that are in non-interacting binaries. In the case of radio pulsars, five of the systems are double neutron star systems, and the masses of each of these neutron stars (distinguished by (1) or (2) in the name) are shown. (Data compiled by S. Clark (University College London))

This type of analysis has been carried out for a number of neutron star binaries. Some of these are interacting binaries which contain X-ray pulsars and others are non-interacting binaries that contain radio pulsars. Figure 9.21 shows values of neutron star masses that have been obtained by such studies, and illustrates that they all lie close to a value of $1.4M_\odot$. Only a small number of masses have been measured in this way and future observations may reveal the existence of neutron stars with masses up to about $3M_\odot$, but equally, it may be found that the limiting mass is much closer to $1.4M_\odot$ than is predicted by current theories.

9.5.4 Black hole candidates

Observationally, the binary systems that are believed to harbour black holes are similar in some respects to interacting systems that contain neutron stars; they are bright X-ray sources that show substantial variability in their luminosity. The challenge to astronomers is to produce convincing evidence that such systems, which are called **black hole candidates**, cannot be explained by accretion onto a neutron star. Any source that exhibits regular pulsation is likely to be a neutron star: black holes have no magnetic field and so cannot produce modulated emission. This does *not* mean that any source that does not show pulsation must be a black hole; it is quite likely that some neutron star binaries do not produce pulsed emission.

■ What property of a stellar remnant in an interacting binary *could* be used to rule out the possibility that it is a neutron star?

❑ Its mass: if the mass of a stellar remnant exceeded the maximum mass of a neutron star, then it could be concluded that the object is a black hole.

Since astronomers require overwhelming evidence that the remnant is not a neutron star, it is usual to accept the highest theoretical upper limit to the mass of a neutron star as the minimum mass that would prove the existence of a black hole. For this purpose, the mass limit is taken to be $5M_\odot$ although most astrophysicists believe that the true limit is substantially lower than this.

There are immense practical difficulties in carrying out measurements to determine the mass of black hole candidates. The first difficulty is the small number of such systems; only about twenty or so are known. Secondly, the method of determining masses that was described for neutron stars has to be modified for two reasons: there are no pulsations, and, none of the known systems show eclipses. The result of this is that while the mass itself cannot be determined without making certain assumptions about the binary system, a *lower limit* to the mass of the black hole candidate can be found. In many cases, there is no known optical identification of the system, and without optical spectroscopy of the mass-losing star, not even a mass limit can be determined. A final difficulty is that the sources that provide the highest lower limits to the mass of the black hole candidate are so-called 'transient' sources – for most of the time their X-ray emission is at a low level and the source is undetectable. It is only when these sources show outbursts that it is possible to identify them as black hole candidates. These outbursts last for a few weeks or months, but their occurrence is unpredictable, and the source may then return to a quiescent state for a period of many years.

A black hole candidate that has a high lower limit to its mass is a system called V404 Cygni. Outbursts of this source had been observed in the optical waveband by amateur observers in 1938 and 1956, and originally it had been classified as a nova (in fact, it was classified as a 'recurrent nova' – a star that seems to repeat nova-like outbursts over an interval of several years). In 1989 a luminous outburst was detected in the X-ray band which revealed that this was not a system in which accretion was occurring onto a white dwarf, but one in which the stellar remnant must be a neutron star or black hole. After the 1989 outburst had subsided, it was possible to make optical spectroscopic measurements of the mass-losing star, which was found to be a subgiant of spectral type K0. The orbital period of the system was measured as 6.5 days, and this led to a lower limit to the mass of the black hole candidate of $6.3M_\odot$. Since this is substantially above the maximum possible mass for a neutron star, the stellar remnant in V404 Cygni is generally accepted to be a black hole.

The firm lower limits to a range of black hole candidates provide good evidence that black holes of a few solar masses do exist. As mentioned above, the mass of a black hole candidate can be calculated if certain assumptions are made about the binary system. So for instance, the best estimate of the mass of the black hole in V404 Cygni is about $12M_\odot$. A summary of mass measurements of some other black hole candidates is given in Figure 9.22.

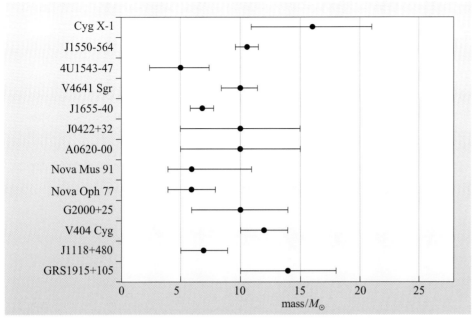

Figure 9.22 A summary of mass determinations, with uncertainties, of a selection of black hole candidates. (Data compiled by S. Clark (University College London))

It is thought that mass transfer in most black hole candidate systems occurs by Roche lobe overflow although there are a few cases in which stellar wind accretion seems a more likely mechanism. Since the black hole has no surface, any X-ray emission that is observed must originate from the accreting material that surrounds the black hole. The outburst behaviour of sources such as V404 Cygni is, in some respects, similar to the outburst behaviour of dwarf novae, and probably arises from an instability in the accretion disc. Although we have concentrated here on the differences between interacting binaries that contain white dwarfs, neutron stars or black holes, it is worthwhile bearing in mind that these systems have many features in common that result from the presence of an accretion disc, and that the process of accretion is in itself a topic of much research effort.

9.5.5 Interacting binaries – effects of binarity on evolution

How did interacting binaries get to their present form, and what will they become next? We assume that both stars were formed at the same time, but that they did not necessarily have the same mass.

- How does the mass of a star affect its rate of evolution?

- The more massive stars take a shorter time to go through their life cycle.

Suppose we have a binary system in which the more massive star has reached the red giant stage while the less massive star is still on the main sequence. As it expands, the envelope of the red giant is held less tightly by the giant's gravity. Meanwhile, part of the envelope is coming closer to the other star and more under the influence of its gravity. Eventually, the envelope may fill the Roche lobe and mass transfer will begin. As the giant star loses mass, its Roche lobe shrinks, but the star's radius changes very little; this enhances the mass transfer. The temperature and luminosity of the giant drop and those of the other star increase.

Following the evolution further, we find that the former giant (we shall call this star 1 from now on) can lose all its outer layers, possibly leaving a naked helium star. The second star is now the more massive and is a bright main sequence star. Nuclear reactions will cease in star 1 and it will become a white dwarf, or perhaps a neutron star. We shall then have a system with an old white dwarf and an *apparently* younger, more massive main sequence star. Figure 9.23 illustrates this sequence.

What happens next? The system remains in this configuration for quite some time, but eventually star 2 will reach the end of its main sequence lifetime; depending on how massive it is, there can be various outcomes. During its red giant or supergiant phase it may fill its Roche lobe and lose mass to star 1 (the white dwarf or neutron star). A number of known cataclysmic variables and X-ray binaries are at this stage of evolution.

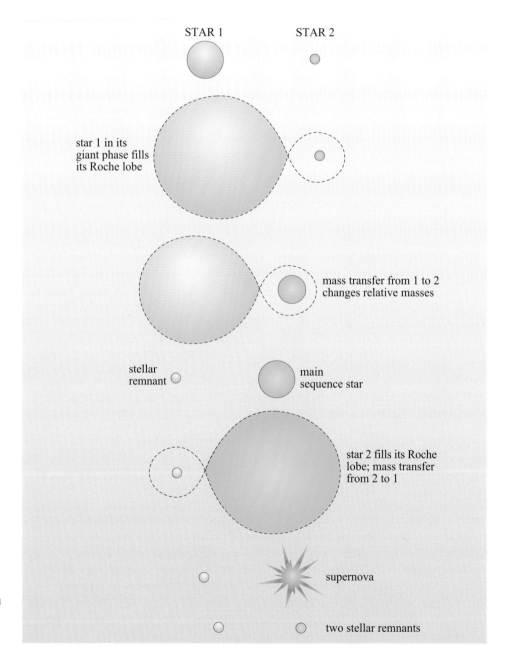

Figure 9.23 The evolution of an interacting binary system. Star 1 is initially the more massive.

If star 2 does not lose too much mass this way, then it may continue its evolution and become a supernova (or even a black hole) so that we can find binary systems that contain a white dwarf and a black hole. Note that there can be variants on this – the origins of many kinds of unusual system can be explained through the evolution of interacting binaries.

9.5.6 Supernovae in binaries

One of the most intriguing effects that is believed to arise in binary systems is a supernova mechanism that is quite distinct from that which is caused by the core collapse of an isolated high-mass star.

▨ Which classes of supernova are believed to arise from the core collapse in a supergiant star?

❏ Type II, Ib and Ic supernovae are all believed to arise from core collapse in high-mass stars (Section 8.3.1).

As has already been noted in Section 8.3.1, Type Ia supernovae do not appear to be associated with sites of star formation and hence cannot arise from processes involving high-mass stars. A plausible model for these type of supernova explosions is one in which mass transfer occurs onto a white dwarf whose mass is close to the Chandrasekhar limit. In this scenario, the white dwarf is composed of carbon and oxygen.

▨ Is it possible to obtain energy from nuclear fusion reactions involving carbon or oxygen?

❏ Yes – fusion reactions are exothermic for all elements with mass numbers lower than those around iron.

Because the material in the white dwarf has not reached the end point of nuclear burning, the Type Ia supernova process is quite different to core collapse in a massive star. As the white dwarf mass approaches the Chandrasekhar limit, the interior temperature increases to the point at which carbon starts to undergo fusion reactions. Once a region is heated by fusion reactions, neighbouring regions are heated by conduction or convection and these too can start to burn. The way in which the zones of nuclear burning propagate through the white dwarf is in many respects similar to the way in which a flame propagates in a chemical fire. Attempts to understand this process by using computer models are an area of active research, and are a challenge to the fastest available computers. An example of a computer model of the propagation of the nuclear 'flame' through a white dwarf is shown in Figure 9.24 (overleaf).

Given the complexity of this rapid thermonuclear burning, it is perhaps not surprising that numerical models are not yet able to predict the exact outcome of such events. Rather the approach taken is to see whether such models can reproduce some of the observed features of Type Ia supernovae. It seems that this type of model can produce an explosion of the correct magnitude and a distribution of elements up to and including the iron group that matches observation.

The successful models have the following features. In the central regions of the supernova, the temperatures reach about 10^{10} K and nuclear reactions proceed to the point where essentially all of the nuclei are converted into iron, cobalt or nickel. In regions further out from the core, temperatures are high enough to form nuclei such as silicon and calcium, but not nuclei of the iron group. The process is rapid, the white dwarf undergoes thermonuclear burning in a matter of seconds, and it is thought that as the zone of nuclear burning moves outwards it accelerates and becomes explosive. The explosion disperses the newly synthesized elements.

■ A large mass of $^{56}_{27}\mathrm{Co}$ is formed as a result of the Type Ia supernova explosion. By analogy with the light curve of Type II supernovae, what would you expect an observational consequence of this to be?

❏ Decay of $^{56}_{27}\mathrm{Co}$ releases energy that emerges as visible light. It is likely that the visible light curve will decay in the same way that $^{56}_{27}\mathrm{Co}$ decays.

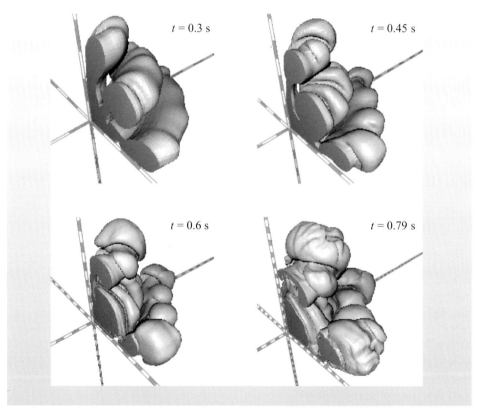

Figure 9.24 A computer simulation of the way in which nuclear burning propagates through a carbon–oxygen white dwarf that is at the Chandrasekhar limit. These plots show one-eighth of the volume of the white dwarf, with the centre of the star at the intersection of the axes. The surface represents the narrow zone in which thermonuclear burning is taking place. The sequence shows how this nuclear 'flame' evolves with time: the nuclear burning starts at the centre of the white dwarf at time $t = 0$ and the snapshots show how it progresses within one second. Note that the tick marks on the axes represent a length of 1×10^5 m and that the scale varies between the images. Thus, the surface of nuclear burning expands outwards through the white dwarf during the sequence. (Reinecke *et al.*, 2002)

This is indeed what is found; the light curves of Type Ia supernovae seem to reflect the radioactive decay of $^{56}_{27}\text{Co}$. The mass of radioactive elements formed in the explosion has been estimated as being about $0.6M_\odot$ in all such supernovae. The energy liberated as kinetic energy and electromagnetic radiation is similar to that released in a Type II supernova explosion.

QUESTION 9.13

Why are Type II and Type Ia supernovae *not* similar in terms of the *total* amount of energy released? What would you expect the total energy released in a Type Ia supernova to be in comparison to the total energy released in a Type II supernova?

A remarkable feature of supernovae of Type Ia is that the peak luminosity of all such explosions is believed to be constant to within about 15%. We have already seen in Chapter 3 (Section 3.3.3) that the distance to a star of known luminosity L can be determined by measuring its flux density F and using Equation 3.12

$$d = \sqrt{\frac{L}{4\pi F}}$$

The same technique can be applied to any source whose intrinsic luminosity is known. This approach to the determination of distance is called a 'standard candle' method, and it turns out that supernovae of Type Ia seem to provide the best standard candle for distances on cosmological scales (i.e. greater than 10^8 pc). Given the importance of Type Ia supernovae to cosmology, they are the subject of much observational and theoretical scrutiny.

Despite intensive research into the nature of Type Ia supernovae, there are some major issues that are yet to be resolved. Firstly, the nature of the mass-donating star in the binary is unknown. There is no case in which the progenitor of a Type Ia supernova has been observed, and given the low frequency of occurrence of supernovae and the probable low luminosity of a pre-supernova binary, it seems unlikely that we will observe such a system in the near future. In some models matter is assumed to be transferred from a main sequence or giant star onto the white dwarf. If this is the case, there is a puzzle as to why the spectrum of the supernova shows no trace of hydrogen or helium. One suggestion is that the transferred material might undergo steady nuclear burning on the surface of the white dwarf such that no significant quantities of hydrogen or helium accumulate. An alternative scenario is that the explosion is the result of the coalescence of two white dwarfs, which would naturally explain the absence of hydrogen features from the observed spectrum. The debate as to which of these models is more appropriate to Type Ia supernovae is on-going.

9.6 Summary of Chapter 9

White dwarfs

- White dwarfs are the remnants of stars which have an initial mass of less than $11M_\odot$.

- Such stars are formed from the cores of giant stars once nuclear burning has ceased. After depletion of all nuclear fuels, the core collapses until degeneracy pressure is able to prevent further collapse.

- With no further nuclear reactions possible, the white dwarf cools and disappears from view.

- White dwarfs cannot have a mass above the Chandrasekhar limit of $1.4M_\odot$.

Neutron stars

- Neutron degeneracy pressure allows the existence of another stable form of stellar matter, more dense than the white dwarf and rich in neutrons. This type of star is known as a neutron star.

- Such a star is formed when the core of a supergiant collapses in a supernova explosion.

- There is a maximum mass that a neutron star can have. This is certainly less than $5M_\odot$ and thought to be around $3M_\odot$. All neutron stars whose mass has been measured are lower than this limit.

- Angular momentum is conserved in the collapse of the core of the supergiant, and so the neutron star will be rotating very rapidly on its axis.

- Magnetic field lines are trapped as the core collapses and consequently the neutron star will have a very strong magnetic field.

- Pulsars are rapidly rotating, highly magnetized neutron stars, which produce beamed radio emission. As the star rotates, the beam is swept around the sky. If the orientation is such that the beam sweeps across the Earth, regular pulses of radio emission can be observed repeating at the pulsar rotation period.

Black holes

- A black hole is believed to be the end-point of the evolution of a star too massive to become a neutron star.

- The Schwarzschild radius of a black hole of mass M is given by

$$R_S = \frac{2GM}{c^2}$$

- When an object collapses to a radius smaller than its Schwarzschild radius then its collapse cannot be halted, and nothing can escape from it.

Interacting binary stars

- Roche lobe overflow and stellar wind accretion are processes that result in the transfer of material in a close binary system.

- When matter accretes onto a white dwarf, a neutron star or a black hole, gravitational potential energy is released, typically leading to emission in the X-ray band.

- The existence of pulsars in binary systems (both interacting and non-interacting) allows the masses of neutron stars to be measured.

- A black hole in a binary-star system may be detectable through the emission of X-rays from an accretion disc. The X-ray source V404 Cygni is almost certainly one such system. Estimates of the mass of the non-luminous star in a binary can help to confirm the existence of a black hole.

- If a star is one of a close binary system, its evolution will be significantly modified because of the transfer of material between the two stars in the system.

- Many unusual binary systems, such as those containing two compact objects, or a compact object and a more normal star, can be accounted for by the evolution of interacting binary systems.

- Supernovae of Type Ia are believed to occur in interacting binaries and arise from the complete nuclear burning of a white dwarf that is close to the Chandrasekhar limit. The nuclear burning results in the formation of a relatively large mass of iron group elements.

Questions

QUESTION 9.14

A malfunctioning satellite, of moment of inertia $2500 \, \text{kg m}^2$, spinning at 0.33 revolutions per second, is to be repaired by a visiting astronaut. The satellite is cylindrical in shape with radius 1 m. The astronaut, who has a mass of 100 kg, latches on to the curved surface of the satellite. At how many revolutions per second will the satellite–astronaut combination rotate? (*Hint*: assume that the astronaut can be considered to be a point mass at a distance of 1 m from the axis of rotation.)

QUESTION 9.15

If the power received at the Earth from a pulsar is $10^{-19} \, \text{W m}^{-2}$ and you have a radio telescope of collecting area $1000 \, \text{m}^2$, compare the power received by the telescope from the pulsar with the power you would use lifting this book (which weighs about 1 kg) through a height of 1 m in 1 second.

QUESTION 9.16

A supergiant, spectral class B, mass $15 M_{\odot}$, is in a non-interacting binary system with a pulsar. The B star has emission lines; describe how the spectrum seen by an optical astronomer changes during one orbital period. What changes does a radio astronomer see during one orbital period? Outline the history and future evolution of the binary system.

CONCLUSION

The stars in the night sky appear to the unaided eye as apparently motionless pinpoints of light. Even with large telescopes and sophisticated detectors, this impression does not change. It is therefore astonishing that astronomers have learned so much about their structure, composition and evolution. Much of this knowledge has been gained as a result of high-resolution (both spectral and spatial) observations, which can be made of one particular star – our Sun. However much detail is visible in images and spectra of the photosphere and outer atmosphere of the Sun, the interior cannot be seen directly and the processes that produce its prodigious energy output must be inferred indirectly. In addition, the Sun may not be typical of many of the stars we see in the night sky. It is the application of the scientific method that has brought this understanding, through

- careful observation,
- construction of models by application of the laws of physics constrained by the observations,
- production of theories to explain the observed properties, and
- testing and refinement (or rejection!) of such theories by prediction and further observation.

Observations of the positions of stars over a long period of time reveal that they are not fixed in the sky. The Doppler shifts of lines in stellar spectra reveal their motions in the line of sight. Extremely precise measurements of their positions over a year reveal parallaxes from which distances can be determined. Observations of the motions of stars in binary systems can be used to determine stellar masses.

Once the distance to a star is known, the absolute brightness (luminosity) can be determined from the apparent brightness. Distances can be derived beyond the limits of parallax measurements by using so-called standard candles – stars for which the luminosity is known independently. One example is Cepheid variables, which have pulsation periods related to luminosity and can be identified from their characteristic light curves.

Stellar temperatures can be derived from the spectra of stars since, to first order, they are approximated by black-body sources. Radii can then be estimated from the luminosity and temperature. Detailed studies of spectra reveal the composition and physical conditions present in the stars' atmospheres.

However, all this information does not in itself provide insight into the internal workings of a star or its evolution. It is the application of the laws of physics to construct models that allows us to infer the extreme conditions that exist in the interiors of stars. Stellar models combine fundamental physical principles such as the balance of forces in hydrostatic equilibrium, the conservation of energy and mass and the processes of energy transfer, with nucleosynthesis – the fusion of atomic nuclei to form heavier elements which provides the power to sustain a star's luminosity.

Such models must always be consistent with the observed properties or they must be refined or rejected. It is vital also to ensure that the limitations of the observations are known and their interpretation is valid. For example, it was long held that the Earth was stationary and located at the centre of the Universe because no stellar parallaxes could be observed – the problem was that parallaxes were much smaller than could be observed at the time. The solar neutrino problem cast doubt on either models of nucleosynthesis, or the observational techniques for neutrino detection, until its resolution with the discovery that neutrinos could change type.

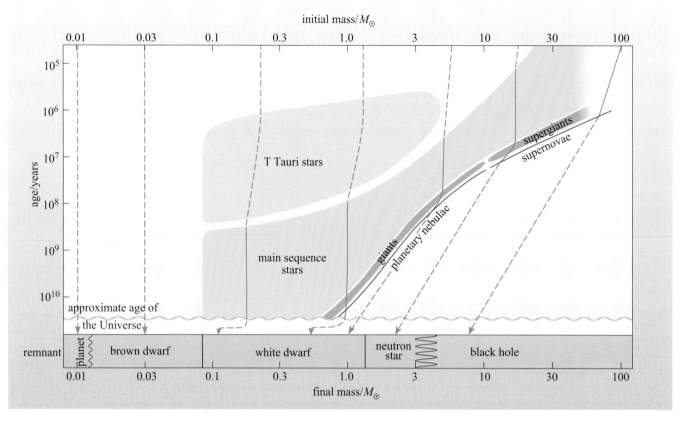

Figure 10.1 A comparison of the time spent in different stages of stellar evolution. The vertical axis indicates the age since formation, on a logarithmic scale. The horizontal axis indicates the mass, also on a logarithmic scale. The coloured zones show the approximate duration of a particular phase of stellar evolution. Note that stars spend the majority of their time on the main sequence. The green lines show the evolution of objects with selected initial masses (solid lines indicate when nucleosynthesis is occurring). The boxes immediately above the lower horizontal axis indicate the eventual end-states of the evolutionary process. Stars with mass rather less than $1 M_\odot$ cannot yet have evolved into the giant stage. Objects with masses less than $0.08 M_\odot$ never attain high enough internal temperatures to sustain nucleosynthesis and become brown dwarfs or planets.

Although it is not usually possible to observe evolutionary processes (the timescales are mostly much too long), you have seen how information on a large number of stars helps us to constrain how the properties of stars evolve. The positions of stars on the Hertzsprung–Russell diagram indicate which combinations of luminosity and temperature are most common. The main sequence, where the majority of stars lie, must represent the properties of stars for a large fraction of their lives. You have learnt that the mass and composition of a star uniquely determine its location on the H–R diagram and that the main sequence is a sequence of mass, the hottest and most luminous stars being the most massive. The red giant, supergiant and white dwarf regions must all also be occupied by some of these stars for some fraction of their lifetimes. Since, for most of the lifetime of most stars, the mass does not change significantly, the change in position on the H–R diagram is largely a result of a change in composition – a natural consequence of the nucleosynthesis that powers the star. Stellar evolution is understood by calculating stellar models with gradually changing composition. The models must predict the distribution of stars that is seen on the H–R diagram today; any largely empty regions on the diagram cannot be occupied by a predicted star except for a very short time.

Figure 10.1 illustrates some of the main features of the process of stellar evolution. The mass of the dense interstellar cloud fragment that forms a star is the critical determinant of its future – or if it becomes a star at all! The mass also dictates the speed with which a star passes through each stage of its evolution and the final remnant. The most massive stars end their lives as spectacular supernovae, often briefly outshining their entire host galaxy of many billions of stars. Stellar remnants become ever more bizarre with increasing mass. White dwarfs have densities equivalent to several hundred thousand times that of this book, whereas neutron stars are a billion times more dense again! Despite their small size and low intrinsic luminosity, they reveal themselves through the influence of their extreme gravitational and magnetic fields. The most massive remnants, black holes, have effectively zero luminosity but are detectable, usually in binary systems by their interaction with matter from their companions.

During the course of stellar evolution, material is being continually reprocessed and returned to the interstellar medium from which the stars originally formed, as shown in Figure 10.2. The ISM is gradually enriched with heavier elements, providing the raw materials for planetary system formation, and potentially, life.

This book has, we hope, provided you with an understanding of the basic properties and processes which shape the structure and evolution of stars. We owe our very existence to the presence of the Sun and perhaps in the future astronomers will find evidence for life on planets orbiting other stars. Galaxies are visible because of the presence of stars within them, so understanding the properties and evolution of stars therefore plays a fundamental role in the study of the structure and evolution of galaxies and ultimately of the entire Universe.

Figure 10.2 Schematic diagram of the processing of interstellar gas and dust. At each stage of star formation and evolution material is recycled back into the interstellar medium. At some stages (stellar winds from evolved stars, planetary nebula formation, and supernovae) the material is enriched with heavier elements.

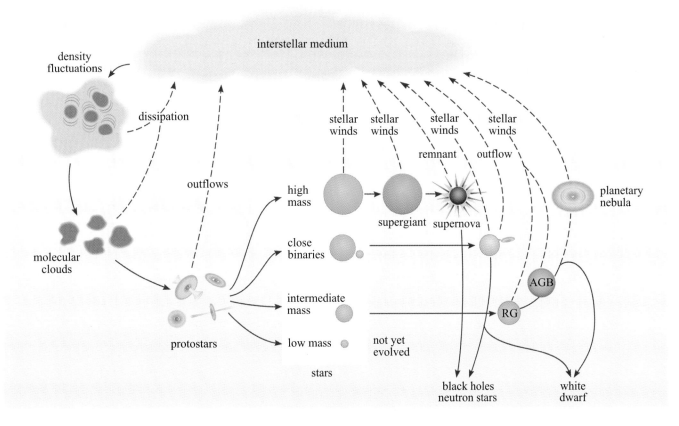

ANSWERS AND COMMENTS

QUESTION 1.1

$1\,\text{W} = 1\,\text{J}\,\text{s}^{-1}$, so the value of $L_{\odot}$ in terms of watts is simply $L_{\odot} = 3.84 \times 10^{26}\,\text{W}$. Therefore the number of typical power stations required to match the Sun's energy output is given by

$$\frac{3.84 \times 10^{26}\,\text{W}}{2.5 \times 10^{9}\,\text{W}} = 1.54 \times 10^{17} \quad \text{to 3 significant figures}$$

So, to 2 significant figures, which correponds to the data in the question, the value is 1.5×10^{17} power stations.

In part, this question is intended to emphasize the immensity of the Sun's energy output, but an equally important purpose is to draw your attention to the use of SI units (the joule (J), the watt (W), etc.) and to the manipulation of powers of ten ($10^{26}/10^{9} = 10^{17}$, etc.) and the importance of working to the relevant number of significant figures.

QUESTION 1.2

It was stated earlier that the diameter of the photosphere is about $1.4 \times 10^{6}\,\text{km}$. The size of the sunspot can be estimated by simply measuring the diameter of the spot's image in Figure 1.1 and comparing this with the size of the image of the photosphere:

$$\text{sunspot diameter} = \frac{d_{\text{si}}}{d_{\text{pi}}} \times d_{\text{p}}$$

where d_{si} is the diameter of the sunspot's image, d_{pi} is the diameter of the photosphere's image, and d_{p} is the diameter of the photosphere. However, in using this formula it is important to remember that the Sun is a spherical body, so the further a sunspot is from the centre of the solar disc, the more it will be 'turned away' from the observer and the more its image will be foreshortened. Although it would be possible to make allowance for this foreshortening, it is much easier just to examine sunspot images near the centre of the disc, where the effect can be ignored. Such sunspot images seem to have a diameter that is about 2% of the diameter of the disc image. It follows that, for the large sunspots near the centre of Figure 1.1,

$$\text{sunspot diameter} \approx \frac{2}{100} \times 1.4 \times 10^{6}\,\text{km} = 2.8 \times 10^{4}\,\text{km}$$

So the estimated sunspot diameter is about $3 \times 10^{4}\,\text{km}$. It follows that large sunspots have diameters that are comparable to the diameter of the Earth ($1.3 \times 10^{4}\,\text{km}$).

QUESTION 1.3

The small-scale structure in Figure 1.3b is partly due to the solar granulation seen in the photosphere. (Sunspots may also contribute.) The granules are bright whereas the lanes that separate them are dark. This variation shows up in the graph.

QUESTION 1.4

The frequencies are given by $f = c/\lambda$ (Equation 1.2, rearranged). So, for $\lambda = 400\,\text{nm}$,

$$f = \frac{3.00 \times 10^8\,\text{m s}^{-1}}{400 \times 10^{-9}\,\text{m}} = 7.50 \times 10^{14}\,\text{Hz}$$

and, for $\lambda = 700\,\text{nm}$,

$$f = \frac{3.00 \times 10^8\,\text{m s}^{-1}}{700 \times 10^{-9}\,\text{m}} = 4.29 \times 10^{14}\,\text{Hz}$$

Thus, the frequency range is from $4.29 \times 10^{14}\,\text{Hz}$ to $7.50 \times 10^{14}\,\text{Hz}$.

QUESTION 1.5

The photon energies are given by $\varepsilon = hf$ (Equation 1.3). So, for $\lambda = 400\,\text{nm}$,

$$\varepsilon = (6.63 \times 10^{-34}\,\text{J s}) \times (7.50 \times 10^{14}\,\text{s}^{-1}) = 4.97 \times 10^{-19}\,\text{J}$$

Since $1\,\text{eV} = 1.602 \times 10^{-19}\,\text{J}$, the photon energy in electronvolts is

$$\varepsilon = (4.97 \times 10^{-19}\,\text{J})/(1.602 \times 10^{-19}\,\text{J eV}^{-1}) = 3.10\,\text{eV}$$

(Alternatively, the energy in electronvolts can be obtained from Equation 1.3 and using a value of the Planck constant of $h = 4.14 \times 10^{-15}\,\text{eV s}$.)

For $\lambda = 700\,\text{nm}$,

$$\varepsilon = (6.63 \times 10^{-34}\,\text{J s}) \times (4.29 \times 10^{14}\,\text{s}^{-1}) = 2.84 \times 10^{-19}\,\text{J}$$

and in electronvolts

$$\varepsilon = (2.84 \times 10^{-19}\,\text{J})/(1.602 \times 10^{-19}\,\text{J eV}^{-1}) = 1.77\,\text{eV}$$

The energy range is from $2.84 \times 10^{-19}\,\text{J}$ to $4.97 \times 10^{-19}\,\text{J}$, or equivalently from $1.77\,\text{eV}$ to $3.10\,\text{eV}$.

QUESTION 1.6

No. In the case of Figure 1.23 all of the sources being measured were *of the same size* and *at the same distance* from the detector. If sources of different sizes had been used, or if the sources had been at different distances from the detector, the results might well have been different. A 3000 K source of sufficient size or located sufficiently close to the detector could certainly provide more energy per second than a 6000 K source that was small or distant.

QUESTION 1.7

Assuming that the heated ball can be treated as a rough approximation to a black-body source of light, it is to be expected that the ball would emit relatively more light with wavelengths towards the blue end of the spectrum (as opposed to the red end) as the temperature increases, as indicated by Figure 1.23. Thus, at relatively low temperatures red light will predominate and the ball will glow 'red-hot'. As the temperature rises the proportion of shorter wavelengths will gradually increase, making the colour progress from red to orange-white to yellowish-white to white – the last being a mix of all the colours.

QUESTION 1.8

It follows from Equation 1.1 that the wavelength of the absorbed electromagnetic radiation is

$$\lambda_{\mathrm{fi}} = \frac{\upsilon}{f_{\mathrm{fi}}}$$

However, the frequency of the electromagnetic wave is related to photon energy by Equation 1.5. Combining these two equations gives,

$$\lambda_{\mathrm{fi}} = \frac{\upsilon h}{E_{\mathrm{f}} - E_{\mathrm{i}}}$$

where υ is the speed of light in the gas containing the atom. If the gas were very thin, then υ would be approximately equal to the speed of light in a vacuum and we could write

$$\lambda_{\mathrm{fi}} = \frac{ch}{E_{\mathrm{f}} - E_{\mathrm{i}}}$$

QUESTION 1.9

On the whole, the chromosphere is very dim compared with the photosphere. Thus, we can expect to see chromospheric emissions at only those wavelengths where the photosphere is relatively dark. These will be the wavelengths at which the solar spectrum exhibits absorption lines, because most of the light at other wavelengths comes from the photosphere. In the case of the $H\alpha$ line, for example, it is the chromosphere that is mainly responsible for the absorption of photospheric light at that wavelength, and it is also the chromosphere that is mainly responsible for the (weak) emitted light.

QUESTION 1.10

This question can be answered with the aid of the relative spectral flux densities (R) given in Figure 1.23. At 6000 K,

$$\frac{R_{6000}\,(400\ \mathrm{nm})}{R_{6000}\,(700\ \mathrm{nm})} = \frac{4.6}{4.0} = 1.2$$

At 5000 K the ratio would have been

$$\frac{R_{5000}\,(400\ \mathrm{nm})}{R_{5000}\,(700\ \mathrm{nm})} = \frac{1.3}{1.9} = 0.7$$

As you can see, the two ratios are very different. When treated properly, ratios of this kind are sufficient to determine the temperature of a black-body source. There is no need to record the entire spectrum or even to determine the precise location of the peak.

QUESTION 1.11

Wien's displacement law (Equation 1.4) needs to be rearranged to make (T/K) the subject,

$$(T/K) = \frac{2.90 \times 10^{-3}}{(\lambda_{\text{peak}}/m)}$$

The peak of the Sun's spectrum seems to be at a wavelength of about 470 nm. Substituting this value in Wien's displacement law gives

$$(T/K) = \frac{2.90 \times 10^{-3}}{(470 \times 10^{-9}\,m/m)} = 6170$$

$$T = 6.2 \times 10^3\,K$$

This value is rather high, but we have taken a crude approach here. In fact, a Planck curve corresponding to a black-body source with a lower temperature (5800 K, say) provides a reasonable approximation to the data in Figure 1.30.

QUESTION 1.12

The completed Table 1.1 is shown below in Table 1.2.

Table 1.2 Completed version of Table 1.1.

Wavelength, λ/m	3×10^{-14}	6×10^{-10}	3×10^{-7}	1×10^{-5}	5×10^{-3}	10
Corresponding frequency, f/Hz	1×10^{22}	5×10^{17}	1×10^{15}	3×10^{13}	6×10^{10}	3×10^{7}
Corresponding photon energy, ε/J	7×10^{-12}	3×10^{-16}	6×10^{-19}	2×10^{-20}	4×10^{-23}	2×10^{-26}
Corresponding photon energy, ε/eV	4×10^{7}	2×10^{3}	4	1×10^{-1}	2×10^{-4}	1×10^{-7}
Temperature, T/K, of a black body that has a peak in its spectrum at this value of λ	1×10^{11}	5×10^{6}	9×10^{3}	3×10^{2}	1	3×10^{-4}
Corresponding part of the electromagnetic spectrum	γ-ray	X-ray	ultraviolet	infrared	microwave	radio wave

QUESTION 1.13

The only way to do this on the basis of the information given in this chapter is to compare the size of the plage in the calcium K image (Figure 1.19) with the size of the solar disc. This is similar to Question 1.2 and suffers from the same problem of foreshortening. At a rough estimate, the longest dimension of the plage is about a quarter of the diameter of the Sun, which implies a (maximum) size of

$$\tfrac{1}{4} \times 1.4 \times 10^6\,km = 3.5 \times 10^5\,km$$

QUESTION 1.14

No. The black-body spectrum depends only on the temperature of the source and not on any other physical characteristic such as its composition.

QUESTION 1.15

(a) It follows from Equation 1.5 that the frequency, f_{32}, of the radiation absorbed during a transition from E_2 to E_3 is $(1/h)(E_3 - E_2)$, which is

$$\frac{1}{4.14 \times 10^{-15}} \left[\left(-\frac{13.6}{3^2} \right) - \left(-\frac{13.6}{2^2} \right) \right] \text{Hz} = 4.56 \times 10^{14} \text{ Hz}$$

Note that because the photon energies are given in terms of electronvolts, the Planck constant must be expressed in terms of eV s, and so the value $h = 4.14 \times 10^{-15}$ eV s is used in the above calculation.

Now, the corresponding wavelength is given by

$$\lambda_{32} = \frac{c}{f_{32}} = \frac{3.00 \times 10^8}{4.56 \times 10^{14}} \text{ m} = 658 \text{ nm}$$

which is close to the 656.3 nm measured for the Hα line.

Since the numerical work in this question has been carried out to three significant figures, the third figure in the final answer is actually somewhat suspect and it would be best to conclude that the given transition will absorb photons of wavelength between 650 nm and 660 nm. Clearly, such a conclusion makes it quite possible for the given transition to be responsible for Hα absorption, but it's hardly firm proof. What it does show is the need to work to much greater precision when dealing with spectroscopic quantities. In fact, more precise calculations *do* confirm that Hα absorption is due to the E_2 to E_3 transition.

(b) Hβ absorption is caused by the E_2 to E_4 transition (which can be discovered by trial and error). Thus

$$\lambda_{42} = \frac{c}{f_{42}} = \frac{c}{(E_4 - E_2)/h} = \frac{ch}{(E_4 - E_2)}$$

$$= \frac{3.00 \times 10^8 \times 4.14 \times 10^{-15}}{\left(-\dfrac{1}{4^2} + \dfrac{1}{2^2} \right) \times 13.6} \text{ m} = 487 \text{ nm}$$

This is to be compared with the wavelength of Hβ absorption, which is 486.1 nm.

(c) Hβ emission is due to the E_4 to E_2 transition (the reverse of the transition causing Hβ absorption).

(d) The transition from E_1 to E_2 produces electromagnetic radiation with a wavelength that is only one-quarter of the Hβ wavelength. Such a wavelength is so small that it is outside the range covered by Figure 1.28. (It is actually in the ultraviolet part of the spectrum.)

QUESTION 1.16

(a) Ca^{14+} and Fe^{9+}.

(b) The energy carried by the photons that contribute to the yellow and red lines must arise from appropriate transitions in the atoms concerned. For the yellow line the relevant energy is

$$\varepsilon = hf = h\frac{c}{\lambda}$$

$$= \frac{(4.14 \times 10^{-15}\text{ eV s}) \times (3.00 \times 10^{8}\text{ m s}^{-1})}{5.69 \times 10^{-7}\text{ m}} = 2.18\,\text{eV}$$

So this must represent the difference in energy between two of the energy levels of Ca^{14+}.

Similarly, in the case of the red line

$$\varepsilon = \frac{(4.14 \times 10^{-15}\text{ eV s}) \times (3.00 \times 10^{8}\text{ m s}^{-1})}{6.37 \times 10^{-7}\text{ m}}$$

$$= 1.95\,\text{eV}$$

So this must represent the difference between two of the energy levels of Fe^{9+}.

QUESTION 1.17

(a) The approach here is the same as was taken in Example 1.1. The typical photon energy is found from Equation 1.6

$$\varepsilon \sim kT$$

Rather than inserting numerical values at this stage, it is better to find an expression for the wavelength. By combining Equations 1.2 and 1.3 as in Example 1.1, the wavelength is given by

$$\lambda = hc/\varepsilon$$

The photon energy is given by Equation 1.6, so we can write

$$\lambda \sim hc/kT$$

Notice that we have replaced the equals sign by '~' because this is now only an approximate relationship.

Now we can insert numerical values, taking care to ensure that the value of h is the appropriate one for the units we are using. The calculation is in SI units (we are not interested in expressing the energy in electronvolts here), so we must use the value $h = 6.63 \times 10^{-34}\,\text{J s}$.

$$\lambda \sim (6.63 \times 10^{-34}\text{ J s}) \times (3.00 \times 10^{8}\text{ m s}^{-1})/[(1.38 \times 10^{-23}\text{ J K}^{-1}) \times (5 \times 10^{6}\text{ K})]$$

$$= 2.88 \times 10^{-9}\,\text{m}$$

So the typical wavelength of emission from this source will be about 3 nm.

(b) If the source acts as a black body, the peak of the spectrum of emission occurs at a wavelength given by Wien's displacement law (Equation 1.4),

$$(\lambda_{\text{peak}}/\text{m}) = \frac{2.90 \times 10^{-3}}{(T/\text{K})} = \frac{2.90 \times 10^{-3}}{5 \times 10^{6}} = 5.8 \times 10^{-10}$$

So the peak of the black-body spectrum occurs at a wavelength of 6×10^{-10} m or 0.6 nm.

(c) At first sight it may appear that the wavelengths found in parts (a) and (b) are quite different, being 3 nm and 0.6 nm, respectively. However, the first calculation is an estimate of the wavelength that a thermal source may emit at, and is in fact, only expected to give an answer that is within a factor of 10 of the true value. Since these answers differ by a factor of 5 (= 3 nm/0.6 nm), the wavelengths calculated in parts (a) and (b) *are* consistent with one another.

QUESTION 1.18

At radio wavelengths, much of the Sun's emission comes from regions of the corona where temperature increases with height. Because (as argued in Section 1.2.2) observations made near the limb sample material that is, on average, at greater altitude than that sampled by observations made near the centre of the solar disc, it follows that at radio wavelengths limb observations will involve higher and hence hotter material than disc observations. Thus, at radio wavelengths the limb will be brighter than the disc centre.

At visible wavelengths the corona is essentially transparent and its own emissions are very faint. Thus there is no chance of observing limb brightening in visible light produced by the corona.

QUESTION 2.1

The volume occupied by the core of the Sun, V_c, is related to the radius of the core, R_c, by

$$V_c = \tfrac{4}{3} \pi R_c^3 \tag{i}$$

Similarly, the volume of the Sun, $V_\odot$, is related to the solar radius, $R_\odot$, by

$$V_\odot = \tfrac{4}{3} \pi R_\odot^3 \tag{ii}$$

To find an expression for the volume of the core as a fraction of the total volume of the Sun, Equation (i) is divided by Equation (ii)

$$\frac{V_c}{V_\odot} = \frac{\tfrac{4}{3} \pi R_c^3}{\tfrac{4}{3} \pi R_\odot^3}$$

The factor of $(4/3)\pi$ is common to both the denominator and the numerator and can so be cancelled to give

$$\frac{V_c}{V_\odot} = \frac{R_c^3}{R_\odot^3} = \left(\frac{R_c}{R_\odot} \right)^3$$

The question states that $R_c/R_\odot = 0.3$, so

$$\frac{V_c}{V_\odot} = (0.3)^3 = 2.7 \times 10^{-2}$$

So, to 1 significant figure, the volume of the core is 3% of the total volume of the Sun.

QUESTION 2.2

Since X departs significantly from a constant value only in the central 30% of the Sun (that is, where $R/R_\odot$ is less than 0.30) it seems pretty clear that the nuclear processes that convert hydrogen into helium must be confined to that inner region. Further, because X falls progressively as the fractional radius decreases in this inner region, conversion of hydrogen into helium must have been most common in the most central parts of the Sun.

QUESTION 2.3

(a) Following the initial reaction between two protons that was considered in Example 2.1, the second and third stages of the ppI chain are:

$$^2_1\text{H} + ^1_1\text{H} \rightarrow ^3_2\text{He} + \gamma \qquad (2)$$

$$2\,^3_2\text{He} \rightarrow ^4_2\text{He} + 2\,^1_1\text{H} \qquad (3)$$

(b) For Reaction 2, the total incoming charge is $2e$ and so is the total outgoing charge.

For Reaction 3, the total incoming charge is that carried by the protons in *two* helium nuclei, so it is $4e$. The total outgoing charge is that carried by *one* helium nucleus and two hydrogen nuclei, $4e$.

Thus, in both reactions, electric charge is conserved.

(c) For both reactions the baryon number entering or leaving is equal to the total number of protons and neutrons entering or leaving, and this total is given by the superscript, the mass number. You can easily see that on both sides of Reaction 2 it is 3, and in Reaction 3 it is 6. The baryon number is thus conserved in both cases.

QUESTION 2.4

(a) The radiant energy eventually resulting from each occurrence of the ppI chain is given by the expression

$$c^2\,[\,(4 \times \text{mass of }^1_1\text{H}) - (\text{mass of }^4_2\text{He}) - (2 \times \text{mass of e}^+) + (2 \times \text{mass of e}^+)$$
$$+\,(2 \times \text{mass of e}^-)\,]$$

$$= 9.00 \times 10^{16}[4 \times 1.673 - 6.645 + 2 \times 0.001] \times 10^{-27}\,\text{J}$$
$$= 4.4 \times 10^{-12}\,\text{J}$$

The accurate value is about 5% lower than this estimate.

(b) If we assume (somewhat incorrectly) that the solar luminosity is entirely provided by the ppI chain and its supplementary reactions, then the number of times per second that the chain is completed is given by

$$\frac{3.84 \times 10^{26}\,\text{J s}^{-1}}{4.4 \times 10^{-12}\,\text{J}} = 8.7 \times 10^{37}\,\text{s}^{-1}$$

This is only an estimate, but it is not too bad.

(c) Each time the chain is completed, four hydrogen nuclei are consumed (and one helium nucleus produced). Since the mass of a hydrogen nucleus is $1.673 \times 10^{-27}\,\text{kg}$, it follows that the rate of hydrogen consumption is roughly

$$4 \times 8.7 \times 10^{37} \times 1.673 \times 10^{-27}\,\text{kg s}^{-1} = 5.8 \times 10^{11}\,\text{kg s}^{-1}$$

Now, the number of seconds in a year is 3.16×10^7. So the annual consumption of hydrogen is

$$3.16 \times 10^7 \times 5.8 \times 10^{11}\,\text{kg} = 1.8 \times 10^{19}\,\text{kg}$$

This is about three millionths of the Earth's mass per year, and about one part in 10^{11} of the Sun's mass per year.

QUESTION 2.5

(a) The wavelength of a photon of a given energy is found by combining Equations 1.2 and 1.3 and rearranging to give

$$\lambda = hc/\varepsilon$$

The energy of one of the γ-rays produced by electron–positron annihilation is $0.51\,\text{MeV}$ or $0.51 \times 10^6\,\text{eV}$. Since energies are being measured here in terms of eV rather joules, it is convenient to use $h = 4.14 \times 10^{-15}\,\text{eV s}$. Hence

$$\lambda = \frac{4.14 \times 10^{-15} \times 3.00 \times 10^8}{5.1 \times 10^5}\,\text{m} = 2.4 \times 10^{-12}\,\text{m}$$

(b) The wavelength of the peak of the black-body curve is given by Equation 1.4,

$$(\lambda_{\text{peak}}/\text{m}) = \frac{2.90 \times 10^{-3}}{(T/\text{K})}$$

The temperature at the core of the Sun is approximately $1.5 \times 10^7\,\text{K}$, so

$$\lambda_{\text{peak}} = \frac{2.90 \times 10^{-3}}{1.5 \times 10^7}\,\text{m} = 1.9 \times 10^{-10}\,\text{m}$$

According to Figure 1.36, this peak lies in the X-ray region of the electromagnetic spectrum.

QUESTION 2.6

Two neutrinos are released each time the ppI chain is executed (Section 2.2.4). Assuming (somewhat unrealistically) that all nuclear reactions apart from those of the ppI chain can be ignored, we know (from Question 2.4) that the ppI chain is completed about 8.71×10^{37} times per second. (This estimate involves various assumptions that are discussed in Question 2.4.) It follows that the rate at which neutrinos are produced in the core of the Sun is about $2 \times 8.71 \times 10^{37}\,\text{s}^{-1}$. Now, assuming (quite realistically) that all these neutrinos escape from the Sun and spread out evenly in all directions, the rate at which they pass through an area of $0.01\,\text{m}^2$ on a spherical surface of radius $1.50 \times 10^{11}\,\text{m}$, centred on the Sun, is

$$(2 \times 8.71 \times 10^{37})\frac{0.01}{4\pi \times (1.50 \times 10^{11})^2}\,\text{s}^{-1} = 6.16 \times 10^{12}\,\text{s}^{-1}$$

Since $0.01\,\text{m}^2$ is roughly the cross-sectional area of a human brain, this calculation justifies the claim made earlier that more than a million million neutrinos pass through your head in the time it takes to read a sentence.

QUESTION 2.7

(a) To estimate the speed of the front of the CME, we need to measure the distance travelled by the front between two intervals. Distances on these images can be roughly estimated from the size of the solar disc that is indicated. The images taken at 19:42 and 21:18 on 5 August 1999 both show the front reasonably well. Between these two times, the front of the CME appears to move a distance (perpendicular to the line of sight) of about 2 solar diameters. The time elapsed is 1 hour 36 minutes, which is equivalent to 96 minutes or 5.760×10^3 s. So the speed is given by

$$\text{speed} = \frac{\text{distance}}{\text{time}} = \frac{2 \times 2 \times 6.96 \times 10^8 \text{ m}}{5.760 \times 10^3 \text{ s}} = 4.83 \times 10^5 \text{ m s}^{-1}$$

The distance travelled can only be roughly estimated from the images, so to an appropriate number of significant figures, the speed is 5×10^5 m s^{-1} or equivalently 500 km s^{-1}.

The key assumptions that are made here are (i) that the speed of the CME does not change dramatically between the times when the two images were taken, and (ii) that the CME is being ejected in a direction that is perpendicular to the line of sight.

(b) The time taken for the CME to reach the radius of the Earth's orbit from the Sun is

$$\text{time} = \frac{\text{distance}}{\text{speed}} = \frac{1.50 \times 10^{11} \text{ m}}{5 \times 10^5 \text{ m s}^{-1}} = 3 \times 10^5 \text{ s}$$

So the CME will reach the radius of the Earth's orbit about 3×10^5 s or about 3.5 days after being ejected from the Sun.

The main assumption made here is that the speed of the CME remains roughly constant as it travels from the Sun to Earth.

A time delay of about 3 days is generally observed between the ejection of CMEs that are heading towards Earth and the onset of geomagnetic storms and auroral activity. So the estimate of the speed and travel time of a CME that has been calculated here corresponds well to the actual behaviour of CMEs.

QUESTION 2.8

The kinetic energy E_k of a body of mass m and speed v is given by

$$E_k = \tfrac{1}{2} m v^2$$

In this case, $m = 5 \times 10^{13}$ kg, and from the answer to Question 2.7(a)

$$v = 5 \times 10^5 \text{ m s}^{-1}$$

so $\quad E_k = \tfrac{1}{2} m v^2 = \tfrac{1}{2}(5 \times 10^{13}) \times (5 \times 10^5)^2 \text{ J} = 6 \times 10^{24} \text{ J}$

So the order of magnitude estimate of the kinetic energy of a large CME is 10^{25} J.

QUESTION 2.9

The average distance from the Sun to Mercury is 5.79×10^{10} m, which can be expressed as $(5.79 \times 10^{10})/(1.50 \times 10^{11})$ AU = 0.386 AU.

The average distance from the Sun to Pluto is 5.90×10^{12} m, which can be expressed as $(5.90 \times 10^{12})/(1.50 \times 10^{11})$ AU = 39.3 AU.

The average distance from the Sun to Mercury can also be expressed in terms of $R_\odot$ as $(5.79 \times 10^{10})/(6.96 \times 10^8)R_\odot = 83.2 R_\odot$.

The average distance from the Sun to the Earth can be expressed in terms of $R_\odot$ as $(1.50 \times 10^{11})/(6.96 \times 10^8)R_\odot = 216 R_\odot$.

QUESTION 2.10

Consider an area of 1 m^2 at a distance from the Sun equal to the radius of the orbit of the Earth around the Sun. This imaginary area is set up so that it directly faces the Sun. By considering how far the solar wind will travel in 1 s after passing this surface, we can work out the volume of solar wind that passes through the surface in every second per 1 m^2. Since the speed of the solar wind is assumed to be 750 km s^{-1}, this volume is

$$1.00 \, \text{m}^2 \times 7.5 \times 10^5 \, \text{m} = 7.5 \times 10^5 \, \text{m}^3$$

The mass of material lost per 1 m^2 per second is this volume multiplied by the density of the solar wind near the Earth. (Recall that the mass of a proton is $m_p = 1.67 \times 10^{-27}$ kg.)

$$\text{volume} \times \text{density} = 7.5 \times 10^5 \, \text{m}^3 \times 7 \times 10^6 \, \text{m}^{-3} \times m_p$$
$$= 5.25 \times 10^{12} \times 1.67 \times 10^{-27} \, \text{kg} = 8.77 \times 10^{-15} \, \text{kg}$$

This is a mass loss rate of 8.77×10^{-15} kg s^{-1} m^{-2}, where we have taken care to ensure that the units reflect the fact that this is the mass lost per unit time and per unit area. To find the total mass loss rate, this rate per unit area needs to be multiplied by the area of the sphere with radius equal to the radius of the Earth's orbit. The area of this sphere is

$$4\pi R^2 = 4 \times \pi \times (1.50 \times 10^{11})^2 \, \text{m}^2 = 2.83 \times 10^{23} \, \text{m}^2$$

So the mass loss rate over this sphere is

$$2.83 \times 10^{23} \, \text{m}^2 \times 8.77 \times 10^{-15} \, \text{kg s}^{-1} \, \text{m}^{-2} = 2.48 \times 10^9 \, \text{kg s}^{-1}$$

In terms of solar masses per year this is a rate of

$$\frac{2.48 \times 10^9 \, \text{kg s}^{-1} \times (365 \times 24 \times 60 \times 60 \, \text{s})}{1.99 \times 10^{30} \, \text{kg}} = 3.93 \times 10^{-14} M_\odot \, \text{yr}^{-1}$$

So the mass loss rate due to the solar wind is, to 1 significant figure, 2×10^9 kg s^{-1} or $4 \times 10^{-14} M_\odot$ yr^{-1}.

QUESTION 2.11

The auroral emission arises from electronic transitions in neutral atoms or ions of those elements that are common in the Earth's atmosphere. (Note also that electronic transitions in molecules, such as N_2 are also possible.) At the Earth's surface the atmosphere comprises 78% molecular nitrogen and 20% molecular oxygen. In the high-energy environment of the upper atmosphere it is likely that molecular oxygen and nitrogen could be split up into atomic form and possibly ionized. Hence it is likely that the aurora would show spectral lines due to atomic oxygen (a) and ionized nitrogen (c). There is a negligible amount of iron in the Earth's atmosphere, so a spectral line due to atomic iron (b) is unlikely to be seen.

QUESTION 2.12

The speed of the fast component of the solar wind is $750 \, \mathrm{km \, s^{-1}}$, and the distance from the Sun to the termination shock is estimated to be 85 AU.

The travel time t, is given by

$$t = \frac{\text{distance}}{\text{speed}}$$

$$\text{distance} = 85 \, \mathrm{AU} = 85 \times 1.50 \times 10^{11} \, \mathrm{m} = 1.275 \times 10^{13} \, \mathrm{m}$$

$$\text{speed} = 750 \, \mathrm{km \, s^{-1}} = 7.50 \times 10^5 \, \mathrm{m \, s^{-1}}$$

$$t = \frac{1.275 \times 10^{13} \, \mathrm{m}}{7.50 \times 10^5 \, \mathrm{m \, s^{-1}}} = 1.7 \times 10^7 \, \mathrm{s}$$

In terms of days this is

$$t = \frac{1.7 \times 10^7 \, \mathrm{s}}{60 \times 60 \times 24 \, \mathrm{s}} = 1.97 \times 10^2 \, \text{days}$$

So the travel time from the Sun to the termination shock is about 200 days.

QUESTION 2.13

The composition of the Sun is roughly 73% hydrogen and 25% helium, by mass. Thus, for every 73 kg of hydrogen in the Sun there are 25 kg of helium and 2 kg of everything else. Since the mass of a hydrogen nucleus, $^1_1\mathrm{H}$, is 1.673×10^{-27} kg and that of a helium nucleus, $^4_2\mathrm{He}$, is 6.645×10^{-27} kg (these figures were given in Question 2.4), it follows that 100 kg of typical solar material will contain roughly

$$\frac{73}{1.673 \times 10^{-27}} = 4.36 \times 10^{28} \text{ hydrogen nuclei}$$

and $\quad \dfrac{25}{6.645 \times 10^{-27}} = 3.76 \times 10^{27}$ helium nuclei

This is a ratio by *number* of nuclei of roughly $100 : 8.6$, hydrogen to helium.

Now, the total mass of the Sun was given in Section 2.2.4 as 1.99×10^{30} kg. Thus, the total number of hydrogen nuclei in the Sun will be roughly

$$\frac{1.99 \times 10^{30}}{100} \times 4.36 \times 10^{28} = 8.7 \times 10^{56}$$

and the number of helium nuclei will be roughly

$$\frac{1.99 \times 10^{30}}{100} \times 3.76 \times 10^{27} = 7.5 \times 10^{55}$$

QUESTION 2.14

According to Question 2.4, the mass of hydrogen consumed every year is about 1.84×10^{19} kg. Since the total mass of the Sun is 1.99×10^{30} kg, and about 73% of this is hydrogen, it follows that the maximum duration of the conversion of hydrogen into helium at the present rate is

$$\frac{0.73 \times 1.99 \times 10^{30}}{1.84 \times 10^{19}} \text{ years} = 7.9 \times 10^{10} \text{ years}$$

Of course, this is a crude estimate because it assumes that all of the hydrogen currently in the Sun will undergo conversion and it also makes use of a result (from Question 2.4) that is itself only an estimate. Nonetheless, the final answer is quite reasonable. More refined estimates of the Sun's hydrogen-fuelled lifetime provide figures of about 10^{10} years. Since the Sun is currently thought to be about 4.5×10^{9} years old, it is usual to regard the Sun as a middle-aged star.

QUESTION 2.15

(a) The nucleus $^{5}_{3}\text{He}$ does not exist. Every helium nucleus *must* have an atomic number of 2, denoting two protons in the nucleus. The atomic number of 3 corresponds to the element lithium.

(b) This reaction does not conserve electric charge.

(c) This reaction also fails to conserve electric charge.

QUESTION 2.16

The solar luminosity was given earlier as 3.84×10^{26} J s^{-1}. Since there are approximately 3.16×10^{7} seconds in one year, it follows that the annual energy output of the Sun, E, is about 1.21×10^{34} J. Assuming that this energy is entirely supplied by the loss of mass from the Sun's core, it follows that the mass lost per year is

$$m = \frac{E}{c^2} = \frac{1.21 \times 10^{34} \text{ J}}{(3.00 \times 10^8 \text{ m s}^{-1})^2} = 1.34 \times 10^{17} \text{ kg}$$

It is worth emphasizing that this means the Sun is losing mass at the amazing rate of 4.25×10^{9} kilograms per second, and will lose about 10^{27} kg in its hydrogen-burning lifetime, which is about 0.05% of its mass.

QUESTION 2.17

From Question 2.4(a) the energy generated in a single ppI reaction is approximately 4×10^{-12} J. Given that this question asks for estimated values, it is appropriate to use a value given to 1 significant figure.

To estimate the number of photospheric photons that eventually carry this energy away, it is necessary to calculate the average energy of a photon that is emitted by the photosphere. A reasonable estimate of this can be made by assuming that the average photon energy corresponds to the peak of the Planck curve at the photospheric temperature. Using Equation 1.4 and a photospheric temperature of 6×10^3 K (i.e. to 1 significant figure),

$$(\lambda_{\text{peak}}/\text{m}) = \frac{2.90 \times 10^{-3}}{(T/\text{K})} = \frac{2.90 \times 10^{-3}}{6000} = 4.8 \times 10^{-7}$$

The corresponding photon energy is found by combining Equations 1.2 and 1.3

$$\varepsilon = hc/\lambda$$

$$= 6.6 \times 10^{-34} \,\text{J s} \times 3.0 \times 10^8 \,\text{m s}^{-1}/4.8 \times 10^{-7} \,\text{m} = 4.1 \times 10^{-19} \,\text{J}$$

(Note that energies are given in joules, so the appropriate value of Planck's constant is $h = 6.6 \times 10^{-34}$ J s.)

So the number of these photons that would be required to carry away the energy released in a single ppI reaction is

$$\text{number of photons} = 4 \times 10^{-12} \,\text{J}/4.1 \times 10^{-19} \,\text{J} = 9.8 \times 10^6$$

So, to 1 significant figure, the number of photospheric photons that carry away the energy generated by a single ppI reaction is 1×10^7.

QUESTION 2.18

The field pattern that is proposed in Figure 2.40 is a loop which is modified by the fact that at the highest part of the loop there is a small dip in the magnetic field lines. Plasma can only move along field lines, so any plasma that lies within the dip will move under the influence of gravity to the lowest point of the dip. It is only in this position that the field lines can provide stable support against gravity. Any plasma that lies outside of the dip would also move along the field lines because of the gravitational force acting on it, but in this case it would not be supported – it would simply follow the field lines down to the chromosphere.

QUESTION 2.19

Energy that is stored in the magnetic field is converted into the thermal energy of the plasma at the site of reconnection and into the kinetic energy of particles in the plasma as field lines move away from this site.

QUESTION 3.1

With a parallax of 0.287 arcsec, the distance to 61 Cygni in parsecs is given by Equation 3.7 as

$$d/\text{pc} = 1/0.287 = 3.48$$

Thus $d = 3.48$ pc. This is $3.48 \times 206\,265$ AU $= 7.2 \times 10^5$ AU, and $7.2 \times 10^5 \times 1.50 \times 10^{11}$ m $= 1.1 \times 10^{17}$ m. The distance to the Sun is 1 AU, so 61 Cygni is 720 000 times further away.

QUESTION 3.2

The angular diameter of Betelgeuse is given in Section 3.3.1 as 0.050 arcsec, which is $0.050/206\,265 = 2.42 \times 10^{-7}$ radians. Thus, from Equation 3.8, the radius of Betelgeuse is given by

$$R = [(\alpha/2)/\text{radians}] \times d$$
$$= (2.42 \times 10^{-7}/2) \times 131\,\text{pc} = 1.59 \times 10^{-5}\,\text{pc} = 4.91 \times 10^{11}\,\text{m}$$

So the radius is 4.9×10^{11} m (to 2 significant figures).

$$1\ \text{solar radius} = 6.96 \times 10^8\,\text{m, so } R = 706 R_\odot.$$

So the radius of Betelgeuse is about 700 times that of the Sun.

QUESTION 3.3

From Figure 3.23 we see that the Balmer lines are weak at temperatures above approximately 40 000 K and below about 5000 K. The corresponding spectral classes (Table 3.2) are O, K and M (G is a marginal case). Figures 3.25 and 3.26 confirm that Balmer lines are strong in spectral types B, A, F and G, but weak in the other spectral classes.

QUESTION 3.4

From Figure 1.38, the more transparent bands in the Earth's atmosphere are: the visible region; some bands in the near infrared region; most of the microwave and radio wave regions.

QUESTION 3.5

(a) Using your calculator, $\log 5 = 0.699$.

For the remaining parts of this question you do not need a calculator.

(b) 50 is 10 times larger than 5, so $\log 50 = (\log 5) + 1 = 1.699$

(c) 5000 000 is 10^6 times larger than 5, so $\log 5000\,000 = (\log 5) + 6 = 6.699$

(d) 0.5 is 10 times *smaller* than 5, so $\log 0.5 = (\log 5) - 1 = -0.301$

(e) 5×10^{-7} is 10^{-7} times smaller than 5, so $\log(5 \times 10^{-7}) = (\log 5) - 7 = -6.301$.

QUESTION 3.6

Using Equation 3.15, with star 1 as α Centauri A:

$$(0 - m(\alpha\ \text{Cen B})) = -2.5 \log(2.5), \text{ so } m(\alpha\ \text{Cen B}) = 1.0$$
$$(0 - m(\alpha\ \text{Cen C})) = -2.5 \log(25\,000), \text{ so } m(\alpha\ \text{Cen C}) = 11.0$$

You may have noticed that in this case it isn't necessary to use Equation 3.15 as you have learnt that a ratio of approximately 2.5 in flux density corresponds to a magnitude difference of 1. (More precisely, a ratio of $2.512 = 1$ magnitude, which is not to be confused with the factor of 2.5 in Equation 3.15.) Also, a ratio of 25 000 is $2.5 \times 100 \times 100$, i.e. a magnitude difference of approximately $1 + 5 + 5 = 11$.

QUESTION 3.7

Using Equation 3.16, with magnitudes in the visual band,

$$M_V = m_V - 5 \log d + 5$$

For α Centauri: $M_V = 0.01 - 5 \log(1.35) + 5 = 4.36$

For β Centauri: $M_V = 0.61 - 5 \log(161) + 5 = -5.42$

α Centauri is of similar absolute visual magnitude to the Sun ($M_V = 4.8$) and therefore similar luminosity. β Centauri has a numerically much smaller absolute visual magnitude and is therefore much more luminous than the Sun.

QUESTION 3.8

From Equation 3.14

$$d = [6.1 \times 10^{30}\,W/(4\pi \times 4.4 \times 10^{-10}\,W\,m^{-2})]^{1/2}$$

$$= 3.32 \times 10^{19}\,m$$

$$= 1070\,pc$$

So, to 2 significant figures, the distance to ι^1 Sco is 1100 pc.

The value of L_V for ι^1 Sco is 1.4×10^5 times that of the Sun (4.44×10^{25} W), so its intrinsic visual brightness is very high. It is thus its distance that makes it seem not very bright. *Note:* ι^1 Sco is in fact both hotter and bigger than the Sun.

QUESTION 3.9

In obtaining temperature, we compare the strengths of absorption lines from *different* elements, whereas in obtaining composition we compare the strengths of different absorption lines from a *single* element.

QUESTION 3.10

From Appendix A5, the ten most abundant elements in the material from which the Solar System was formed are as follows, in descending order of abundance:

Order	By number of nuclei	By mass
1	hydrogen	hydrogen
2	helium	helium
3	oxygen	oxygen
4	carbon	carbon
5	neon	neon
6	nitrogen	iron
7	magnesium	nitrogen
8	silicon	silicon
9	iron	magnesium
10	sulfur	sulfur

QUESTION 3.11

Squaring both sides of Equation 3.18 we get

$$P^2 = \frac{4\pi^2 r^3}{G(M + m)}$$

We then divide both sides by P^2, and multiply both sides by $(M + m)$, to obtain the desired result:

$$M + m = \frac{4\pi^2 r^3}{GP^2}$$

QUESTION 3.12

If $M/m = 3$, $M = 3m$. Therefore we can substitute $3m$ for M into the equation for the sum of the masses, i.e. $M + m = 12M_\odot$, to obtain

$$(3m + m) = 12M_\odot$$

so $\qquad m = 3M_\odot$

Therefore, with $M/m = 3$,

$$M = 9M_\odot$$

QUESTION 3.13

(a) The angular diameter of the Moon (0.5°) is 1800 arcsec. So, at the rate of 1.259 arcsec yr^{-1}, it will take Procyon A 1800/1.259 = 1430 years to travel this angular distance.

(b) The transverse speed is obtained from the proper motion by means of Equation 3.1. Thus we must first obtain the distance d to Procyon. This is obtained using Equation 3.7:

$$d/\mathrm{pc} = 1/0.286$$

Thus $\qquad d = 3.50\,\mathrm{pc} = 1.08 \times 10^{14}\,\mathrm{km}$

Secondly, the proper motion is given as 1.259 arcsec yr^{-1}. There are 206 265 arcsec in a radian (Section 3.2.2), and 3.16×10^7 seconds in a year, so this proper motion corresponds to

$$\left(\frac{1.259}{206\,265}\right)\left(\frac{1}{3.16 \times 10^7}\right) \text{radians per second}$$

that is 1.93×10^{-13} radians s^{-1}. Thus, from Equation 3.1,

$$v_\mathrm{t} = d \times (\mu/\mathrm{radians})$$
$$= (1.08 \times 10^{14}\,\mathrm{km}) \times (1.93 \times 10^{-13}\,\mathrm{s}^{-1}) = 20.8\,\mathrm{km\,s}^{-1}$$

(c) Its spectral lines are blue-shifted, so its radial velocity is directed towards us.

(d) The large proper motion suggests that Procyon A is a nearby star, since from Equation 3.1 we see that nearby stars will on average have large proper motions. If Procyon A is nearby then it will have a large parallax.

QUESTION 3.14

(a) Using Equation 3.15, set star 1 as Sirius A and star 2 as Sirius B.

Then $b_1/b_2 = 7600/1$

and $(-1.46 - m_2) = -2.5 \log(7600)$

so V magnitude of Sirius B is $m_2 = -1.46 + 2.5 \log(7600) = 8.24$

(b) Since both stars are at the same distance, the difference in absolute visual magnitudes is the same as the difference in apparent visual magnitudes $(m_2 - m_1 = 9.70)$.

(c) Using Equation 3.16, the absolute visual magnitude of Sirius A is given by
$$M_V = m_V - 5 \log d + 5 = -1.46 - 5 \log(2.64) + 5 = 1.43$$

(d) The Sun's absolute visual magnitude is 4.8. Sirius A has a numerically smaller absolute visual magnitude so its intrinsic brightness and hence its luminosity is higher than the Sun's.

QUESTION 3.15

(a) From Table 3.2 we see that 13 000 K is a reasonable temperature for a B8 star. At 13 000 K, the Balmer lines are strong compared with the helium lines, and very much stronger than the lines of ionized helium and ionized calcium (Figure 3.23). Remember that the actual procedure runs the other way: from the observed line strengths we establish the spectral type.

(b) A far less luminous star of the same temperature would have broader spectral lines, and weaker lines from certain ionized atoms.

(c) Following the method in Example 3.2, we get
$$R/R_\odot \approx (1.4 \times 10^5)^{1/2} \times (5770/13\,000)^2 \approx 74$$

Therefore $R = 74 R_\odot$.

(d) From Figures 3.18 and 3.28, a rough estimate is that about 10% of its luminosity lies in the V waveband. Any value between about 5% and 15% is reasonable. Thus, with $L = 1.4 \times 10^5 L_\odot$, we get
$$L_V \approx 0.1 \times 1.4 \times 10^5 \times 3.84 \times 10^{26}\,\text{W} = 5.4 \times 10^{30}\,\text{W}$$

Thus, the approximate value of L_V is 5×10^{30} W.

To find the distance we use Equation 3.14
$$d \approx [5.4 \times 10^{30}\,\text{W}/(4\pi \times 3.0 \times 10^{-9}\,\text{W m}^{-2})]^{1/2}$$
$$= 1.20 \times 10^{19}\,\text{m} = 388\,\text{pc}$$

Thus the estimated distance to Rigel is $d \sim 400$ pc.

Note that the actual distance to Rigel A is about 240 pc.

(e) The angular diameter of Rigel A, from the Earth, is given by a straightforward rearrangement of Equation 3.8. Thus, using the values of radius R and distance d obtained earlier in this question, we get

$$\alpha = (2 \times 74 \times 6.96 \times 10^8 \, \text{m})/(1.2 \times 10^{19} \, \text{m}) \text{ radians}$$

$$= 8.6 \times 10^{-9} \text{ radians}$$

$$= 8.6 \times 10^{-9} \times 206 \, 265 \text{ arcsec}$$

$$= 0.002 \text{ arcsec}$$

This is measurable: at present the smallest measured values are about 0.0004 arcsec (Section 3.3.1).

QUESTION 3.16

From Appendix A5 we obtain Figure 3.42 for the heavy elements ($Z > 2$) as far as $Z = 30$. There is no simple trend with Z. Instead, there are some notably large values, as follows (values of Z in parentheses): carbon (6), nitrogen (7), oxygen (8), neon (10), magnesium (12), silicon (14), sulfur (16) and iron (26).

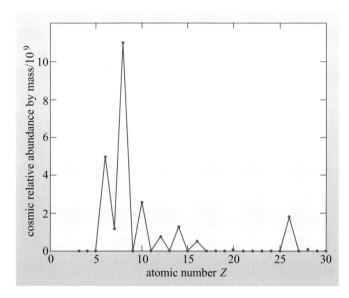

Figure 3.42 Elemental abundances in the Solar System by mass, with respect to $H = 10^{12}$, for the elements $2 < Z < 31$.

The reasons for this relationship with Z will be clearer to you by the end of the book.

QUESTION 3.17

From Equation 3.20 the sum of masses is

$$M + m = 4\pi^2 \frac{(20 \times 1.50 \times 10^{11} \, \text{m})^3}{(6.67 \times 10^{-11} \, \text{N m}^2 \, \text{kg}^{-1}) \times (50 \times 3.16 \times 10^7 \, \text{s})^2}$$

$$= 6.4 \times 10^{30} \text{ kg}$$

From Figure 3.37

$$d_m/d_M = 2.3$$

This ratio is not influenced by our projected view of the orbit.

Thus, from Equation 3.21

$$M/m = 2.3$$

We thus have

$$(2.3m + m) = 6.4 \times 10^{30}\,\text{kg}$$

and so $m = 1.9 \times 10^{30}\,\text{kg} = 1.0M_\odot$

Thus $M = 2.3 \times 1.0M_\odot = 2.3M_\odot$

Note also that if we measure P in years, and a in AU, then we can write Equation 3.20 as

$$(M + m)/M_\odot = (a/\text{AU})^3/(P/\text{yr})^2$$

This numerical simplification arises from the Earth–Sun system, for which $M = M_\odot$, $M_\odot \gg M_\text{E}$ (Earth's mass), $P = 1$ year and $a = 1$ AU. You may want to try answering this question using this version of Equation 3.20.

QUESTION 4.1

The positions in the H–R diagram of these types of star are as follows:

hot, high luminosity stars   top left
hot, low luminosity stars   bottom left
cool, low luminosity stars   bottom right
cool, high luminosity stars   top right

QUESTION 4.2

From Figure 4.5, we see that white dwarfs have radii of order $0.01R_\odot$, which is about the radius of the Earth. Likewise, we see that red giants have radii of order $30R_\odot$, which is about 3000 times the Earth's radius, or about a tenth of the distance of the Earth from the Sun. Note that the *ranges* of radii for white dwarfs, and particularly for red giants, are large.

QUESTION 4.3

Such stars would move diagonally to the right and downwards, the luminosity as well as the temperature decreasing. You will see later that many stars do indeed end their lives in this way.

QUESTION 4.4

(a) The H–R diagram of M67 in Figure 4.10b is notable for the absence from the main sequence of all but the low-mass stars (Figure 4.8), and the presence of considerable numbers of stars between the main sequence and the red giant region, which could represent the higher masses missing from the main sequence. This suggests that the more massive a star, the sooner it leaves the main sequence, and that most stars that have left the main sequence go on to become red giants. Supergiants are absent in M67, and this could be because massive main sequence stars, which are their precursors, are rare. Also, if, as it seems, massive stars evolve rapidly, then any supergiants could have become Type II supernovae, and have thus vanished from the H–R diagram. The absence of white dwarfs is presumably because they are too faint to detect. Thus, the H–R diagram for M67 is consistent with the model of stellar evolution in Figure 4.12.

(b) Assuming the model is right, we can conclude that M67 is older than the Pleiades, because in the Pleiades the main sequence is populated to higher stellar masses than the main sequence in M67 (Figure 4.10). This occurs because the more massive the star the sooner it leaves the main sequence. In M67 there has been enough time for all but the low mass stars to leave the main sequence, whereas the Pleiades is too young for this to have happened.

QUESTION 4.5

From Figure 4.16 we see that, for CO, the difference in energy ε between the lowest electronic level and the one above it is 5.94 eV ($= 9.52 \times 10^{-19}$ J).

ε corresponds to a photon wavelength given by

$$\lambda = hc/\varepsilon$$

$$= (6.63 \times 10^{-34}\,\text{J s}) \times (3.00 \times 10^8\,\text{m s}^{-1})/(9.52 \times 10^{-19}\,\text{J})$$

$$= 2.09 \times 10^{-7}\,\text{m} = 209\,\text{nm}$$

This is the maximum photon wavelength (minimum energy) for this excitation.

(b) From Equation 4.3, the minimum temperature is given by

$$T = 2\varepsilon/(3k)$$

$$= 2 \times 9.52 \times 10^{-19}\,\text{J}/(3 \times 1.38 \times 10^{-23}\,\text{J K}^{-1}) = 4.60 \times 10^4\,\text{K}$$

So typically a temperature of at least 5×10^4 K is required for the collisional excitation of this electronic state in CO molecules.

QUESTION 4.6

Equation 3.16 does not take account of interstellar extinction, A, as in Equation 4.1:

$$M = m - 5 \log d + 5 - A$$

The derived absolute visual magnitude will therefore be too faint (M numerically too large). Since interstellar dust also causes reddening, the $B - V$ colour will be redder and therefore the derived temperature will be too low. Examination of the axes of the H–R diagram in Figure 4.5 shows that the star will appear below and to the right of its correct position.

If a spectrum is observed, the temperature can be derived from its spectral type (based on the strengths of certain spectral lines, Section 3.3.2) and therefore not affected by interstellar reddening. Its luminosity can also be inferred directly from its spectrum (Section 3.3.4) and hence its true position on the H–R diagram can be determined.

QUESTION 4.7

The distance to a star does not influence its position on the H–R diagram: photospheric temperature and luminosity are intrinsic properties of a star. However, we sometimes have to apply corrections to our observations, because of interstellar extinction, in order to obtain intrinsic properties.

QUESTION 4.8

(a) Figure 4.22 shows the five stars plotted on an H–R diagram. By comparison with Figure 4.5, we can make the following assignments given in Table 4.2.

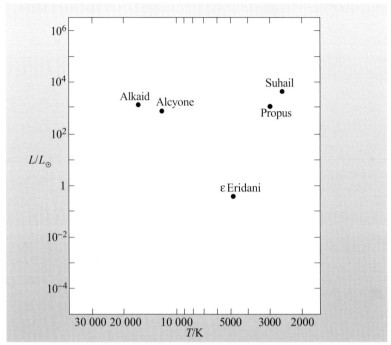

Figure 4.22 An H–R diagram with five stars, from Question 4.8.

Table 4.2

Star	Main stellar class	Comment
Alkaid	main sequence	
Alcyone	giant	
ε Eridani	main sequence	a nearby (3.2 pc), solar-type star
Propus	red giant	
Suhail	red giant/supergiant	some stars fall between the four main classes

(b) Low luminosity stars can have a large apparent visual brightness only if they are particularly close to us. Thus, low-luminosity stars will be under-represented in an H–R diagram that contains only the stars with the greatest apparent visual brightness. (There is always a likelihood that low-luminosity stars will be under-represented, simply because any detection system will fail to detect stars that give too low a flux density at the detector. This is an example of a *selection effect*.)

QUESTION 4.9

From Figure 4.5 we see that the Sun's photospheric temperature, $T_\odot$, luminosity, $L_\odot$, and radius, $R_\odot$, have the following relationships to the temperature T, luminosity L and radius R of stars near the upper and lower ends of the main sequence:

upper end: $T \sim 6T_\odot$, $L \sim 10^6 L_\odot$, $R \sim 10R_\odot$

lower end: $T \sim 0.3T_\odot$, $L \sim 10^{-4}L_\odot$, $R \sim 0.03R_\odot$

Thus, the Sun is a very modest main sequence star.

QUESTION 4.10

The region where T Tauri stars are found on the H–R diagram is shown in Figure 4.7. Since the question states that T Tauri stars approach the main sequence with little change in temperature, they must follow vertical paths from this region to the main sequence. By inspection of Figure 4.5 it can be seen that by following a vertical track from the T Tauri region to the main sequence a small reduction in radius must occur.

QUESTION 4.11

To evolve into a supergiant, the red giant would have to acquire a lot of mass: see Figure 4.8. However, observations suggest that red giants *lose* mass (in the form of stellar winds), and so we can rule out the scenario in which a red giant evolves into a supergiant.

QUESTION 5.1

Note: We choose nitrogen to represent the whole of the Earth's atmosphere as it is the most abundant gas (its mass is very similar to that of oxygen, which is the other major component).

The common isotope of nitrogen, $^{14}_{7}N$, has an atomic mass of $14 \times (1.67 \times 10^{-27}\,\text{kg})$. Nitrogen is present in its molecular form in the Earth's atmosphere so the number n of molecules per cubic metre is given by

$$n \approx \frac{1\,\text{kg m}^{-3}}{14 \times 2 \times (1.67 \times 10^{-27}\,\text{kg})} \approx 2 \times 10^{25}\,\text{m}^{-3}$$

Comparison with Figure 5.1 shows that the atmosphere at the surface of the Earth is about 10^9 times more dense than the densest part of the ISM.

The surface temperature, about 300 K, is not much greater than that of the warmer diffuse clouds.

QUESTION 5.2

For an object that is emitting as a black-body source at temperature T, the wavelength, λ_{peak}, at which the maximum energy is radiated is given by Wien's displacement law, (Equation 1.4):

$$\left(\frac{\lambda_{\text{peak}}}{\text{m}}\right) = \frac{2.90 \times 10^{-3}}{(T/K)}$$

For a dense cloud collapsing on its way to becoming a protostar, we expect the temperature to still be very low, probably still only a few hundred kelvin – let us assume 300 K. In this case

$$\lambda_{peak} = \frac{2.90 \times 10^{-3} \text{ m}}{300} \approx 10^{-5} \text{ m}$$

This is in the infrared part of the spectrum (see Figure 1.36).

If we had assumed a temperature different by a factor of 10, larger or smaller, our answer would still have been in the infrared!

QUESTION 5.3

The smaller the Jeans mass, the more likely gravitational contraction is to occur. From Equation 5.1, we can see that a small value for the Jeans mass requires that n, the number density, be as large, and T, the temperature, be as low, as possible. The same conclusion could be derived from Figure 5.9.

QUESTION 5.4

The number of stars observed in a given phase of evolution should be roughly proportional to the time an individual star spends in that phase. When evolution is rapid, the phase of evolution is soon over and the probability of our observing objects in that phase is low. (One exception to this is a supernova – but that will be discussed later.)

QUESTION 5.5

In bipolar outflow, the flow is highly directed. This means that observations of the Doppler shift will tend to show two distinct values (except for the case where the outflow is perpendicular to the line of sight) corresponding to flow in two opposite directions. For a T Tauri star, outflow is in all directions, so a range of Doppler shifts can be observed. In addition, the outflow from T Tauri stars tends to have a higher velocity, leading to a larger Doppler shift. Furthermore, T Tauri stars sometimes show variable outflow.

QUESTION 5.6

The mass of a hydrogen molecule is $2 \times 1.67 \times 10^{-27}$ kg.

Thus the total mass of the cloud is

mass = volume × density

$$= [\tfrac{4}{3} \pi \times (3 \times 3.09 \times 10^{16} \text{ m})^3] \times [10^9 \text{ m}^{-3} \times (2 \times 1.67 \times 10^{-27} \text{ kg})]$$

$$= 1.1 \times 10^{34} \text{ kg}$$

If we assume, for the present calculation, that all stars have a mass equal to that of the Sun (2×10^{30} kg), we can work out how many stars could be formed from the cloud.

number of stars = $(1.1 \times 10^{34} \text{ kg})/(2 \times 10^{30} \text{ kg}) \approx 5.5 \times 10^3$

If we assume therefore that all the material in the original cloud goes into making stars, we find that there is plenty of material to make a large number of stars.

In Chapter 3, we learnt that stars generally have masses in the range from about $0.08M_\odot$ to about $50M_\odot$, so our conclusion wouldn't change whatever mass we assumed for the stars that formed from this cloud.

QUESTION 5.7

We calculate the Jeans mass in each case using Equation 5.1:

$$M_J = \frac{9}{4} \times \left(\frac{1}{2\pi n}\right)^{1/2} \times \frac{1}{m^2} \times \left(\frac{kT}{G}\right)^{3/2}$$

Using the parameters for the two cases from the question and $m = 2 \times 1.67 \times 10^{-27}\,\text{kg}$ for the mass of the hydrogen molecule, we find, in the uncompressed case

$$M_J = \frac{9}{4}\left[\frac{1}{2\pi \times 5 \times 10^9\,\text{m}^{-3}}\right]^{1/2} \times \frac{1}{(3.34 \times 10^{-27}\,\text{kg})^2} \times \left[\frac{1.38 \times 10^{-23}\,\text{J K}^{-1} \times 10\,\text{K}}{6.67 \times 10^{-11}\,\text{N m}^2\,\text{kg}^{-2}}\right]^{3/2}$$

$$= \frac{9}{4} \times (5.6 \times 10^{-6}\,\text{m}^{3/2}) \times (8.9 \times 10^{52}\,\text{kg}^{-2}) \times (3.0 \times 10^{-18}\,\text{m}^{-3/2}\,\text{kg}^3)$$

$$= 3.4 \times 10^{30}\,\text{kg}$$

Thus, in solar masses ($M_\odot = 2 \times 10^{30}\,\text{kg}$) (and rounding to 1 significant figure, consistent with the information given in the question) $M_J = 2M_\odot$.

Therefore, the core, mass $M_\odot$, has a mass *less* than its Jeans mass, so it is unlikely to contract, particularly if rotation or magnetic fields hinder contraction. In the compressed case, a similar calculation yields $M_J = 0.5M_\odot$.

Thus, the core mass is now *greater* than its Jeans mass, so it is likely to contract unless supported by rotation or magnetic fields.

QUESTION 6.1

The pressure is assumed to be uniform over the whole area of contact between the two hemispheres. In reality it will vary, being highest in the centre.

The two hemispheres are assumed to be point masses. The mass is distributed throughout the hemispheres which are very close together.

(In real stellar models the material is treated as a series of thin concentric spherical shells.)

QUESTION 6.2

The composition of the core could change during the main sequence through (i) convection and (ii) fusion reactions. Figure 6.5 shows that, in stars of low mass, convection is confined to a thin outer envelope, so convective mixing of core material with the material around it won't occur. Therefore the only way in which the composition of the core will change during the main sequence lifetime is through fusion reactions. As in the Sun, these result in a progressive depletion of hydrogen and enrichment of helium in the core.

QUESTION 6.3

The time for which the Sun's present luminosity could be maintained is given by (energy available)/(rate of energy radiation), where the rate of energy radiated is the Sun's luminosity. The energy available from chemical energy is given by the Sun's mass multiplied by the energy available per unit mass:

$$(2 \times 10^{30} \, \text{kg}) \times (3.5 \times 10^{7} \, \text{J kg}^{-1}) = 7 \times 10^{37} \, \text{J}$$

Thus, the time for which the Sun's present luminosity (of 4×10^{26} W) could be maintained is

$$(7 \times 10^{37} \, \text{J})/(4 \times 10^{26} \, \text{W}) = 2 \times 10^{11} \, \text{s, i.e. } {\sim}6000 \text{ years!}$$

Evidence shows that the Sun has been shining at (approximately) its present luminosity for about 5×10^9 years. Therefore the generation of energy by chemical means does not appear to be the likely explanation of the Sun's luminosity.

QUESTION 6.4

The loss of gravitational energy is given by

$$\frac{-GM_\odot^2}{R_\odot} - \frac{-GM_\odot^2}{0.1R_\odot} = \frac{GM_\odot^2}{R_\odot}\left(\frac{1}{0.1} - 1\right)$$

$$= \frac{9GM_\odot^2}{R_\odot}$$

$$= \frac{9 \times (6.67 \times 10^{-11} \, \text{N m}^2 \, \text{kg}^{-2}) \times (2 \times 10^{30} \, \text{kg})^2}{7 \times 10^8 \, \text{m}}$$

$$= 3 \times 10^{42} \, \text{J}$$

Assuming that this energy is all converted into power for the Sun, the time for which the Sun's present luminosity could be maintained is given by

$$\frac{3 \times 10^{42} \, \text{J}}{4 \times 10^{26} \, \text{W}} = 8 \times 10^{15} \, \text{s} = 3 \times 10^8 \text{ years}$$

Although much more promising than chemical energy (see Question 6.3) as an energy source, gravitational energy still can't provide sufficient energy to power the Sun for its known lifetime.

QUESTION 6.5

The electrical potential energy is found using Equation 6.8,

$$E_e = 8.99 \times 10^9 \, \text{N m}^2 \, \text{C}^{-2} \, (1 \times 1.6 \times 10^{-19} \, \text{C})(1 \times 1.6 \times 10^{-19} \, \text{C})/(10^{-15} \, \text{m})$$

$$= 2.3 \times 10^{-13} \, \text{J}$$

The kinetic energy due to thermal motion is found using Equation 4.2

$$E_k = 3 \, (1.38 \times 10^{-23} \, \text{J K}^{-1})(1.6 \times 10^7 \, \text{K})/2$$

$$= 3.3 \times 10^{-16} \, \text{J}$$

The average kinetic energy is therefore $\sim$1000 times smaller than the electrical potential energy. So it can be concluded that the thermal motion of the particles is insufficient to overcome the electrostatic forces that act between protons.

QUESTION 6.6

The ratio of radiation pressure to gas pressure can be calculated from Equations 6.11 and 6.12 as

$$\frac{P_{\text{rad}}}{P_{\text{gas}}} = \frac{\alpha T^3}{3nk}$$

For the Sun, the average mass density is

$$\rho = \frac{1.99 \times 10^{30}\ \text{kg}}{\frac{4}{3}\pi(6.96 \times 10^8\ \text{m})^3} = 1.41 \times 10^3\ \text{kg m}^{-3}$$

The mean number density of hydrogen nuclei is

$$\frac{\text{mass density}}{\text{mass per particle}} = \frac{1.41 \times 10^3\ \text{kg m}^{-3}}{1.67 \times 10^{-27}\ \text{kg}}$$

$$= 8.44 \times 10^{29}\ \text{m}^{-3}$$

In the ionized solar interior, there are as many free electrons as there are hydrogen nuclei, so n (the number of particles per m^3) is double this value. Thus, using $T = 1.5 \times 10^7$ K for the centre of the Sun,

$$\frac{P_{\text{rad}}}{P_{\text{gas}}} = \frac{(7.55 \times 10^{-16}\ \text{kg m}^{-1}\ \text{s}^{-2}\ \text{K}^{-4}) \times (1.6 \times 10^7\ \text{K})^3}{3 \times (2 \times 8.44 \times 10^{29}\ \text{m}^{-3}) \times (1.38 \times 10^{-23}\ \text{kg m}^2\ \text{s}^{-2}\ \text{K}^{-1})}$$

$$= 4.4 \times 10^{-2}$$

So, according to this calculation, at the centre of the Sun radiation pressure is not very significant, amounting to about 4% of that due to gas pressure.

In Figure 2.1, we see that the mass density at the centre of the Sun is predicted to be about 1.5×10^5 kg m^{-3} compared with our average value of 1.4×10^3 kg m^{-3}. So, we have underestimated the number density by about a factor of 100. This means that the contribution of radiation pressure is even less than calculated here (by a factor of 100).

QUESTION 6.7

Centre to surface of the Sun: radiation in the deep interior, then convection in the outer regions. (Conduction doesn't play a significant part in energy transport in most stars.)

Surface of the Sun to the top of the Earth's atmosphere: radiation, because this region is occupied by a vacuum.

Top of the atmosphere to the surface of the Earth: radiation (the atmosphere is partially transparent to some of the Sun's radiation). Although convection is an important process in energy transport in the atmosphere, it does not play a part in the initial transfer of energy to the surface as heat cannot be convected downwards.

QUESTION 6.8

We see from Table 6.1 that upper main sequence stars have main sequence lifetimes much less than the present age of the Earth. In the table, a lifetime of 4.5×10^9 years falls between the values for stars of mass $1M_\odot$ and $1.5M_\odot$, at a mass of approximately $1.3M_\odot$. Any star more massive than this, and that is now on the main sequence, cannot have been on the main sequence when the Earth formed.

QUESTION 6.9

(a) The reaction obeys the laws of conservation of electric charge and of baryon number. It will obey the law of conservation of energy provided that the sum of the rest energies and kinetic energies of the carbon nuclei equals this sum for the magnesium nucleus plus the energy of the γ-ray. The reaction is thus possible.

(b) Figure 6.6 shows that the rest energy *per nucleon* for $A = 12$ exceeds that for $A = 24$. Denoting these values by E_{12} and E_{24}, respectively, for the reactants we have a rest energy of $2 \times 12 \times E_{12}$, and for the product $24 \times E_{24}$. Thus, with $E_{12} > E_{24}$, the rest energy of the reactants exceeds that of the product, and so the reaction is exothermic. However, it is not an appreciable source of energy compared with the pp and CNO cycles, for main sequence stars because (i) with $Z = 6$ the electrical repulsion between the carbon nuclei results in a much lower reaction rate, and (ii) carbon is not nearly as abundant as hydrogen (or as helium). (This reaction is important however, in the post main sequence evolution of stars more massive than about 8–$10M_\odot$.)

QUESTION 7.1

The luminosity of a star is related to its radius, R, and surface temperature, T, by Equation 3.9, i.e.

$$L \approx 4\pi R^2 \sigma T^4$$

where σ is the Stefan–Boltzmann constant. For the star we are considering here, we know that the star has expanded (i.e. R has increased) whereas the luminosity hasn't changed significantly (L is approximately constant). Thus, referring to Equation 3.9, it is required that T *decreases*, i.e. the surface of the star cools.

QUESTION 7.2

To answer this question it is necessary to determine whether the spectral types and luminosities of the stars in Table 7.1 correspond to positions on the H–R diagram that lie within the instability strip. The H–R diagram that shows the instability strip (Figure 7.6) is given in terms of temperature and luminosity. Thus the given spectral types and absolute magnitudes have to be first converted into temperatures and luminosities. This can be done by referring back to Table 3.2 and Figure 4.5. Table 7.2 shows the temperatures and luminosities that correspond to the spectral types and luminosities in Table 7.1. (Remember that the correspondence between absolute visual magnitude and luminosity is not exact.)

By plotting the values of luminosity and temperature on Figure 7.6, it can be seen that stars X and Y lie within the instability strip, whereas Z lies well outside it. Thus X and Y may be pulsating variables, whereas Z is unlikely to show large amplitude stellar pulsations.

Table 7.2 The spectral types and absolute magnitudes from Table 7.1 and the corresponding values of temperature and luminosity.

Star	Spectral type	Temperature/K	M_V	Luminosity/$L_\odot$
X	F0	7400	3	10
Y	K5	4100	−6	10^4
Z	A0	9900	1	10^2

QUESTION 7.3

In going to the right in the H–R diagram the surface temperature drops (considerably). The luminosity will rise a little if the star's evolution lifts it a little higher in the H–R diagram. (See, for example, the evolutionary track of a $15M_\odot$ star in Figure 7.2.)

QUESTION 7.4

Temperature and radius are related by Equation 3.9, $L \approx 4\pi R^2 \sigma T^4$.
Since the luminosity L is constant, and so are 4, π and σ, we can rearrange Equation 3.9 to give

$$R = \frac{1}{T^2}\left(\frac{L}{4\pi\sigma}\right)^{1/2} = \frac{\text{constant}}{T^2}$$

Therefore $\quad \dfrac{R_\text{final}}{R_\text{initial}} = \dfrac{(T_\text{initial})^2}{(T_\text{final})^2} = \left(\dfrac{25\,000}{5000}\right)^2$

$$= 5^2 = 25$$

Therefore $\quad R_\text{final} = 25 R_\text{initial}$

$$= 25 \times 10 R_\odot = 250 R_\odot$$

In units of AU,

$$R_\text{final} = \frac{250 R_\odot \times 7 \times 10^8 \text{ m } R_\odot^{-1}}{1.5 \times 10^{11} \text{ m AU}^{-1}}$$

$$= 1.2\,\text{AU}$$

QUESTION 7.5

One reason for there being so few stars in this area is that massive stars evolve very quickly from being upper main sequence stars to being supergiant stars, so the chances of catching a star in between are slender.

QUESTION 7.6

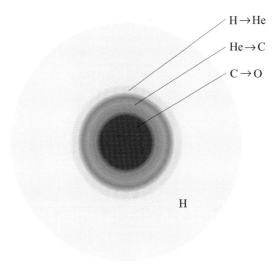

H → He

He → C

C → O

H

The answer is shown in Figure 7.14.

Figure 7.14   The structure of a star at the carbon-burning stage (not to scale).

QUESTION 7.7

In such a high-temperature environment, it is likely that any protons or α-particles would very rapidly undergo fusion reactions with the heavier nuclei present in the core. A reaction that may (and does) occur when α-particles are mixed with carbon at such extreme temperatures is:

$$^{12}_{6}C + {}^{4}_{2}He \rightarrow {}^{16}_{8}O + \gamma$$

Since these light nuclei would undergo reactions very quickly, no significant quantities of hydrogen or helium would build up in the core.

QUESTION 7.8

The wavelength of the peak of the black-body spectrum is given by Wien's displacement law (Equation 1.4):

$$\lambda_{peak} = \frac{2.90 \times 10^{-3}}{(T/K)}\,m = \frac{2.90 \times 10^{-3}}{(10^{9})}\,m = 2.90 \times 10^{-12}\,m$$

So the wavelength of the peak of the black-body spectrum is at about 3×10^{-12} m. From Figure 1.36 it can be seen that this corresponds to the γ-ray part of the electromagnetic spectrum.

QUESTION 7.9

Initially, the fusion of hydrogen to form helium takes place in the core of the star. When the amount of available hydrogen there is appreciably reduced the reaction stops in the core. With the rise in temperature it can continue where there is hydrogen available at the edge of the core. So we find next that hydrogen is being fused to helium in a thin shell around the core. Eventually there is a shortage of hydrogen here too, and when the temperature rises again the site of the reaction moves outwards again to a fresh supply. Then the reaction is found in the next layer out, i.e. in a shell of slightly larger radius surrounding both the previous shell and the core. This process is repeated over and over again, with the site of the hydrogen fusion reaction steadily moving outwards to shells of larger radius.

QUESTION 7.10

The mass lost from a solar mass main sequence star in 10^{10} years is simply

$$10^{10} \text{ years} \times 10^{-14} M_{\odot} \text{ year}^{-1} = 10^{-4} M_{\odot}$$

The fractional mass loss is therefore $10^{-4}/1$ or 0.01%.

For the $50 M_{\odot}$ main sequence star the mass lost is

$$10^{6} \text{ years} \times 10^{-6} M_{\odot} \text{ year}^{-1} = 1 M_{\odot}$$

The fractional mass loss is therefore $1/50$ or 2%.

Despite its short main sequence lifetime, the $50 M_{\odot}$ star loses a larger fraction of its mass.

QUESTION 7.11

Your diagram should resemble Figure 7.15.

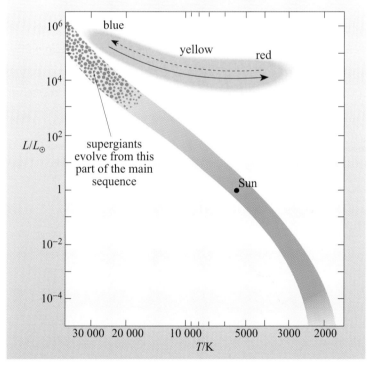

Figure 7.15 H–R diagram showing the positions of supergiants and an indication of their evolutionary tracks from blue to red (solid line) and back again (dashed line).

QUESTION 7.12

(a) For a $25M_\odot$ star, the correct chronological order of core nuclear burning processes is: hydrogen burning, helium burning, carbon burning, neon burning, oxygen burning and silicon burning.

(b) The processes that involve the fusion of two identical nuclei are: hydrogen, carbon and oxygen burning, whereas helium burning involves the fusion of three identical nuclei. Photodisintegration plays an important role in neon burning and silicon burning, as it provides a source of α-particles that can fuse with other nuclei.

QUESTION 7.13

There are three reasons why a massive star goes through its life cycle at an ever-increasing pace:

1   As the star evolves its central temperature rises; the nuclear reaction rates increase markedly as this happens.

2   As the reaction rates increase there is an increase in the emission of neutrinos; the neutrinos carry away energy that is then lost to the star and has to be replaced by further nuclear reactions; the star must contract and heat up so that the reaction rate increases to compensate for this increasing energy loss.

3   Heavier nuclei are produced as the star ages; the fusion of heavier nuclei produces less energy per kilogram than the fusion of lighter nuclei and so more of the heavy nuclei undergo fusion per second to maintain the necessary energy supply.

QUESTION 8.1

The density is equal to the mass divided by the volume, and so

$$3 \times 10^{17} \text{ kg m}^{-3} = \frac{6 \times 10^{24} \text{ kg}}{\text{volume}}$$

Therefore $\text{volume} = \dfrac{6 \times 10^{24}}{3 \times 10^{17}} \text{ m}^3 = 2 \times 10^7 \text{ m}^3$

But volume $= \frac{4}{3}\pi r^3$, therefore

$$r = \left(\frac{3}{4\pi} \times 2 \times 10^7\right)^{1/3} \text{ m} = (4.8 \times 10^6)^{1/3} \text{ m} = 168 \text{ m (!)}$$

So the radius would be 2×10^2 m (to one significant figure).

QUESTION 8.2

A supergiant star might have a radius of 100 times the radius of the Sun (Section 4.2.1), that is $100 \times 7 \times 10^5$ km, which is 7×10^7 km. Our expanding supernova has a radius of 2×10^{10} km which is about 2×10^{10} km$/7 \times 10^7$ km $\approx 3 \times 10^3$ times bigger than such a supergiant. Pluto, the outermost planet in our Solar System, is approximately 6×10^9 km from the Sun; taking this as the radius of the Solar System, we see that the supernova has expanded to about three times this size. The distance to the nearest star is 1.30 parsecs, which is roughly 4×10^{13} km; the supernova has (so far) expanded to only one two-thousandth of this.

QUESTION 8.3

Neutrinos travel readily through the Earth, so it is not necessary for the source to be above the horizon. In this case one would suspect that some, if not all, of the neutrinos had travelled through some of the Earth before detection because (i) Japan and Ohio are approximately on opposite sides of the world, so it is unlikely that a source that they detect simultaneously would be above both their horizons at that time, and (ii) the Large Magellanic Cloud is in the southern sky and these detectors are in the northern hemisphere, so it might well always be below both their horizons. (In fact, this is case – the Large Magellanic Cloud never rises above the horizon for observers in Ohio or Japan.)

QUESTION 8.4

If there are two γ-rays of energy 1.3×10^{-13} J and 1.9×10^{-13} J, then the energy output per decay $= 3.2 \times 10^{-13}$ J. To provide a luminosity of 10^{33} W, i.e. 10^{33} J s^{-1},

$$\text{the rate of decay} = \frac{10^{33} \text{ J s}^{-1}}{3.2 \times 10^{-13} \text{ J}} = 3 \times 10^{45} \text{ decays s}^{-1}$$

Each $^{56}_{27}\text{Co}$ nucleus weighs about 56 times the mass of a neutron or proton, which is $56 \times 1.7 \times 10^{-27}$ kg. Therefore

$$\begin{aligned}
\text{mass decaying per second} &= (56 \times 1.7 \times 10^{-27} \text{ kg}) \times (3 \times 10^{45} \text{ s}^{-1}) \\
&= 3 \times 10^{20} \text{ kg s}^{-1} \\
&= 1.5 \times 10^{-10} M_\odot \text{ s}^{-1}
\end{aligned}$$

The cobalt γ-rays illuminate the remnant for hundreds of days, so clearly a large mass of nickel (which decayed into the cobalt) was created. It is estimated that, in SN 1987A, $0.07 M_\odot$ of nickel was formed.

QUESTION 8.5

The kinetic energy of the shell is given by

$$E_k = \tfrac{1}{2} M v^2$$

where M is the mass of the shell and v is the required speed. Thus

$$v = (2 E_k / M)^{1/2}$$

With $E_k \sim 10^{44}$ J, and $M \sim 0.25 M_\odot \sim 5 \times 10^{29}$ kg, we get

$$v \approx \left(\frac{2 \times 10^{44} \text{ J}}{5 \times 10^{29} \text{ kg}} \right)^{1/2}$$

Replacing J by the equivalent units $\text{kg m}^2 \text{ s}^{-2}$, we get

$$\begin{aligned}
v &\approx \left(\frac{2 \times 10^{44} \text{ kg m}^2 \text{ s}^{-2}}{5 \times 10^{29} \text{ kg}} \right)^{1/2} \\
&\approx (4 \times 10^{14} \text{ m}^2 \text{ s}^{-2})^{1/2} \\
&\approx 2 \times 10^7 \text{ m s}^{-1} \approx 0.07c
\end{aligned}$$

QUESTION 8.6

(a) For a thermal source of radiation, a rough estimate of the photon energy at which the emission will be relatively strong can be made using the relation (Equation 1.6):

$$\varepsilon \sim kT$$

$$\varepsilon \sim 1.38 \times 10^{-23} \, \text{J K}^{-1} \times 10^6 \, \text{K} = 1.38 \times 10^{-17} \, \text{J}$$

Converting into electronvolts,

$$\varepsilon \sim (1.38 \times 10^{-17} \, \text{J})/(1.60 \times 10^{-19} \, \text{J eV}^{-1}) = 86.3 \, \text{eV}$$

Thus the expected photon energy is about 90 eV. (Remember that this equation gives only a rough estimate of the expected photon energy, and that the answer should only be quoted to 1 significant figure.)

(b) The wavelength corresponding to a photon of this energy is found by combining Equations 1.2 and 1.3 (see Example 1.1)

$$\lambda = hc/\varepsilon$$

$$\lambda = (6.63 \times 10^{-34} \, \text{J s}) \, (3.00 \times 10^8 \, \text{m s}^{-1})/(1.38 \times 10^{-17} \, \text{J})$$

$$= 1.44 \times 10^{-8} \, \text{m}$$

So the wavelength at which strong thermal emission would be expected is $1 \times 10^{-8} \, \text{m}$. (Again, the answer is a rough estimate and should only be quoted to 1 significant figure.)

From Figure 1.36, it can be seen that such a wavelength corresponds to the X-ray regime of the electromagnetic spectrum.

QUESTION 8.7

If it is assumed that the outflow speed has remained constant since the planetary nebula formed, then the age can be calculated from

$$\text{age} = \frac{\text{radius}}{\text{expansion speed}} = \frac{(0.4/2) \times 3.09 \times 10^{16} \, \text{m}}{20 \times 10^3 \, \text{m s}^{-1}}$$

$$= 3.09 \times 10^{11} \, \text{s}$$

Converting this into years

$$\text{age} = (3.09 \times 10^{11})/(365 \times 24 \times 60 \times 60) \, \text{years} = 9.80 \times 10^3 \, \text{years}$$

So to 1 significant figure, the age of the planetary nebula is 1×10^4 years.

QUESTION 8.8

The flux density F depends on luminosity L and distance d as given by Equation 3.10:

$$F = \frac{L}{4\pi d^2}$$

If the flux density from the supernova equals that from the Sun, then

$$\frac{L_\odot}{4\pi d_\odot^2} = \frac{L_{\text{SN}}}{4\pi d_{\text{SN}}^2}$$

where SN denotes the supernova. Thus

$$d_{SN}^2 = \left(\frac{L_{SN}}{L_\odot} \right) d_\odot^2$$

Now, we are given that $(L_{SN}/L_\odot) = 5 \times 10^9$, and we know that

$$d_\odot = 1.0\,\text{AU} = 4.9 \times 10^{-6}\,\text{pc}$$

Therefore $d_{SN} = (5 \times 10^9)^{1/2} \times 1.0\,\text{AU}$

$$\approx 7 \times 10^4\,\text{AU}$$

or $\quad d_{SN} = (5 \times 10^9)^{1/2} \times 4.9 \times 10^{-6}\,\text{pc}$

$$\approx 0.3\,\text{pc}$$

There are no supergiants that close! The nearest known star (of any sort) after the Sun is 1.3 parsecs away (Section 3.2.2).

QUESTION 8.9

The answer is a diagram – Figure 8.20.

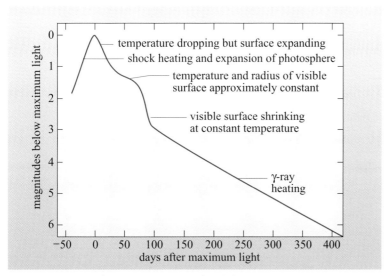

Figure 8.20 For the answer to Question 8.9.

QUESTION 8.10

It can be seen from Table 8.2, that addition of neutrons by the s-process to the stable isotopes of palladium with mass numbers 104, 105 and 106 will successively form isotopes with mass numbers 105, 106 and 107. Table 8.2 shows that $^{107}_{46}\text{Pd}$ is unstable, and so it is necessary to consider whether the nuclei of this isotope will decay before the next neutron capture event. The half-life for the β^--decay is 6.5×10^6 years. This is far longer than the typical interval between s-process neutron capture events, which is about 10^4 years. So the nuclei of $^{107}_{46}\text{Pd}$ are much more likely to capture another neutron than to undergo β^--decay. This means that the isotope of palladium with mass number 108 *is* likely to be formed by the s-process. This is a stable isotope and so can also capture a neutron by the s-process to form $^{109}_{46}$Pd. This is an unstable isotope with a half-life of only 13.7 hours, and so nuclei of $^{109}_{46}$Pd will undergo β⁻-decay (to form an isotope of silver) before neutron capture can occur. Thus, it is impossible to form the isotope $^{110}_{46}$Pd by the s-process. Thus the isotopes of palladium with mass numbers 105 to 109 can be formed by the s-process, whereas the isotope with mass number 110 cannot.

QUESTION 8.11

Stellar winds and the diffusion of old planetary nebulae would be the main ways in which material was returned to the interstellar medium. Both these, however, are from the outermost layers of the star, and are predominantly hydrogen. Small amounts of dust (silicates and graphite) would continue to diffuse into the interstellar medium from the dust shells found surrounding some stars. A few stars have deep convection currents that pull material up to the surface, but in most cases the elements created in stars would remain locked in the stars. Thus there would be a lower abundance of elements heavier than helium. Elements that can be created only by the r-process would not exist.

QUESTION 9.1

The density ρ is calculated from ρ = mass/volume. For a spherical body, with mass M and radius R,

$$\rho = \frac{M}{\frac{4}{3}\pi R^3}$$

For a white dwarf with

$$M = 0.6M_\odot = 0.6 \times 2.0 \times 10^{30}\,\text{kg} = 1.2 \times 10^{30}\,\text{kg}$$

and $\quad R = 7000\,\text{km} = 7 \times 10^6\,\text{m}$

the density ρ_{WD} is given by

$$\rho_{WD} = \frac{2.0 \times 10^{30}\,\text{kg}}{\frac{4}{3}\pi(7 \times 10^6\,\text{m})^3} = 8.4 \times 10^8\,\text{kg m}^{-3}$$

To one significant figure the density of the white dwarf is $8 \times 10^8\,\text{kg m}^{-3}$.

For the Sun, the average density $\rho_\odot$ is given by

$$\rho_\odot = \frac{2.0 \times 10^{30}\,\text{kg}}{\frac{4}{3}\pi(7 \times 10^8\,\text{m})^3} = 1.4 \times 10^3\,\text{kg m}^{-3}$$

Thus, the white dwarf has a density that is about 6×10^5 times greater than that of the Sun.

QUESTION 9.2

From Equation 3.9, we know that the luminosity of a star is given by

$$L = 4\pi R^2 \sigma T^4$$

We can write, for the case of two different luminosities at constant radius

$$L_1 = 4\pi R^2 \sigma T_1^4 \qquad \text{(i)}$$

$$L_2 = 4\pi R^2 \sigma T_2^4 \qquad \text{(ii)}$$

Therefore, by dividing Equation (i) by Equation (ii),

$$\left(\frac{L_1}{L_2}\right) = \left(\frac{T_1}{T_2}\right)^4$$

In this case, $L_2 = 10^{-4} L_1$, (i.e. $L_1/L_2 = 10^4$)

$$10^4 = \left(\frac{T_1}{T_2}\right)^4$$

and so $T_1/T_2 = 10$

In other words, the surface temperature has dropped by a factor of 10. We can check this by referring to Figure 9.1. If we follow the evolutionary track shown for a white dwarf, a drop in luminosity from, say, $1L_\odot$ to $10^{-4} L_\odot$ (in other words, a reduction by a factor of 10^4), corresponds to a change in surface temperature from about 60 000 K to about 6000 K. This confirms that a drop in luminosity by a factor of 10^4 corresponds to a drop in surface temperature by a factor of 10.

QUESTION 9.3

It is best to express the star's mass in kilograms:

$$1.5M_\odot = 3.0 \times 10^{30}\,\text{kg}$$

and the radius in metres:

radius $(r) = 10\,\text{km} = 10^4\,\text{m}$

From the radius, we can get the volume:

volume $= \frac{4}{3}\pi r^3$ (assuming the star is spherical)

$$= \frac{4}{3}\pi(10^4)^3\,\text{m}^3 = 4.2 \times 10^{12}\,\text{m}^3$$

density $=$ mass/volume

Therefore

$$= \frac{3.0 \times 10^{30}}{4.2 \times 10^{12}}\,\text{kg m}^{-3} = 7.1 \times 10^{17}\,\text{kg m}^{-3}$$

We can take the volume of a thimble as $1\,\text{cm}^3 = 10^{-6}\,\text{m}^3$, so the mass of a thimble-full, M_t, is

$$M_t = 10^{-6} \times 7.1 \times 10^{17}\,\text{kg} = 7.1 \times 10^{11}\,\text{kg}$$
$$= 7.1 \times 10^8\,\text{tonnes}$$

So a thimble-full of neutron star material will have a mass of 7×10^8 tonnes.

QUESTION 9.4

(a) The acceleration due to gravity at the surface of the Earth is given by

$$g_E = \frac{GM_E}{R_E^2}$$

where M_E is the mass of the Earth and R_E its radius. Using the values for mass and radius given in the question, (mass = 5.98×10^{24} kg, radius = 6378 km),

$$g_E = \frac{(6.67 \times 10^{-11} \text{ N m}^2 \text{ kg}^{-2}) \times (5.98 \times 10^{24} \text{ kg})}{(6.378 \times 10^6 \text{ m})^2}$$

$$= 9.805 \text{ m s}^{-2}$$

So the acceleration due to gravity at the surface of the Earth is 9.81 m s^{-2} (to 3 significant figures).

(b) The acceleration due to gravity at the surface of a neutron star is

$$g_n = \frac{GM_n}{R_n^2}$$

so, with $M_n = 1.5 M_\odot = 3.0 \times 10^{30}$ kg and $R_n = 10$ km $= 10^4$ m,

$$g_n = \frac{(6.67 \times 10^{-11} \text{ N m}^2 \text{ kg}^{-2}) \times (3.0 \times 10^{30} \text{ kg})}{(10^4 \text{ m})^2}$$

$$= 2.00 \times 10^{12} \text{ m s}^{-2}$$

The acceleration due to gravity at the surface of this neutron star is 2.0×10^{12} m s^{-2} (to 2 significant figures).

QUESTION 9.5

The speed v of an object that is initially at rest, if it has undergone constant acceleration a for a time t is

$$v = at$$

In this case, $t = 1.0 \times 10^{-5}$ s, and the accelerations are those found in Question 9.4

(a) For an object that is dropped near the surface of the Earth,

$$v = 9.81 \times 1.0 \times 10^{-5} \text{ m s}^{-1}$$

The speed of the object after 1.0×10^{-5} s is 9.8×10^{-5} m s^{-1} (to 2 significant figures).

(b) For an object that is dropped near the surface of the neutron star,

$$v = 2.0 \times 10^{12} \times 1.0 \times 10^{-5} \text{ m s}^{-1}$$

Therefore the speed of the object after 1.0×10^{-5} s is 2.0×10^7 m s^{-1} (to 2 significant figures).

This is about 7% of the speed of light, and is an indication that a full description of gravitational effects near a neutron star require the use of Einstein's General Theory of Relativity rather than Newton's theory of gravity.

QUESTION 9.6

(a) A full rotation corresponds to an angle of 2π radians. Hence a body that rotates through a half turn in one second, turns through an angle of π radians per second. The angular speed is then π rad s^{-1} (or equivalently 3.14 rad s^{-1}).

(b) The period of rotation is 0.25 s, so using Equation 9.1, the angular speed is

$$\omega = 2\pi/T = 2\pi/0.25 = 8\pi \text{ rad s}^{-1} \text{ (or equivalently } \omega = 25.1 \text{ rad s}^{-1})$$

(c) The period of rotation is 4.0 s, so again using Equation 9.1, the angular speed is

$$\omega = 2\pi/T = 2\pi/4.0 \text{ rad s}^{-1} = \pi/2 \text{ rad s}^{-1} \text{ (or equivalently } \omega = 1.57 \text{ rad s}^{-1})$$

(d) The frequency is 4.0 Hz. Using Equation 9.2 the angular speed is

$$\omega = 2\pi f = 2\pi \times 4.0 \text{ rad s}^{-1} = 8\pi \text{ rad s}^{-1} \text{ (or equivalently } \omega = 25.1 \text{ rad s}^{-1}).$$

Note that in all these cases, the answer may be expressed in terms of π. This is often a convenient way of expressing an angular speed. For instance, in converting from an angular speed to a frequency, we have to divide the angular speed by 2π; this calculation is very straightforward if angular speed is expressed as a multiple of π.

QUESTION 9.7

This calculation is analogous to calculations about the brightness of stars at various distances, and similarly we have to remember that the flux density received diminishes as $1/d^2$, where d is the distance to the source. We need to compare the ratio of radio luminosity to d^2 for the transmitter and the pulsar.

$$\text{flux density from transmitter} = \frac{L_t}{4\pi d_t^2}$$

$$\text{flux density from pulsar} = \frac{L_p}{4\pi d_p^2}$$

Therefore

$$\frac{\text{flux density from transmitter}}{\text{flux density from pulsar}} = \frac{L_t}{L_p}\left(\frac{d_p}{d_t}\right)^2$$

$$= \frac{10^5 \text{ W}}{10^{20} \text{ W}}\left(\frac{10^4 \times 3 \times 10^{16} \text{ m}}{10^5 \text{ m}}\right)^2$$

$$= 10^{-15} \times 9 \times 10^{30} = 9 \times 10^{15}$$

$$\approx 10^{16}$$

Therefore the flux density from the radio transmitter is 10^{16} times stronger than that from the pulsar.

This highlights one of the problems of radio astronomy – the cosmic signals are weak and can easily be swamped by terrestrial radio signals if they stray onto the channels reserved for radio astronomy.

QUESTION 9.8

An accuracy of 1 part in 10^{14} means, for example, that the accuracy can be expressed as 1 second in 10^{14} seconds.

$$1 \text{ century} = 100 \text{ years} = 100 \times 3.2 \times 10^7 \text{ seconds} = 3.2 \times 10^9 \text{ s}$$

Therefore the number of seconds accuracy per century

$$= 3.2 \times 10^9/10^{14} \text{ s} = 3.2 \times 10^{-5} \text{ s} = 32 \text{ microseconds}$$

QUESTION 9.9

(a) The wavelength of the peak of emission of the black-body spectrum is given by Wien's displacement law (Equation 1.4). In this case $T = 5 \times 10^5$ K, so

$$(\lambda_{\text{peak}}/\text{m}) = \left(\frac{2.90 \times 10^{-3}}{5 \times 10^5} \right)$$

$$\lambda_{\text{peak}} = 5.80 \times 10^{-9} \text{ m}$$

The peak of the emission occurs at about 6×10^{-9} m, and according to Figure 1.36, this corresponds to a band between X-ray and ultraviolet regimes of the electromagnetic spectrum. This regime is sometimes referred to as a soft X-ray band or extreme-UV.

(b) The luminosity of the neutron star is given by Equation 3.9

$$L = 4\pi R^2 \sigma T^4$$

$$= 4\pi \, (1 \times 10^4 \text{ m})^2 \times (5.67 \times 10^{-8} \text{ W m}^{-2} \text{ K}^{-4}) \times (5 \times 10^5 \text{ K})^4$$

$$= 4.45 \times 10^{24} \text{ W}$$

In terms of the solar luminosity, this is

$$L/L_\odot = (4.45 \times 10^{24} \text{ W})/(3.84 \times 10^{26} \text{ W}) = 1.2 \times 10^{-2}$$

So the luminosity of the neutron star will be about 1% of the solar luminosity.

QUESTION 9.10

The magnitude of the escape speed is given by

$$v_{\text{esc}} = \sqrt{2GM/R}$$

The escape speed is equal to c at $R = R_S$, so

$$c = \sqrt{2GM/R_S}$$

$$c^2 = 2GM/R_S$$

Therefore

$$R_S = \frac{2GM}{c^2}$$

This is an important equation, which gives the Schwarzschild radius of a black hole of mass M. We have derived it without using relativity, so the derivation is somewhat artificial. However, it does give the right result, and it does illustrate some of the physical processes that we have to consider.

QUESTION 9.11

Using the relationship derived in Question 9.10,

$$R_S = \frac{2 \times (6.7 \times 10^{-11} \text{ N m}^2 \text{ kg}^{-2}) \times (2.0 \times 10^{30} \text{ kg})}{(3.0 \times 10^8 \text{ m s}^{-1})^2}$$

$$= 3.0 \times 10^3 \text{ N kg}^{-1} \text{ s}^2$$

$$= 3.0 \times 10^3 \text{ kg m s}^{-2} \text{ kg}^{-1} \text{s}^2 = 3.0 \times 10^3 \text{ m}$$

$$= 3.0 \text{ km}$$

QUESTION 9.12

(a) The answer is a diagram. See Figure 9.25.

In comparing the magnitudes of the two gravitational forces we shall work with the masses of the objects in $M_\odot$ and distances in $R_\odot$:

$$F_g = \frac{GMm}{R^2}$$

Gravitational force due to the main sequence star:

$$F_{MS} = \frac{G(0.70M_\odot)m_{blob}}{(0.96R_\odot)^2}$$

Gravitational force due to the white dwarf:

$$F_{WD} = \frac{G(1.20M_\odot)m_{blob}}{(2.18R_\odot - 0.96R_\odot)^2}$$

$$\frac{F_{WD}}{F_{MS}} = \frac{G(1.20M_\odot)m_{blob}}{(2.18R_\odot - 0.96R_\odot)^2} \times \frac{(0.96R_\odot)^2}{G(0.70M_\odot)m_{blob}}$$

$$= \left(\frac{1.20}{0.70}\right) \times \left(\frac{0.96}{2.18 - 0.96}\right)^2 = 1.06$$

and so F_{WD} is slightly larger than F_{MS}.

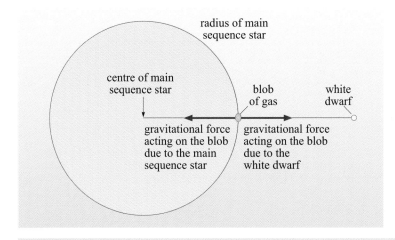

Figure 9.25 A diagram of the binary system described in Question 9.12, showing the direction of forces acting on the blob of gas under consideration.

QUESTION 9.13

Most of the energy released in a Type II supernova is in the form of neutrinos that accompany the collapse of the core of the star. This is about a factor of 100 greater than the energy released as kinetic energy of the ejecta and in electromagnetic radiation. Type Ia supernovae do not undergo core collapse, and there is no corresponding loss of energy by neutrinos. Hence the total energy released in a Type Ia supernova is only about 1% of the energy released in a Type II supernova.

QUESTION 9.14

If angular momentum is conserved, then $L_1 = L_2$, where the subscript 1 refers to the satellite before the astronaut arrives, and subscript 2 to the satellite–astronaut combination. Considering first the satellite alone:

$$L_1 = I_1 \omega_1$$

and $I_1 = 2500 \, \text{kg m}^2$. The satellite spins at 0.33 revolutions per second, therefore the angular speed is (Equation 9.2)

$$\omega_1 = 2\pi \times 0.33 \, \text{rad s}^{-1} = 0.66 \times \pi \, \text{rad s}^{-1}$$

(The factor π is left as a symbol, since later on, we will be required to divide by a factor 2π to get an answer in terms of the frequency.)

$$L_1 = (2500 \, \text{kg m}^2) \times (0.66 \times \pi \, \text{rad s}^{-1})$$
$$= 1.65 \times 10^3 \ \times \pi \, \text{kg m}^2 \, \text{rad s}^{-1}$$

The astronaut has a moment of inertia about the spin axis of $100 \, \text{kg} \times (1 \, \text{m})^2 = 100 \, \text{kg m}^2$, so the satellite–astronaut combination has a moment of inertia of $(2500 + 100) \, \text{kg m}^2$; therefore

$$I_2 = 2600 \, \text{kg m}^2$$

Now, since

$$L_2 = I_2 \, \omega_2$$

$$\omega_2 = \frac{L_2}{I_2}$$

But $\quad L_2 = L_1$ so

$$\omega_2 = \frac{L_1}{I_2}$$
$$= \frac{1.65 \times 10^3 \times \pi \, \text{kg m}^2 \, \text{rad s}^{-1}}{2600 \, \text{kg m}^2}$$
$$= 0.635 \times \pi \, \text{rad s}^{-1}$$

This can be converted from an angular speed to an angular frequency by rearranging Equation 9.2

$$f = \frac{\omega_2}{2\pi} = \frac{0.635 \times \pi}{2\pi} = 0.318 \, \text{s}^{-1}$$

So the frequency of rotation drops to 0.32 revolutions per second.

QUESTION 9.15

The power received from the pulsar is

$$10^{-19}\,\text{W m}^{-2} \times 1000\,\text{m}^2 = 1 \times 10^{-16}\,\text{W}$$

The change in gravitational potential energy when lifting the book is

$$\Delta E_g = mg\Delta h$$
$$= 1\,\text{kg} \times 10\,\text{m s}^{-2} \times 1\,\text{m}$$
$$= 10\,\text{J}$$

$$\text{power used} = \Delta E_g/\text{time taken} = 10\,\text{J}/1\,\text{s} = 10\,\text{W}$$

Therefore $\dfrac{\text{power used in lifting book}}{\text{power received from pulsar}} = \dfrac{10\,\text{W}}{1 \times 10^{-16}\,\text{W}} = 10^{17}$

It can be concluded that pulsar signals are weak!

QUESTION 9.16

The B star and the pulsar will be moving around each other. The optical astronomer will therefore expect there to be changing Doppler shifts in the observed frequencies of the emission lines from the B star. The optical astronomer will not see the pulsar.

The radio astronomer will see similar changes in the pulsar period: owing to Doppler shifts the pulse period will be shorter than average for part of the orbit, and for part of the orbit it will be longer. The radio astronomer will not see the B star.

Assuming these two stars were formed as a binary pair, then the star that is now the pulsar (the neutron star) must originally have been the more massive and be more rapidly evolving. To become a neutron star it must have passed through its main sequence stage and its supergiant stage and then exploded as a supernova. Its core became the pulsar. Meanwhile the other star was more slowly going through the main sequence stages, and is now presumably forming heavier elements in its core. There cannot at the moment be much transfer of material from the B star to the pulsar (because if there were, this material would blanket the pulsar and prevent the radio emission).

We are not given any information about the size of the binary system. It may be that the stars are sufficiently far apart that they have not significantly affected each other. Or it could be that the B star received material from its companion when it was a supergiant. It could also be that the mass transfer from the B star to the neutron star has not yet begun because the B star has yet to swell sufficiently to fill its Roche lobe. If mass transfer does start, the radio emission will cease and X-ray emission begin. The rotation rate of the neutron star may be increased by the transferred material.

If there is a lot of mass transferred from the B star its evolution could be seriously affected, e.g. it might lose all its outer envelope. If the mass transfer is limited the B star will probably still pass through the supernova stage, perhaps with its core becoming another neutron star.

APPENDICES
APPENDIX A1 USEFUL QUANTITIES

Quantity	Symbol	Value[a]
Physical constants		
speed of light in a vacuum	c	$3.00 \times 10^8 \, \text{m s}^{-1}$
Planck constant	h	$6.63 \times 10^{-34} \, \text{J s}$
Boltzmann constant	k	$1.38 \times 10^{-23} \, \text{J K}^{-1}$
gravitational constant	G	$6.67 \times 10^{-11} \, \text{N m}^2 \text{kg}^{-2}$
Stefan–Boltzmann constant	σ	$5.67 \times 10^{-8} \, \text{W m}^{-2} \text{K}^{-4}$
charge of electron	$-e$	$-1.60 \times 10^{-19} \, \text{C}$
mass of hydrogen atom	m_{H}	$1.67 \times 10^{-27} \, \text{kg}$
mass of electron	m_{e}	$9.11 \times 10^{-31} \, \text{kg}$
Astronomical constants		
mass of the Earth	M_{E}	$5.98 \times 10^{24} \, \text{kg}$
radius (equatorial) of the Earth	R_{E}	$6.38 \times 10^6 \, \text{m}$
mass of the Sun	$M_{\odot}$	$1.99 \times 10^{30} \, \text{kg}$
radius of the Sun	$R_{\odot}$	$6.96 \times 10^8 \, \text{m}$
photospheric temperature of the Sun	$T_{\odot}$	5770K
luminosity of the Sun	$L_{\odot}$	$3.84 \times 10^{26} \, \text{W}$

[a]Values are given to 3 significant figures. Many of these are known more accurately.

Units

Quantity	SI Unit	Other units	In SI	Alternative SI units
length	metre, m	Astronomical unit, AU	$1.50 \times 10^{11} \text{m}$	
		parsec, pc	$3.09 \times 10^{16} \text{m}$	
time	second, s	year, yr	$3.16 \times 10^7 \text{s}$	
frequency	hertz, Hz			s^{-1}
force	newton, N			kg m s^{-2}
pressure	pascal, Pa			$\text{kg m}^{-1}\text{s}^{-2}, \text{N m}^{-2}$
temperature	kelvin, K	°C	$(\text{kelvin} - 273)$	
energy	joule, J	electronvolt, eV	$1.60 \times 10^{-19} \text{J}$	$\text{kg m}^2\text{s}^{-2}$
power	watt, W			$\text{kg m}^2\text{s}^{-3}, \text{J s}^{-1}$
angle	radian, rad	degree, °	$1/57.3 \, \text{rad}$	
		$1° = 60 \, \text{arcmin} = 3600 \, \text{arcsec}$		
		arcsec, ″	$1/206265 \, \text{rad}$	

APPENDIX A2
STELLAR NOMENCLATURE

The brightest stars in a constellation are referred to by Greek letters (usually, but not always) in order of brightness followed by a constellation designation shortened to three letters. For example, Betelgeuse in the constellation of Orion is referred to as α Ori (alpha Orionis).

The Greek alphabet

Name	Pronounced	Lower case	Upper case	Name	Pronounced	Lower case	Upper case
alpha		α	A	nu	new	ν	N
beta	bee-ta	β	B	xi	cs-eye	ξ	Ξ
gamma		γ	Γ	omicron		o	O
delta		δ	Δ	pi	pie	π	Π
epsilon		ε	E	rho	roe	ρ	P
zeta	zee-ta	ζ	Z	sigma		σ	Σ
eta	ee-ta	η	H	tau		τ	T
theta	thee-ta (th as in theatre)	θ	Θ	upsilon		υ	Y
iota	eye-owe-ta	ι	I	phi	fie	φ	Φ
kappa		κ	K	chi	kie	χ	X
lambda	lam-da	λ	Λ	psi	ps-eye	ψ	Ψ
mu	mew	μ	M	omega	owe-me-ga	ω	Ω

The constellations

No.	Abbr.	Name	Genitive	English name	Region[a]
1	And	Andromeda	Andromedae	Andromeda	N
2	Ant	Antlia	Antliae	Air Pump	S
3	Aps	Apus	Apodis	Bird of Paradise	S
4	Aqr	Aquarius	Aquarii	Water Carrier	Eq
5	Aql	Aquila	Aquilae	Eagle	Eq
6	Ara	Ara	Arae	Alter	S
7	Ari	Aries	Arietis	Ram	E
8	Aur	Auriga	Aurigae	Charioteer	N
9	Boo	Bootes	Bootis	Herdsman	NEq
10	Cae	Caelum	Caeli	Chisel	S
11	Cam	Camelopardalis[b]	Camelopardalis[b]	Giraffe	N

No.	Abbr.	Name	Genitive	English name	Region[a]
12	Cnc	Cancer	Cancri	Crab	Eq
13	CVn	Canes Venatici	Canes Venaticorum	Hunting Dogs	N
14	CMa	Canis Major	Canis Majoris	Great Dog	Eq
15	CMi	Canis Minor	Canis Minoris	Small Dog	Eq
16	Cap	Capricornus	Capricorni	Goat (Capricorn)	Eq
17	Car	Carina[c]	Carinae	Keel	S
18	Cas	Cassiopeia	Cassiopeiae	Cassiopeia	N
19	Cen	Centaurus	Centauri	Centaur	S
20	Cep	Cepheus	Cephei	Cepheus	N
21	Cet	Cetus	Ceti	Whale	Eq
22	Cha	Chameleon	Chameleontis	Chamelion	S
23	Cir	Circinus	Circini	Compasses	S
24	Col	Columba	Columbae	Dove	S
25	Com	Coma Berenices	Comae Berenices	Berenice's Hair	Eq
26	CrA	Corona Australis	Coronae Australis	Southern Crown	S
27	CrB	Corona Borealis	Coronae Borealis	Northern Crown	N
28	Crv	Corvus	Corvi	Crow	Eq
29	Crt	Crater	Crateris	Cup	Eq
30	Cru	Crux	Crucis	Southern Cross	S
31	Cyg	Cygnus	Cygni	Swan	N
32	Del	Delphinus	Delphini	Dolphin	Eq
33	Dor	Dorado	Doradus	Goldfish	S
34	Dra	Draco	Draconis	Dragon	N
35	Equ	Equuleus	Equulei	Small Horse	Eq
36	Eri	Eridanus	Eridani	River	EqS
37	For	Fornax	Fornacis	Furnace	EqS
38	Gem	Gemini	Geminorum	Twins	Eq
39	Gru	Grus	Gruis	Crane	S
40	Her	Hercules	Herculis	Hercules	NEq
41	Hor	Horologium	Horologii	Clock	S
42	Hya	Hydra	Hydrae	Water Monster	Eq
43	Hyi	Hydrus	Hydri	Sea Serpent	S

No.	Abbr.	Name	Genitive	English name	Region[a]
44	Ind	Indus	Indi	Indian	S
45	Lac	Lacerta	Lacertae	Lizard	N
46	Leo	Leo	Leonis	Lion	Eq
47	LMi	Leo Minor	Leonis Minoris	Small Lion	N
48	Lep	Lepus	Leporis	Hare	Eq
49	Lib	Libra	Librae	Scales	Eq
50	Lup	Lupus	Lupi	Wolf	S
51	Lyn	Lynx	Lyncis	Lynx	N
52	Lyr	Lyra	Lyrae	Lyre	N
53	Men	Mensa	Mensae	Table (mountain)	S
54	Mic	Microscopium	Microscopii	Microscope	S
55	Mon	Monoceros	Monocerotis	Unicorn	Eq
56	Mus	Musca	Muscae	Fly	S
57	Nor	Norma	Normae	Square	S
58	Oct	Octans	Octantis	Octant	S
60	Ori	Orion	Orionis	Orion (Hunter)	Eq
61	Pav	Pavo	Pavonis	Peacock	S
62	Peg	Pegasus	Pegasi	Winged Horse	Eq
63	Per	Perseus	Persei	Perseus	N
64	Phe	Phoenix	Phoenicis	Phoenix	S
65	Pic	Pictor	Pictoris	Easel	S
66	Psc	Pisces	Piscium	Fishes	Eq
67	PsA	Pisces Austrinus	Pisces Austrini	Southern Fish	EqS
68	Pup	Puppis[c]	Puppis	Stern	S
69	Pyx	Pyxis	Pyxidis	Compass	EqS
70	Ret	Reticulum	Reticuli	Net	S
71	Sge	Sagitta	Sagittae	Arrow	Eq
72	Sgr	Sagittarius	Sagittarii	Archer	EqS
73	Sco	Scorpius	Scorpii	Scorpion	EqS
74	Scl	Sculptor	Sculptoris	Sculptor	EqS
75	Sct	Scutum	Scuti	Shield	Eq

No.	Abbr.	Name	Genitive	English name	Region[a]
76	Ser	Serpens[d]	Serpentis	Serpent[d]	Eq
77	Sex	Sextans	Sextantis	Sextant	Eq
78	Tau	Taurus	Tauri	Bull	Eq
79	Tel	Telescopium	Telescopii	Telescope	S
80	Tri	Triangulum	Trianguli	Triangle	N
81	TrA	Triangulum Australe	Trianguli Australis	Southern Triangle	S
82	Tuc	Toucana	Toucanae	Toucan	S
83	UMa	Ursa Major	Ursae Majoris	Great Bear	N
84	UMi	Ursa Minor	Ursae Minoris	Little Bear	N
85	Vel	Vela[c]	Velorum	Sails	S
86	Vir	Virgo	Virginis	Virgin	Eq
87	Vol	Volans	Volantis	Flying Fish	S
88	Vul	Vulpecula	Vulpeculae	Fox	Eq

[a]N = Northern sky, Eq = Equatorial sky, S = Southern sky.

[b]Also known as Camelopardus (Camelopardi).

[c]Carina, Vela and Puppis were originally one constellation, Argo.

[d]In two separate parts: Serpens Caput (Head) and Serpens Cauda (Tail).

APPENDIX A3
THE 100 CLOSEST STARS TO THE SUN

The data are taken from the RECONS survey of objects beyond the Solar System within 10 parsecs and are correct at mid July 2002. The positions of objects in the list may change as more accurate measurements are made or faint objects are discovered. Most of the stars are so faint they do not have names and are known by catalogue numbers or their positions. The missing numbers in the first column indicate stars in binary or multiple systems.

	Name	HIP^a	RA^b/degrees	Dec^b/degrees	V^c	$M_V{}^d$	Sp type^e	d/pc^f
1	α Cen C, Proxima	70890	217.43	−62.68	11.09	15.53	M5.5 V	1.295 ± 0.004
	α Cen A, Rigel Kent	71683	219.90	−60.83	0.01	4.38	G2 V	1.338 ± 0.001
	α Cen B, Rigel Kent	71681	219.90	−60.84	1.34	5.71	K0 V	1.338 ± 0.001
4	Barnard's star	87937	269.45	4.67	9.53	13.22	M4 V	1.828 ± 0.003
5	Wolf 359		164.12	7.02	13.44	16.55	M6 V	2.386 ± 0.012
6	Lalande 21185	54035	165.84	35.98	7.47	10.44	M2 V	2.542 ± 0.005
7	α CMa A, Sirius A	32349	101.29	−16.71	−1.43	1.47	A1 V	2.631 ± 0.009
	α CMa B, Sirius B		101.29	−16.71	8.44	11.34	WD	2.631 ± 0.009
9	−^{g}		24.76	−17.95	12.54	15.40	M5.5 V	2.68 ± 0.02
	−^{g}		24.76	−17.95	12.99	15.85	M6 V	2.68 ± 0.02
11	−^{g}	92403	282.45	−23.85	10.43	13.07	M3.5 V	2.968 ± 0.016
12	−^{g}		355.48	44.18	12.29	14.79	M5.5 V	3.165 ± 0.011
13	ε Eri	16537	53.24	−9.46	3.73	6.18	K2 V	3.226 ± 0.008
14	−^{g}	114046	346.45	−35.86	7.34	9.75	M1.5 V	3.293 ± 0.009
15	−^{g}	57548	176.93	0.81	11.13	13.51	M4 V	3.348 ± 0.015
16	−^{g}		339.64	−15.30	13.33	15.64	M5 V	3.45 ± 0.05
	−^{g}		339.64	−15.30	13.27	15.58		3.45 ± 0.05
	−^{g}		339.64	−15.30	14.03	16.34		3.45 ± 0.05
19	α CMi A, Procyon A	37279	114.83	5.23	0.38	2.66	F5 IV–V	3.496 ± 0.010
	α CMi B, Procyon B		114.83	5.23	10.70	12.98	WD	3.496 ± 0.010
21	61 Cyg A	104214	316.71	38.74	5.21	7.49	K5 V	3.496 ± 0.007
	61 Cyg B	104217	316.71	38.74	6.03	8.31	K7 V	3.496 ± 0.007
23	−^{g}	91768	280.70	59.63	8.90	11.16	M3 V	3.53 ± 0.02
	−^{g}	91772	280.70	59.63	9.69	11.95	M3.5 V	3.53 ± 0.02

	Name	HIP[a]	RA[b]/degrees	Dec[b]/degrees	V^c	M_V^d	Sp type[e]	d/pc[f]
25	–[g]	1475	4.59	44.02	8.08	10.32	M1.5 V	3.563 ± 0.012
	–[g]		4.59	44.02	11.06	13.30	M3.5 V	3.563 ± 0.012
27	ε Ind	108870	330.82	−56.78	4.69	6.89	K5 V	3.625 ± 0.009
28	–[g]		127.46	26.78	14.78	16.98	M6.5 V	3.63 ± 0.04
29	τ Cet	8102	26.02	−15.94	3.49	5.68	G8 V	3.644 ± 0.010
30	–[g]		54.00	−44.51	13.03	15.21	M5.5 V	3.68 ± 0.02
31	–[g]	5643	18.13	−17.00	12.02	14.17	M4.5 V	3.72 ± 0.04
32	Luyten's star	36208	111.85	5.24	9.86	11.97	M3.5 V	3.79 ± 0.02
33	Kapteyn's star	24186	77.90	−45.00	8.84	10.87	M1.5 V	3.917 ± 0.013
34	–[g]	105090	319.32	−38.87	6.67	8.69	M0 V	3.95 ± 0.02
35	Kruger 60	110893	337.00	57.70	9.79	11.76	M3 V	4.03 ± 0.02
	–[g]		337.00	57.70	11.41	13.38	M4 V	4.03 ± 0.02
37	–[g]	30920	97.35	−2.81	11.15	13.09	M4.5 V	4.09 ± 0.03
	–[g]		97.35	−2.81	14.23	16.17		4.09 ± 0.03
39	–[g]	80824	247.58	−12.66	10.07	11.93	M3 V	4.24 ± 0.03
40	Van Maanen 2	3829	12.29	5.40	12.38	14.21	WD	4.31 ± 0.03
41	–[g]	439	1.34	−37.35	8.55	10.35	M3 V	4.36 ± 0.02
42	–[g]		188.32	9.02	13.18	14.97	M5.5 V	4.39 ± 0.09
	–[g]		188.32	9.02	13.17	14.96		4.39 ± 0.09
44	–[g]		30.06	13.05	12.27	14.03	M4.5 V	4.45 ± 0.06
45	–[g]		161.13	−61.19	13.92	15.66	M5.5 V	4.5 ± 0.2
46	–[g]	86162	264.11	68.34	9.17	10.89	M3 V	4.54 ± 0.02
47	–[g]		162.05	−11.34	15.60	17.32	M6.5 V	4.54 ± 0.07
48	–[g]	85523	262.17	−46.89	9.38	11.09	M3 V	4.54 ± 0.03
49	–[g]		298.48	44.42	13.46	15.17	M5.5 V	4.54 ± 0.02
	–[g]		298.48	44.42	14.01	15.72	M6 V	4.54 ± 0.02
	–[g]		298.48	44.42	16.75	18.46		4.54 ± 0.02
52	–[g]	57367	176.43	−64.84	11.5	13.18	WD	4.62 ± 0.04
53	–[g]		1.68	−7.54	13.76	15.40	M5.5 V	4.69 ± 0.08
54	–[g]	113020	343.32	−14.26	10.17	11.81	M3.5 V	4.70 ± 0.04
	–[g]		343.32	−14.26			planet?	4.70 ± 0.04

	Name	HIP[a]	RA[b]/degrees	Dec[b]/degrees	V[c]	M$_V$[d]	Sp type[e]	d/pc[f]
56	_[g]	54211	166.38	43.52	8.77	10.34	M1 V	4.85 ± 0.03
	_[g]		166.38	43.52	14.48	16.05	M5.5 V	4.85 ± 0.03
58	_[g]	49908	152.85	49.46	6.59	8.16	K7 V	4.86 ± 0.02
59	_[g]		154.90	19.87	9.32	10.87	M3 V	4.89 ± 0.07
60	_[g]	106440	323.39	−49.01	8.66	10.20	M3 V	4.93 ± 0.03
61	_[g]		54.90	−35.43	18.50	20.02	M9 V	4.97 ± 0.10
62	_[g]	86214	264.27	−44.32	10.95	12.45	M4.5 V	5.01 ± 0.06
63	_[g]	19849	63.82	−7.65	4.43	5.92	K1 V	5.03 ± 0.02
	_[g]		63.82	−7.65	9.52	11.01	WD	5.03 ± 0.02
	_[g]		63.82	−7.65	11.19	12.68	M4.5 V	5.03 ± 0.02
66	_[g]	112460	341.71	44.34	10.22	11.70	M3.5 V	5.05 ± 0.04
67	70 Oph	88601	271.36	2.50	4.20	5.66	K0 V	5.10 ± 0.02
	_[g]		271.36	2.50	6.05	7.51	K5 V	5.10 ± 0.02
69	α Aql	97649	297.70	8.87	0.77	2.22	A7 IV–V	5.13 ± 0.02
70	_[g]		134.56	19.76	14.06	15.47	M5.5 V	5.23 ± 0.07
	_[g]		134.56	19.76	14.92	16.33		5.23 ± 0.07
72	_[g]		90.02	2.71	11.33	12.68	M3.5 V	5.37 ± 0.3
73	_[g]	57544	176.91	78.69	10.79	12.14	M3.5 V	5.38 ± 0.04
74	_[g]	1242	3.87	−16.13	11.58	12.93	M4 V	5.38 ± 0.09
	_[g]		3.87	−16.13	14.33	15.68		5.38 ± 0.09
76	_[g]	67155	206.43	14.90	8.46	9.79	M1.5 V	5.43 ± 0.03
	_[g]		75.49	−6.95	12.15	13.46	M3.5 V	5.47 ± 0.10
78	_[g]	103039	313.14	−16.98	11.41	12.71		5.49 ± 0.11
79	_[g]	21088	67.79	58.98	11.04	12.32	M4 V	5.54 ± 0.02
	_[g]		67.79	58.98	12.44	13.72	WD	5.54 ± 0.02
81	_[g]	33226	103.71	33.27	10.02	11.29	M3 V	5.57 ± 0.05
82	_[g]		290.20	−45.56	12.23	13.45	M4.5 V	5.71 ± 0.3
83	_[g]	25878	82.86	−3.67	7.95	9.17	M1.5 V	5.71 ± 0.03
84	σ Dra	96100	293.09	69.67	4.68	5.88	K0 V	5.761 ± 0.014
85	_[g]	29295	92.65	−21.86	8.12	9.31	M1 V	5.77 ± 0.04

	Name	HIP[a]	RA[b]/degrees	Dec[b]/degrees	V[c]	M_V[d]	Sp type[e]	d/pc[f]
	–[g]		92.65	−21.86			BD	5.77 ± 0.04
87	–[g]	86990	266.65	−57.32	10.75	11.92	M4 V	5.82 ± 0.07
88	–[g]	94761	289.23	5.17	9.11	10.28	M3 V	5.85 ± 0.02
	–[g]		289.23	5.17	17.50	18.67	M8 V	5.85 ± 0.02
90	–[g]	26857	85.53	12.49	11.51	12.67	M4 V	5.87 ± 0.11
	–[g]		123.17	−21.55	12.07	13.22	M3.5 V	5.89 ± 0.5
92	–[g]	73184	224.36	−21.41	5.75	6.90	K5 V	5.89 ± 0.03
	–[g]		224.36	−21.41	8.28	9.43	M1 V	5.89 ± 0.03
	–[g]		224.36	−21.41	10.05	11.20		5.89 ± 0.03
	–[g]		224.36	−21.41			BD	5.89 ± 0.03
96	–[g]	117473	357.30	2.40	8.99	10.12	M1 V	5.93 ± 0.05
97	η Cas	3821	12.27	57.82	3.45	4.58	G3 V	5.94 ± 0.02
	–[g]		12.27	57.82	7.51	8.64	K7 V	5.94 ± 0.02
99	–[g]	76074	233.06	−41.27	9.31	10.44	M3 V	5.94 ± 0.05
100	–[g]		116.17	3.55	11.19	12.31	M4 V	5.97 ± 0.08

[a]Hipparcos catalogue number for those stars detected by Hipparcos.

[b]Coordinate system fixed with respect to the stars. Right Ascension and Declination at the beginning of the year 2000. RA usually has units of hours, minutes and seconds but degrees are often used in computer databases.

[c]Apparent visual magnitude (Section 3.3.3).

[d]Absolute visual magnitude (Section 3.3.3).

[e]Spectral type (Section 3.3.2): WD = White dwarf, BD = Brown dwarf.

[f]Distance from the Sun in parsecs (Section 3.2.2).

[g]Unnamed star.

Data adapted from RECONS: Research Consortium on Nearby Stars
http://www.chara.gsu.edu/~thenry/RECONS/
and ESA, The Hipparcos Space Astrometry Mission
http://astro.estec.esa.nl/Hipparcos

Appendix A4
The 100 brightest stars visible from Earth

This list gives the 100 'stars' with the lowest apparent visual magnitudes as measured by Hipparcos. In many cases these stars are binary or multiple systems which cannot be separated with the unaided eye. Often one star is much brighter than the other(s) so the apparent magnitude of this star is indistinguishable from that of the combined system and the spectrum is dominated by the brightest component. The binary designation is only shown when the stars have been detected individually by Hipparcos. The values of apparent visual magnitude V, (and hence position in the list), absolute visual magnitude M_V and distance as measured by Hipparcos differ in detail from those in other catalogues and Appendix A3. These differences illustrate the uncertainties in measurements obtained from different sources. In particular, if the distance is highly uncertain then the absolute magnitude derived using it may differ significantly from that expected for the particular spectral type. No account has been taken of interstellar absorption in calculation of the absolute magnitudes.

	Name	HIP[a]	RA[b]/degrees	Dec[b]/degrees	Bin[c]	V[d]	M_V[e]	Sp type[f]	d/pc[g]
1	α CMa Sirius	32349	101.29	−16.71	1	−1.44	1.45	A1 V	2.64 ± 0.01
2	α Car Canopus	30438	95.99	−52.70		−0.62	−5.53	F0 II	96 ± 5
3	α Boo Arcturus	69673	213.92	19.19		−0.05	−0.31	K1.5 III	11.26 ± 0.09
4	α Cen A Rigel Kent	71683	219.92	−60.84	1	−0.01	4.34	G2 V	1.347 ± 0.003
5	α Lyr Vega	91262	279.23	38.78	1	0.03	0.58	A0 V	7.76 ± 0.03
6	α Aur Capella	24608	79.17	46.00	1+2	0.08	−0.48	G5 III + G0 III	12.9 ± 0.2
7	β Ori Rigel	24436	78.63	−8.20	1	0.18	−6.69	B8 I	237 ± 50
8	α CMi Procyon	37279	114.83	5.23	1	0.40	2.68	F5 IV–V	3.50 ± 0.01
9	α Ori Betelgeuse	27989	88.79	7.41	1 *	0.45	−5.14	M2 I	131 ± 30
10	α Eri Achernar	7588	24.43	−57.24		0.45	−2.77	B3 V	44 ± 1
11	β Cen Hadar	68702	210.96	−60.37	1	0.61	−5.42	B1 III	161 ± 14
12	α Aql Altair	97649	297.70	8.87	1	0.76	2.20	A7 V	5.14 ± 0.03
13	α Cru Acrux	60718	186.65	−63.10	1+2	0.77	−4.19	B0.5 IV + B1 V	98 ± 7
14	α Tau Aldebaran	21421	68.98	16.51	1 *	0.87	−0.63	K5 III	20 ± 0.4
15	α Vir Spica	65474	201.30	−11.16	1	0.98	−3.55	B1 III	80 ± 6
16	α Sco Antares	80763	247.35	−26.43	1+2 *	1.06	−5.28	M1.5 I + B4 V	185 ± 60
17	β Gem Pollux	37826	116.33	28.03	1	1.16	1.09	K0 III	10.34 ± 0.09
18	α PsA Fomalhaut	113368	344.41	−29.62		1.17	1.74	A3 V	7.69 ± 0.05
19	α Cyg Deneb	102098	310.36	45.28	1	1.25	−8.73	A2 I	990 ± 560
20	β Cru Mimosa	62434	191.93	−59.69	1	1.25	−3.92	B0.5 III	108 ± 7
21	α Cen B Rigel Kent	71681	219.91	−60.84	2	1.35	5.70	K1 V	1.347 ± 0.003

	Name	HIP[a]	RA[b]/degrees	Dec[b]/degrees	Bin[c]	V[d]	$M_V^{\,e}$	Sp type[f]	d/pc[g]
22	α Leo Regulus	49669	152.09	11.97	1	1.36	−0.52	B7 V	23.8 ± 0.5
23	ε CMa Adhara	33579	104.66	−28.97	1	1.50	−4.10	B2 II	132 ± 10
24	α Gem Castor	36850	113.65	31.89	1+2	1.58	0.59	A1 V + A2 V	15.8 ± 0.3
25	γ Cru Gacrux	61084	187.79	−57.11	1	1.59	−0.56	M3.5 III	27.0 ± 0.5
26	λ Sco Shaula	85927	263.40	−37.10	1	1.62	−5.05	B2 IV	216 ± 40
27	γ Ori Bellatrix	25336	81.28	6.35	1	1.64	−2.72	B2 III	75 ± 5
28	β Tau Alnath	25428	81.57	28.61	1	1.65	−1.37	B7 III	40 ± 1
29	β Car Miaplacidus	45238	138.30	−69.72		1.67	−0.99	A2 IV	34.1 ± 0.6
30	ε Ori Alnilam	26311	84.05	−1.20	1	1.69	−6.38	B0 I	410 ± 150
31	α Gru Alnair	109268	332.06	−46.96	1	1.73	−0.73	B7 IV	31.1 ± 0.8
32	ζ Ori Alnitak	26727	85.19	−1.94	1	1.74	−5.26	O9 I	251 ± 50
33	γ Vel Regor	39953	122.38	−47.34	1+2 *	1.75	−5.31	W + O7.5	258 ± 35
34	ε UMa Alioth	62956	193.51	55.96		1.76	−0.21	A0	24.8 ± 0.4
35	ε Sgr Kaus Australis	90185	276.04	−34.38	1	1.79	−1.44	B9.5 III	44 ± 2
36	α Per Mirphak	15863	51.08	49.86	1	1.79	−4.50	F5 I	181 ± 20
37	α UMa Dubhe	54061	165.93	61.75	1	1.81	−1.08	K0 III	37.9 ± 0.7
38	δ CMa Al Wazor	34444	107.10	−26.39		1.83	−6.87	F8 I	550 ± 170
39	η UMa Alkaid	67301	206.89	49.31		1.85	−0.60	B3 V	30.9 ± 0.7
40	θ Sco Sargas	86228	264.33	−43.00		1.86	−2.75	F1 II	83 ± 6
41	ε Car She	41037	125.63	−59.51	1+2	1.86	−4.58	K3 III + B2 V	194 ± 20
42	β Aur Menkalinam	28360	89.88	44.95	1	1.90	−0.10	A2 IV	25.2 ± 0.5
43	α Tr Ras Al Muthallath	82273	252.17	−69.03		1.91	−3.62	K2 II–III	127 ± 10
44	δ Vel Koo She	42913	131.18	−54.71	1	1.93	−0.01	A1 V	24.5 ± 0.2
45	γ Gem Alhena	31681	99.43	16.40	1	1.93	−0.60	A0 IV	32 ± 2
46	α Pav Joo Tseo	100751	306.41	−56.74	1	1.94	−1.81	B2 IV	56 ± 2
47	α UMi Polaris	11767	37.95	89.26	1 *	1.97	−3.64	F7 I	132 ± 8
48	β CMa Murzim	30324	95.68	−17.96	1	1.98	−3.95	B1 II–III	153 ± 15
49	α Hya Alphard	46390	141.90	−8.66	1	1.99	−1.69	K3II-III	54 ± 2
50	γ Leo Algieba	50583	154.99	19.84	1+2	2.01	−0.92	K1 III +G7 III	38.5 ± 1.2
51	α Ari Hamal	9884	31.79	23.46		2.01	0.48	K2 III	20.2 ± 0.4
52	β Cet Diphda	3419	10.90	−17.99		2.04	−0.30	G9.5 III	29.4 ± 0.7
53	σ Sgr Nunki	92855	283.82	−26.30	1	2.05	−2.14	B2.5 V	69 ± 4

	Name	HIP[a]	RA[b]/degrees	Dec[b]/degrees	Bin[c]	V[d]	M_V[e]	Sp type[f]	d/pc[g]
54	θ Cen Haratan	68933	211.67	−36.37	1	2.06	0.70	K0 III	18.7 ± 0.3
55	κ Ori Saiph	27366	86.94	−9.67		2.07	−4.65	B0.5 I	221 ± 40
56	α And (δ Peg) Alpheratz	677	2.10	29.09	1	2.07	−0.30	B8 IV	29.8 ± 0.7
57	β Gru Al Dhanab	112122	340.67	−46.89	*	2.07	−1.52	M5 III	52 ± 2
58	β And Mirach	5447	17.43	35.62	1	2.07	−1.86	M0 III	61 ± 3
59	β UMi Kocab	72607	222.68	74.16	1	2.07	−0.87	K4 III	38.8 ± 0.8
60	α Oph Rasalhague	86032	263.73	12.56		2.08	1.30	A5 III	14.3 ± 0.2
61	β Per Algol	14576	47.04	40.96	*	2.09	−0.18	B8 V	28.5 ± 0.7
62	γ And A Alamach	9640	30.98	42.33	1	2.10	−3.08	K3 II	109 ± 9
63	β Leo Denebola	57632	177.27	14.57	1	2.14	1.92	A3 V	11.1 ± 0.1
64	γ Cas Cih	4427	14.18	60.72	1 *	2.15	−4.22	B0 IV	188 ± 20
65	γ Cen Koo Low	61932	190.38	−48.96	1	2.20	−0.81	A1 IV	40 ± 2
66	ζ Pup Suhail Hadar	39429	120.90	−40.00		2.21	−5.95	05	429 ± 90
67	ι Car Tureis	45556	139.27	−59.28		2.21	−4.42	A8 I	212 ± 21
68	α CrB Alphekka	76267	233.67	26.72	1+2 *	2.22	0.42	A0 V + G5 V	22.9 ± 0.4
69	λ Vel Suhail	44816	137.00	−43.43	1	2.23	−3.99	K4.5 I-II	175 ± 16
70	γ Cyg Sadr	100453	305.56	40.26	1	2.23	−6.12	F8 I	470 ± 110
71	ζ UMa Mizar	65378	200.98	54.93	1	2.23	0.33	A1 V	24.0 ± 0.4
72	α Cas Shedir	3179	10.13	56.54	1	2.24	−1.99	K0 III	70 ± 3
73	γ Dra Etamin	87833	269.15	51.49	1	2.24	−1.04	K5 III	45.2 ± 1.0
74	δ Ori Mintaka	25930	83.00	−0.30	1	2.25	−4.99	O9.5 II	281 ± 65
75	β Cas Caph	746	2.29	59.15	1	2.28	1.17	F2 III-IV	16.7 ± 0.2
76	δ Sco Dschubba	78401	240.08	−22.62		2.29	−3.16	B0.3 IV	123 ± 13
77	ε Sco Wei	82396	252.54	−34.29		2.29	0.78	K2.5 III	20.1 ± 0.3
78	ε Cen	66657	204.97	−53.47	1	2.29	−3.02	B1 III	115 ± 10
79	α Lup	71860	220.48	−47.39	1	2.30	−3.83	B1.5 III	168 ± 20
80	η Cen	71352	218.88	−42.16	*	2.33	−2.55	B1.5 V	95 ± 7
81	β UMa Merak	53910	165.46	56.38		2.34	0.41	A1 V	24.3 ± 0.4
82	ε Boo Izar	72105	221.25	27.07	1+2	2.35	−1.69	K0 II-III + A0 V	64 ± 3
83	ε Peg Enif	107315	326.05	9.88	*	2.38	−4.19	K2 I	206 ± 35
84	κ Sco	86670	265.62	−39.03		2.39	−3.38	B1.5 III	142 ± 15

	Name	HIP[a]	RA[b]/degrees	Dec[b]/degrees	Bin[c]	V[d]	M_V[e]	Sp type[f]	d/pc[g]
85	α Phe Ankaa	2081	6.57	−42.31		2.40	0.52	K0 III	23.7 ± 0.5
86	γ UMa Phad	58001	178.46	53.70		2.41	0.36	A0 V	25.6 ± 0.4
87	η Oph	84012	257.59	−15.73	1	2.43	0.37	A0 V	25.8 ± 0.6
88	β Peg Scheat	113881	345.94	28.08	1 *	2.44	−1.49	M2.5 II-III	61 ± 3
89	α Cep Alderamin	105199	319.64	62.59	1	2.45	1.58	A7 V	15.0 ± 0.1
90	η CMa Aludra	35904	111.02	−29.30	1	2.45	−7.51	B5 I	980 ± 550
91	κ Vel Cih	45941	140.53	−55.01		2.47	−3.62	B2 IV-V	165 ± 13
92	ε Cyg	102488	311.55	33.97	1	2.48	0.76	K0 III	22.1 ± 0.3
93	α Peg Markab	113963	346.19	15.21		2.49	−0.67	B9 V	42.8 ± 1.4
94	ζ Oph	81377	249.29	−10.57		2.54	−3.20	O9.5 V	140 ± 14
95	α Cet Menkar	14135	45.57	4.09		2.54	−1.61	M1.5 III	67 ± 4
96	ζ Cen	68002	208.89	−47.29		2.55	−2.81	B2.5 IV	118 ± 10
97	β Sco A	78820	241.36	−19.81	1	2.56	−3.50	B1 V	163 ± 30
98	δ Leo	54872	168.53	20.52	1	2.56	1.32	A4 V	17.7 ± 0.3
99	δ Cen	59196	182.09	−50.72		2.58	−2.84	B2 IV	121 ± 11
100	α Lep Arneb	25985	83.18	−17.82	1	2.58	−5.40	F0 I	390 ± 110

[a]Hipparcos catalogue number.

[b]Coordinate system fixed with respect to the stars. Right Ascension and Declination at the beginning of the year 2000.

[c]Binary or Variable star: 1 = data for brightest component only or brightest component dominates total radiation;
2 = data for fainter component only; 1+2 = data for combined star; * = variable.

[d]Apparent visual magnitude (Section 3.3.3).

[e]Absolute visual magnitude (Section 3.3.3).

[f]Spectral type (Section 3.3.2): W = Wolf–Rayet star.

[g]Distance from the Sun in parsecs.

Data adapted from ESA, The Hipparcos Space Astrometry Mission
http://astro.estec.esa.nl/Hipparcos

APPENDIX A5
THE CHEMICAL ELEMENTS AND THEIR ABUNDANCES

The relative abundance for hydrogen is arbitrarily set at 10^{12}. The relative atomic mass, A_r, is the average mass of the atoms of the element as it occurs on Earth. It is thus an average over all the isotopes of the element. The scale is fixed by giving the carbon isotope $^{12}_{6}C$ a relative atomic mass of 12.0.

Atomic number, Z	Name	Chemical symbol	Relative atomic mass, A_r	Solar System abundance by number	by mass
1	hydrogen	H	1.01	1.0×10^{12}	1.0×10^{12}
2	helium	He	4.00	9.8×10^{10}	3.9×10^{11}
3	lithium	Li	6.94	2.0×10^{3}	1.4×10^{4}
4	beryllium	Be	9.01	26	2.4×10^{2}
5	boron	B	10.81	6.3×10^{2}	6.8×10^{3}
6	carbon	C	12.01	3.6×10^{8}	4.4×10^{9}
7	nitrogen	N	14.01	1.1×10^{8}	1.6×10^{9}
8	oxygen	O	16.00	8.5×10^{8}	1.4×10^{10}
9	fluorine	F	19.00	3.0×10^{4}	5.7×10^{5}
10	neon	Ne	20.18	1.2×10^{8}	2.5×10^{9}
11	sodium	Na	22.99	2.0×10^{6}	4.7×10^{7}
12	magnesium	Mg	24.31	3.8×10^{7}	9.2×10^{8}
13	aluminium	Al	26.98	3.0×10^{6}	8.1×10^{7}
14	silicon	Si	28.09	3.5×10^{7}	1.0×10^{9}
15	phosphorus	P	30.97	3.7×10^{5}	1.2×10^{7}
16	sulfur	S	32.07	1.9×10^{7}	6.0×10^{8}
17	chlorine	Cl	35.45	1.9×10^{5}	6.6×10^{6}
18	argon	Ar	39.95	3.6×10^{6}	1.5×10^{8}
19	potassium	K	39.10	1.3×10^{5}	5.2×10^{6}
20	calcium	Ca	40.08	2.2×10^{6}	8.8×10^{7}
21	scandium	Sc	44.96	1.2×10^{3}	5.5×10^{4}
22	titanium	Ti	47.88	8.5×10^{4}	4.1×10^{6}
23	vanadium	V	50.94	1.0×10^{4}	5.3×10^{5}
24	chromium	Cr	52.00	4.8×10^{5}	2.5×10^{7}
25	manganese	Mn	54.94	3.4×10^{5}	1.9×10^{7}
26	iron	Fe	55.85	3.2×10^{7}	1.8×10^{9}

Atomic number, Z	Name	Chemical symbol	Relative atomic mass, A_r	Solar System abundance by number	by mass
27	cobalt	Co	58.93	8.1×10^4	4.8×10^6
28	nickel	Ni	58.69	1.8×10^6	1.0×10^8
29	copper	Cu	63.55	1.9×10^4	1.2×10^6
30	zinc	Zn	65.39	4.5×10^4	2.9×10^6
31	gallium	Ga	69.72	1.3×10^3	9.4×10^4
32	germanium	Ge	72.61	4.3×10^3	3.1×10^5
33	arsenic	As	74.92	2.3×10^2	1.8×10^4
34	selenium	Se	78.96	2.2×10^3	1.8×10^5
35	bromine	Br	79.90	4.3×10^2	3.4×10^4
36	krypton	Kr	83.80	1.7×10^3	1.4×10^5
37	rubidium	Rb	85.47	2.5×10^2	2.1×10^4
38	strontium	Sr	87.62	8.5×10^2	7.5×10^4
39	yttrium	Y	88.91	1.7×10^2	1.5×10^4
40	zirconium	Zr	91.22	4.1×10^2	3.7×10^4
41	niobium	Nb	92.91	25	2.3×10^3
42	molybdenum	Mo	95.94	91	8.7×10^3
43	technetium	Tc[a]	98.91	—[b]	—[b]
44	ruthenium	Ru	101.07	66	6.8×10^3
45	rhodium	Rh	102.91	12	1.3×10^3
46	palladium	Pd	106.42	50	5.3×10^3
47	silver	Ag	107.87	17	1.9×10^3
48	cadmium	Cd	112.41	58	6.5×10^3
49	indium	In	114.82	6.6	7.6×10^2
50	tin	Sn	118.71	140	1.6×10^4
51	antimony	Sb	121.76	11	1.3×10^3
52	tellurium	Te	127.60	170	2.2×10^4
53	iodine	I	126.90	32	4.1×10^3
54	xenon	Xe	131.29	170	2.2×10^4
55	caesium	Cs	132.91	13	1.8×10^3
56	barium	Ba	137.33	160	2.2×10^4
57	lanthanum	La	138.91	16	2.2×10^3
58	cerium	Ce	140.12	41	5.7×10^3
59	praseodymium	Pr	140.91	6.0	8.5×10^2

Atomic number, Z	Name	Chemical symbol	Relative atomic mass, A_r	Solar System abundance by number	by mass
60	neodymium	Nd	144.24	30	4.3×10^3
61	promethium	Pm[a]	146.92	–[c]	–[c]
62	samarium	Sm	150.36	9.3	1.4×10^3
63	europium	Eu	151.96	3.5	5.3×10^2
64	gadolinium	Gd	157.25	12	1.8×10^3
65	terbium	Tb	158.93	2.1	3.4×10^2
66	dysprosium	Dy	162.50	14	2.3×10^3
67	holmium	Ho	164.93	3.2	5.2×10^2
68	erbium	Er	167.26	8.9	1.5×10^3
69	thulium	Tm	168.93	1.3	2.3×10^2
70	ytterbium	Yb	170.04	8.9	1.5×10^3
71	lutetium	Lu	174.97	1.3	2.3×10^2
72	hafnium	Hf	178.49	5.3	9.6×10^2
73	tantalum	Ta	180.95	1.3	2.4×10^2
74	tungsten	W	183.85	4.8	8.8×10^2
75	rhenium	Re	186.21	1.9	3.5×10^2
76	osmium	Os	190.2	24	4.6×10^3
77	iridium	Ir	192.22	23	4.5×10^3
78	platinum	Pt	195.08	48	9.3×10^3
79	gold	Au	196.97	6.8	1.3×10^3
80	mercury	Hg	200.59	12	2.5×10^3
81	thallium	Tl	204.38	6.6	1.4×10^3
82	lead	Pb	207.2	110	2.3×10^4
83	bismuth	Bi	208.98	5.1	1.1×10^3
84	polonium	Po[a]	209.98	–[c]	–[c]
85	astatine	At[a]	209.99	–[c]	–[c]
86	radon	Rn[a]	222.02	–[c]	–[c]
87	francium	Fr[a]	223.02	–[c]	–[c]
88	radium	Ra[a]	226.03	–[c]	–[c]
89	actinium	Ac[a]	227.03	–[c]	–[c]
90	thorium	Th[a]	232.04	1.2	2.8×10^2

Atomic number, Z	Name	Chemical symbol	Relative atomic mass, A_r	Solar System abundance by number	by mass
91	protoactinium	Pa[a]	231.04	–[c]	–[c]
92	uranium	U[a]	238.03	0.32	7.7×10^{1}
93	neptunium	Np[a]	237.05	–[c]	–[c]
94	plutonium	Pu[a]	239.05	–[c]	–[c]
95	americium	Am[a]	241.06	–[c]	–[c]
96	curium	Cm[a]	244.06	–[c]	–[c]
97	berkelium	Bk[a]	249.08	–[c]	–[c]
98	californium	Cf[a]	252.08	–[c]	–[c]
99	einsteinium	Es[a]	252.08	–[c]	–[c]
100	fermium	Fm[a]	257.10	–[c]	–[c]
101	mendelevium	Md[a]	258.10	–[c]	–[c]
102	nobelium	No[a]	259.10	–[c]	–[c]
103	lawrencium	Lr[a]	262.11	–[c]	–[c]
104	rutherfordium	Rf[a]	261	–[c]	–[c]
105	dubnium	Db[a]	262	–[c]	–[c]
106	seaborgium	Sg[a]	266	–[c]	–[c]
107	bohrium	Bh[a]	264	–[c]	–[c]
108	hassium	Hs[a]	269	–[c]	–[c]
109	meitnerium	Mt[a]	268	–[c]	–[c]
110	ununnilium	Uun[a]	271	–[c]	–[c]
111	unununium	Uuu[a]	272	–[c]	–[c]
112	ununbium	Uub[a]	285	–[c]	–[c]

[a] No stable isotopes.

[b] Detected in spectra of rare evolved stars but has a half-life too short for survival in the interstellar medium.

[c] Far too scarce to have been detected beyond the Earth, and probably very scarce.

Data adapted from Däppen, 2000, from the original data of Ander and Grevesse, 1989, and Grevesse and Noels, 1993.

GLOSSARY

3α process *See* triple alpha process.

aberration An effect that causes the apparent direction of a star (as observed from the Earth) to differ from its true direction in a way that changes systematically throughout the year. The effect arises from the combination of the finite speed of light and the movement of the Earth as it orbits the Sun, and requires that a telescope being used to observe a star must be pointed slightly away from the true direction of the star.

absolute visual magnitude A numerical measure of the intrinsic brightness of a star, equal to the apparent visual magnitude the star would have if observed from a standard distance of 10 parsecs, in the absence of interstellar absorption.

absorption A process that leads to a decrease in the intensity (of flux density) of radiation as it passes through a medium (e.g. the interstellar medium).

absorption lines Narrow wavelength or frequency ranges in a spectrum where the spectral flux density is less than at adjacent wavelengths (or frequencies).

absorption spectrum A spectrum in which absorption lines are prominent.

accretion Any process by which material is added to an astronomical body. This may, for example, occur during the formation of a protostar or during the transfer of matter in an interacting binary system.

accretion disc A disc of gas which forms around massive objects such as the accreting star in an interacting binary system. Material spirals inwards within the disc and falls onto the central object from the inner edge of the disc.

active regions Areas on the photosphere of the Sun where magnetic field lines are concentrated. These generate a number of different phenomena such as sunspots and flares.

angular momentum A measure of the momentum associated with rotation. The magnitude (L) of a body's angular momentum at any time is found by multiplying its moment of inertia by its angular speed at that time. The SI unit of angular momentum is the $kg\,m^2\,s^{-1}$.

angular speed (ω) The rate at which an object (e.g. a wheel) turns, or at which a body (e.g. a satellite) orbits another body. Defined as the angle turned through in unit time. The SI unit of angular speed is the radian per second ($rad\,s^{-1}$).

apparent visual magnitude A numerical measure of the apparent brightness of a body. For a star, it is determined by the flux density received in the V band, i.e. a band that approximates the response of human vision versus wavelength.

asteroseismology The study of stellar interiors from observations of global oscillations of their photospheres. Analogous to helioseismology but using only the combined light from the whole surface of a star.

astronomical unit (AU) The average distance from the Earth to the Sun. More precisely, the semimajor axis of the Earth's elliptical orbit around the Sun. (The size of the Earth's orbit varies slightly, so for precise work the informal definitions given above have been superseded by a more technical and exact definition.)

asymptotic giant branch (AGB) A region on the Hertzsprung–Russell diagram occupied by large cool stars after helium core burning has been completed. On the H–R diagram, stars of different masses approach this region from the left and evolve upwards so that the evolutionary tracks approach each other.

aurora A luminous atmospheric phenomenon observed in high latitudes. It is due to visible light emitted from atoms and molecules in the Earth's atmosphere that have been excited by collisions with electrons from the magnetosphere.

Balmer absorption lines A set of absorption lines due to photoexcitation of electronic transitions in hydrogen atoms, in which the electron makes a transition from a state corresponding to the energy level with $n = 2$, to a higher energy level.

baryon An elementary particle that comprises three quarks. Protons and neutrons are baryons.

big bang model The generally accepted scientific account of the evolution of the Universe from the earliest phases of its expansion. A feature of this model is that early in the history of the Universe, all matter was in a very hot and dense state, and the temperature of the matter dropped as time progressed.

binary star A system of two stars in orbit around each other.

bipolar outflow The phenomenon in which material ejected from a stellar system takes the form of two opposing jets. Some such jets appear to be associated with the early stages of stellar evolution.

black hole A region of space from which electromagnetic radiation and matter are unable to escape due to the action of gravity. Such regions are expected to be created by the catastrophic collapse of some massive stars.

black hole candidate An object believed, through its observed properties, to be a possible black hole.

black-body radiation curve *See* black-body spectrum.

black-body spectrum The spectrum of a black-body source (or ideal thermal source). This is a continuous spectrum with a characteristic shape, the peak wavelength depending on the temperature of the source, in accord with Wien's displacement law. A characteristic of sources that produce spectra that are close to the black-body form is that there is a degree of interaction between electromagnetic radiation and the material that makes up the source. (This leads to the formal definition of a black-body source as one which has the property that it absorbs perfectly any electromagnetic radiation that is incident on it and emits a black-body spectrum.)

blue-shift The decrease in the observed wavelength of radiation relative to the wavelength at which the radiation was emitted (the frequency is correspondingly increased). A common cause of blue-shift is the relative movement of the source towards the observer (an example of the Doppler effect).

Boltzmann constant The constant that relates kinetic energy and temperature in a gas (or pressure and temperature in a fixed volume and quantity of gas). It has the value $k = 1.38 \times 10^{-23}\,\mathrm{J\,K^{-1}}$.

bow shock The boundary of the Earth's magnetosphere where the solar wind is deflected. More generally it refers to any boundary between the flow of fast moving and stationary gas or plasma.

broadband spectrum A spectrum covering a wide range of wavelengths or frequencies, which represents the energy distribution of a source. It does not generally show narrow features such as absorption lines.

brown dwarf A stellar object of mass less than about $0.08M_\odot$, in which the core temperature is too low for hydrogen burning to be initiated.

carbon burning Nuclear fusion reactions in which two carbon nuclei fuse to form a heavier nucleus with a mass number of around 20 (e.g. sodium, neon or magnesium). Carbon burning occurs in massive post main sequence stars. The term is sometimes also used to refer to any nuclear fusion reaction where carbon is a reactant, for example when a carbon nucleus combines with a helium nucleus to produce an oxygen nucleus.

cataclysmic variable A form of interacting binary star in which material is transferred from a star which fills its Roche Lobe onto a white dwarf. They exhibit dramatic changes in luminosity that are most prominent in the optical, ultraviolet and X-ray parts of the spectrum.

centre of mass A point in a body, or amidst a set of bodies, that moves through space in the same way that the whole mass of the system would move were it concentrated at that point.

Cepheid A type of regular variable star, which pulsates with a period in the range from about a day to about 100 days. The changes in radius, temperature, and hence luminosity, arise from instabilities in the envelopes of such evolved giant or supergiant stars.

Chandrasekhar limit The theoretical upper limit to the mass of a white dwarf; about $1.4M_\odot$. Above this limit, electron degeneracy pressure cannot support the star and it collapses further.

chromosphere The region of the Sun just above the photosphere. It is the lowest layer of the Sun's atmosphere, characterized by a reddish hue, and by an increase of temperature with altitude through all but its lower reaches.

circumstellar disc A disc (or torus) of material surrounding a protostar. Possibly because of its rotational motion, it has not collapsed into the protostar.

circumstellar shell A shell of material, rich in molecules and dust, formed near a cool giant/supergiant from matter ejected by the star.

CNO cycle A cycle of nuclear reactions that convert hydrogen into helium using the elements carbon, nitrogen and oxygen as catalysts. This cycle predominates in upper main sequence stars.

cocoon nebula A compact, dense cloud of warm dust surrounding a protostar, which absorbs electromagnetic radiation from the hot protostar and re-radiates the energy in the infrared part of the spectrum.

collisional excitation A process that raises an ion, atom or molecule into a higher energy level when it collides with another particle.

colour index The difference, measured in magnitudes, of the brightness of an object in two specified wavebands (e.g. in the blue 'B' and visual 'V' wavebands of the UBV system, in which case the difference is denoted $m_B - m_V$ or simply $B - V$). It is a measure of the colour and hence the temperature of the object.

conduction A process of energy transfer in which atoms or molecules pass energy to their neighbours through collisions. In such a process energy is transferred from regions of relatively high temperature to regions of relatively low temperature.

conservation of angular momentum The principle that the angular momentum of a body, measured about any point, will not change provided no external force acts on that body. For an isolated system of several bodies, the total angular momentum, about any point, of the whole system will not change.

continuous spectrum A spectrum that is broad and smooth, i.e. the spectral flux density exhibits no sharp changes with wavelength.

convection A process of energy transfer in which a fluid in a gravitational field is heated from below to the point where the hotter, less dense fluid rises upwards, displacing the cooler, denser fluid downwards.

convection cells Localized regions of fluid flow set up by convection. A cell consists of a region where warm fluid is rising in a surrounding region where cool fluid is descending.

convection current A current of fluid in motion because of convection.

convective core The core of a star in which convection is the dominant process of energy transfer. For a star of mass greater than about $1.5M_\odot$, the temperature gradient in the core is sufficiently high to set up energy transfer by convection.

convective envelope The regions of a star outside the core in which convection is the dominant process of energy transfer. For a star of mass less than about $1.5M_\odot$, the core is non-convective but convection can take place further out in the form of a convective shell or envelope.

convective zone The region in the Sun (or in any other body) in which heat transfer is predominantly by convection.

core The central region of a body, distinguished on the basis of chemical composition, or physical properties (such as temperature), or physical processes (such as nuclear fusion).

corona The outer region of the Sun's atmosphere; it is very extensive, has a very low density, and is extremely hot.

coronagraph An optical device that enables the Sun's corona to be investigated. It blocks the far greater amounts of light that come from the photosphere, thus enabling the corona to be seen.

coronal hole Regions of the Sun's corona where the solar magnetic field opens outward to interplanetary space rather than looping back on to the Sun. Such holes are thought to be a major source of the solar wind.

coronal loop Closed loops of magnetic field lines, extending from the Sun's photosphere into the corona, which contain plasma at high temperature. They are typically observed through ultraviolet or X-ray emission from the hot plasma.

degeneracy A phenomenon described by quantum mechanics, where, at sufficiently low temperatures/high densities, certain particles (such as electrons or neutrons) are forced into higher energy levels because all the lower energy levels are fully occupied. Degeneracy gives rise to degeneracy pressure, which can support a star against gravitational collapse even if there is no nuclear burning to provide a gas pressure gradient.

dense clouds One of the coldest and densest components of the interstellar medium, rich in molecules. Dense clouds give birth to stars, mainly in the form of open clusters.

differential rotation The phenomenon whereby the rotation period of one part of a fluid body may differ from that of another. In the case of the Sun, for example, the rotation period varies with latitude and with depth.

diffuse cloud Interstellar clouds that have temperatures that are comparable to dense clouds, but which are much less dense, and not so rich in molecules.

dipole field The magnetic field produced by a dipole, such as a bar magnet, which has two poles, north and south.

dissociation The process of separation of two or more atoms in a molecule so that they are no longer bound together in the molecule.

Doppler effect The effect whereby the observed frequency of waves received from a source depends on the motion of the source relative to the observer. There is a corresponding change in the observed wavelength.

Doppler shift The change in wavelength (or frequency) at an observer's position, with respect to that emitted by the source, due to the motion of the source relative to the observer.

dwarf novae Cataclysmic variable stars that exhibit erratic outbursts in their optical emission of 2–5 magnitudes in a few days, followed by a slower decline to quiescent levels. The outbursts are not periodic, but recur over a timescale of weeks to months.

eccentricity A numerical measure of the extent to which an ellipse (such as an elliptical orbit) differs from a circle. It is half the distance between the two foci, divided by the semimajor axis.

eclipsing binary A binary star in which the plane of the orbit of the two components is sufficiently close to the observer's line of sight that one star can pass in front of (or behind) the other, causing a change in the perceived brightness of the binary system.

electric field The quantity, specified throughout some region of space, that determines the electric force that would act on a particle of given electric charge at any point within that region. At each point in the region the electric field possesses a strength and a direction, and these two quantities are given by the electric force per unit charge that would act at that point. The electric field in a region may be produced by electric charges, but is deemed to exist irrespective of whether there are any other electric charges present to 'feel' its effect.

electromagnetic radiation A form of radiation in which the transfer of energy from place to place may be attributed to the passage of wave-like disturbances in the electric and magnetic fields between those places (i.e. electromagnetic waves), or to the passage of photons from one place to the other. The wavelength or frequency of the waves (or the energy of the photons) may be used to assign particular kinds of electromagnetic radiation to specific parts of the electromagnetic spectrum, resulting in their classification as light, radio waves, X-rays, etc.

electromagnetic spectrum A collective term used to describe the various wavelength ranges of electromagnetic radiation. In order of increasing wavelength, these ranges are gamma (γ) rays, X-rays, ultraviolet radiation, visible light, infrared radiation, microwaves and radio waves. *See also* spectrum.

electromagnetic wave model of light A model of visible light or any other form of electromagnetic radiation, in which the behaviour of light is described entirely in terms of electromagnetic waves. Some properties of electromagnetic radiation cannot be explained by this model. *See* quantum theory.

electromagnetic waves Fluctuating patterns of electric field and magnetic field in which the two fields are at right angles to each other and to the direction in which the waves travel. The speed at which they travel in a vacuum (to 3 significant figures) is $3.00 \times 10^8 \, \mathrm{m \, s^{-1}}$; this is the speed of light in a vacuum.

electron degeneracy pressure Pressure due to electron degeneracy. It occurs in the cores of red giants, and supports white dwarfs.

electronvolt (eV) A unit of energy often used for specifying the energy of photons or in energy-level diagrams. $1 \, \mathrm{eV} = 1.602 \times 10^{-19} \, \mathrm{J}$.

elliptical orbit The path followed by a celestial body, or spacecraft, in orbit about a more massive body (or of one object about the centre of mass of a binary system). In the absence of any perturbing forces, this path, or orbit, has the shape of an ellipse.

emission lines Narrow wavelength or frequency ranges in a spectrum where the spectral flux density is greater than at adjacent wavelengths (or frequencies).

emission spectrum A spectrum in which emission lines are prominent.

endothermic A term used to describe a process or reaction that requires a net input of energy for it to be sustained.

energy-level diagram A diagram showing the energies of the allowed states of a quantum physical system, such as an atom or a molecule, by virtue of the arrangement of its internal constituents. In the diagram energy is shown as increasing in the upward direction, and short horizontal lines are used to denote the energy levels themselves. The gaps between the levels indicate energies that cannot be attained by the system.

equations of stellar structure A set of equations describing the internal state of a star using parameters such as temperature, pressure, thermal conductivity.

equivalent width A measure of the strength of a spectral line. If a spectrum is plotted as spectral flux density against wavelength, it is the width of an area bounded by the spectral flux density of the continuum that has the same area as that between the actual spectral line and the continuum.

escape speed The minimum speed that enables a small body to just escape from the gravitational field of a far more massive body.

excited state Any state of a quantum physical system, such as an atom or molecule, that has a higher energy than the lowest energy state (ground state).

exothermic A term used to describe a process or reaction that produces a net output of energy as it progresses.

extinction The combined effect of the scattering of electromagnetic radiation by a medium, and absorption of such radiation by the medium.

extrasolar planet A planet orbiting a star other than the Sun.

extrinsic variables Variable stars that change in brightness as a result of some external effect rather than physical changes in the stars themselves. An example is an eclipsing binary.

faculae Bright patches on the solar photosphere generally associated with sunspots.

filament A long, winding dark feature, seen in visible light images of the Sun (particularly Hα images), that is caused by huge clouds of relatively cool gas held high above the chromosphere by magnetic forces. Can also be seen as a prominence when at the limb of the Sun's disc.

flux density (F) A quantity describing the rate at which energy transferred by radiation is received from a source, per unit area facing the source. The SI unit of flux density is the $W\,m^{-2}$.

foci (singular focus) Two special points within an ellipse. In the case of an elliptical orbit of body A with respect to body B, body B lies at one focus, the other being empty.

fractional radius The distance from the centre of a spherical body, expressed as a fraction of the body's radius.

fragmentation The process by which a contracting interstellar cloud breaks up into a number of separate contracting cloudlets.

frequency The rate at which wavelengths of a wave pass a fixed point (i.e. the number per second passing the fixed point). The SI unit of frequency is the hertz (Hz), where $1\,Hz = 1\,s^{-1}$.

gamma-rays (γ-rays) Electromagnetic radiation with the highest frequencies, above the highest frequencies of X-rays. The photon energies are consequently also the highest.

geomagnetic disturbances Variations in the magnetic field at the surface of the Earth caused by motions of charged particles in the magnetosphere. In extreme cases these disturbances are known as geomagnetic storms.

geomagnetic storm Intense and rapid variations of the magnetic field at the surface of the Earth caused by motions of charged particles in the magnetosphere. These motions result from large coronal mass ejections striking the magnetosphere.

giant molecular cloud (GMC) complex A large type of cloud in the interstellar medium, consisting mainly of a mixture of diffuse clouds and dense clouds. They are probably the fundamental type of cloud structure in the interstellar medium.

giant star Any star that lies, on a Hertzsprung–Russell diagram, between the main sequence and the supergiants. Such stars are typically between about 5 and 25 times larger than the Sun.

global oscillations *See* helioseismology.

globular clusters Clusters of 10^5 to 10^6 very old stars tightly bound by gravity into a spherical region of space less than about 50 pc in diameter. They are found in a spherical distribution about the centre of our Galaxy and in other galaxies.

granules Bright, small patches on the Sun's photosphere that are the tops of rising columns of hot material. They form a seething pattern called solar granulation.

ground state The lowest energy state of an atom, a molecule or other quantum system.

Hα absorption line An absorption line due to an electronic transition in hydrogen atoms, in which the atom absorbs a photon (photoexcitation), so that the electron makes a transition from a state corresponding to the electronic energy level with $n = 2$, to the one above, with $n = 3$. The photon wavelength (in a vacuum) is 656.3 nm: this is the longest wavelength Balmer absorption line.

Hα image An image made using the emission line from hydrogen atoms corresponding to the reverse transition from that which gives rise to the Hα absorption line.

Harvard Spectral Classification A classification of stellar spectra based on the relative strengths of the absorption lines. The main classes relate to photospheric temperature, and in order of descending temperature are O, B, A, F, G, K, M.

Hayashi track A track on the Hertzsprung–Russell diagram showing the theoretical evolution of a protostar as it approaches the main sequence.

heavy elements All chemical elements except the two that have the atoms with the lowest mass, i.e. all elements except hydrogen and helium.

heliopause The boundary between the region where the solar wind dominates the motion of plasma in the Solar System and beyond which lies the interstellar medium.

helioseismology The study of the solar interior based on the analysis of global oscillations of the Sun. Such oscillations can be studied because they cause observable effects at the photosphere.

heliosphere The volume of space within which the Sun, through the solar wind, influences its environment in the interstellar medium.

helium flash (core helium flash) An explosive release of energy in the core of a star that occurs after the star has left the main sequence and which initiates the process of helium burning (i.e. helium fusion) in the star. In order for helium ignition to take the form of what is, in effect, an internal explosion, the core must be degenerate at the time of ignition, and this implies that the star must be of relatively low mass, probably less than about 2.25 solar masses.

helium fusion The fusion of helium nuclei. Once the temperature in a star is sufficiently high, helium nuclei undergo fusion, via the triple alpha process, to produce carbon nuclei.

helium mass fraction (Y) A numerical measure of the proportion of helium in a sample of material, obtained by dividing the mass of helium in the sample by the total mass of the sample. In the Sun, $Y \approx 0.25$.

hertz (Hz) The SI unit of frequency: the rate at which wavelengths of a wave pass a fixed point (i.e. the number per second passing the fixed point).

Hertzsprung–Russell (H–R) diagram A diagram showing the luminosity and temperature of stars, which is useful for comparing large numbers of stars and for tracking their evolution. Photospheric temperature is shown along the horizontal axis (increasing to the left), and luminosity is shown along the vertical axis. A star appears as a point on the diagram, corresponding to its current temperature and luminosity.

HII regions A hot, luminous component of the interstellar medium, comprising ionized hydrogen gas that is made visible by the presence of a hot, young star or stars. Strong ultraviolet radiation from hot stars ionizes the hydrogen, and the occasional recombination of an electron and proton to form a neutral hydrogen atom results in the emission of light, before the hydrogen is reionized.

horizontal branch A region on the Hertzsprung–Russell diagram occupied by stars of low mass and low metallicity after they have left the red giant branch during core helium burning. It is often seen in H–R diagrams of globular clusters, where many stars have similar luminosity but a wide range of surface temperatures, and hence lie in an approximately horizontal strip.

hot intercloud medium A component of the interstellar medium, characterized by a low density and a very high temperature. It is formed by material from within the cavities of supernova remnants.

hydrogen burning The conversion of hydrogen into helium through nuclear processes (hydrogen fusion).

hydrogen mass fraction (X) A numerical measure of the proportion of hydrogen in a sample of material, obtained by dividing the mass of hydrogen in the sample by the total mass of the sample. In the Sun, $X \approx 0.73$.

hydrostatic equilibrium The condition in which there is a balance between the inward force on a layer of a star (due to the gravitational attraction of material closer to the centre of the star) and the outward force on the layer (due to the difference in the pressure exerted on the layer by the layers above and below).

ideal gas A hypothetical gas that satisfies the ideal gas law. Real gases behave like ideal gases at sufficiently high temperatures and/or low densities.

ideal gas law The assertion that, for an ideal gas,

$$P = \frac{k\rho T}{m}$$

where P is pressure, k is the Boltzmann constant, ρ the density, T the absolute temperature, and m the average mass of the gas particles.

infrared radiation Electromagnetic radiation with frequencies or wavelengths between those of visible light and microwaves.

inner Lagrangian point The Lagrangian point in a binary system, between the two stars, at the centre of mass where the gravitational forces exerted by the two stars are equal in magnitude but act in opposite directions. This is where the two pear-shaped Roche lobes touch and it is through this point that mass transfer in an interacting binary system takes place.

instability strip A roughly vertical region on the Hertzsprung–Russell diagram where the structure of stars is unstable. Any star in this region (e.g. a Cepheid) pulsates and therefore shows variability.

interstellar cirrus Widespread wisps of cool interstellar matter, possibly the result of circumstellar shells as they spread out and become cooler and more diffuse.

interstellar medium (ISM) The thinly distributed matter that fills interstellar space in the Galaxy. It consists of gas (mainly hydrogen), with a trace of dust, and is made up of several different components, such as dense clouds, HII regions and the intercloud medium.

interstellar reddening The effect whereby the observed colour of a star is influenced by the presence of dust in the interstellar medium. The size of the dust grains results in more efficient absorption and scattering of short wavelength (blue) light than of longer wavelength (red) light, so as light from a star encounters more and more dust, the apparent colour of the star becomes progressively redder.

intrinsic variables Variable stars that change in brightness as a result of a change in the properties of the stars themselves. An example is a pulsating variable.

ionization energy The energy required to remove an electron from an atom in its ground state.

iron group Elements with relative atomic masses around $A \sim 56$, such as iron, chromium, manganese, cobalt and nickel, which have the lowest rest energy per nucleon.

irregular variable A star whose luminosity varies with time in an irregular manner. The timescale is *very* short compared with the stellar lifetime.

Jeans mass The critical mass that a uniform, spherical, non-rotating cloud must have before it will collapse under its own gravitational attraction.

kinetic energy The energy that a body has by virtue of its motion.

Lagrangian point One of the five points in a binary system at which a small body can maintain a stable orbit about the system's centre of mass. Viewed from a frame of reference rotating with the binary system, a small body placed at any one of these points feels no net force.

light curve A diagram showing the variation of brightness (e.g. magnitude, flux density or luminosity) with time, for a celestial object such as a variable star or supernova.

light-year (ly) The distance travelled by light (or any other form of electromagnetic radiation) through a vacuum in one year. $1 \, \text{ly} \approx 9.46 \times 10^{15} \, \text{m}$.

limb darkening The phenomenon whereby the extremities of the image of a spherical body are darker than the central regions. In the case of the Sun it arises in visible light images because towards the solar limb we are seeing less deeply into the photosphere, our view being confined to the outer and hence cooler and dimmer part of the photosphere.

line spectra Spectra which exhibit narrow lines due to absorption (absorption lines) or emission (emission lines) of electromagnetic radiation (called, respectively, absorption spectra and emission spectra).

local thermodynamic equilibrium The condition in which the matter in any appropriately localized region within a body may be characterized by a local temperature, and the radiation coming from any such region is in equilibrium with (i.e. has the same local temperature as) the matter in that region. The assumption of local thermodynamic equilibrium is useful in models of stellar envelopes and atmospheres.

lower main sequence star A star with a mass of less than about $1.5M_\odot$, occupying the lower part of the main sequence on the Hertzsprung–Russell diagram. In such stars the pp-chain dominates the production of energy.

luminosity The rate at which energy is carried away from a star by electromagnetic radiation. The SI unit of luminosity is the watt (W), where $1 \, \text{W} = 1 \, \text{J s}^{-1}$.

luminosity class One of a number of classes of stars (determined by means of certain spectral features such as the widths of particular absorption lines) that typically contains stars with a wide range of spectral classes but a rather narrow range of sizes. Amongst the commonly used luminosity classes are Class Ia (highly luminous supergiants), Class III (giants) and Class V (main sequence stars). Each luminosity class can be represented by a line on the Hertzsprung–Russell diagram, with the consequence that specifying both the spectral class and the luminosity class of a star (as in G2 V, for the Sun) permits the luminosity of the star to be approximately determined.

magnetic field The quantity, specified throughout some region of space, that determines the magnetic force that would act on a particle of given electric charge moving with given velocity through any point within that region. At each point in the region, the magnetic field possesses a strength and a direction. The magnetic field in a region may be produced by magnets or by moving charged particles (e.g. electric currents), but is deemed to exist irrespective of whether there are any other moving charged particles or magnets present to 'feel' its effect.

magnetic field line An imaginary directed line (i.e. a line with an arrow head on it) passing through the region occupied by a magnetic field in such a way that it has the same direction as the magnetic field at every point. In diagrams, it is conventional to draw magnetic field lines in such a way that their density (i.e. the number of lines per unit area) indicates the relative strength of the magnetic field. *See* polarity.

magnetic reconnection A process taking place in a plasma, in which neighbouring, oppositely directed, magnetic field lines suddenly part and reconnect in a new configuration. The abrupt change in the magnetic field represented by this process can release large amounts of energy and is thought to be important in explaining solar flares and other phenomena.

magnetogram A map of the magnetic field strength over a surface (usually of the Sun).

magnetograph An instrument which maps the magnetic field strength in the Sun's photosphere by measuring the splitting of certain spectral lines due to the local magnetic field.

magnetosphere The region around the Earth (or another planet) where its magnetic field influences the motion of charged particles.

main sequence The region on the Hertzsprung–Russell diagram where stars spend most of their lives. Most of the observed stars lie on the main sequence where they are powered by hydrogen burning in their cores.

main sequence turn-off The point on the main sequence of the Hertzsprung–Russell diagram of a star cluster above which no stars are present. It corresponds to stars that are just reaching the end of their time on the main sequence, and is therefore an indication of the age of the cluster.

mass–luminosity relationship A relationship between the mass of a main sequence star and its luminosity. It shows a dramatic increase in luminosity with increasing mass.

medium Any material (solid, liquid or gas) through which particles or radiation may pass.

metallicity (Z) A numerical measure of the proportion of heavy elements in a sample of material, obtained by dividing the mass of heavy elements in the sample by the total mass of the sample. In the Sun, $Z \approx 0.02$.

microwaves Electromagnetic radiation with frequencies or wavelengths between those of infrared radiation and radio waves.

Mira variable A type of variable star. Mira variables are giant stars in a late stage of evolution (on the asymptotic giant branch) which exhibit long period (100–1000 days) global pulsations. The periods and amplitudes are subject to variations from cycle to cycle. Mira variables are named after the star Mira (omicron Ceti).

moment of inertia (I) A quantity, expressed relative to a specified axis, that measures the distribution of a system's mass with respect to that axis. The moment of inertia of a rigid body plays an important part in determining how that body will rotate about the specified axis under given conditions. Broadly speaking, the same mass, more widely distributed, has a larger moment of inertia about a specified axis. The SI unit of moment of inertia is the kg m^2.

neon burning A nuclear fusion reaction in which neon nuclei react with helium nuclei (α-particles) to form magnesium. The helium nuclei are themselves formed from the photodisintegration of neon.

neutrino A type of elementary particle that is electrically neutral and has very low mass. Neutrinos generally travel at speeds very close to the speed of light in a vacuum, and interact with other particles extremely weakly. There are three sub-types: electron, muon and tauon neutrinos.

neutron degeneracy pressure Pressure that arises from neutron degeneracy. It supports neutron stars.

neutron star A star of mass between around 1.4 and 3 to $5M_\odot$ with a radius of about 10 km, formed from the collapsed core of a supergiant, and made of material exceptionally rich in neutrons. Most known neutron stars are observed as pulsars.

non-thermal source A source of electromagnetic radiation that emits for reasons other than those relating to its temperature.

nova An outburst in which a star increases its brightness by a factor of about 10^3 in a few days, and then slowly returns to its original brightness. Novae are now associated with stars in certain types of close binary system, where the nova outburst is a consequence of mass being transferred from one star to the other.

OB association A group of young stars containing several stars of spectral type O and B in the Harvard Spectral Classification.

open cluster A cluster of up to a few hundred stars, formed from a cloudlet that has fragmented from a larger dense cloud. The stars are only loosely bound together in an open structure.

oxygen burning A nuclear fusion reaction, occurring in massive post main sequence stars, in which two oxygen nuclei fuse to form silicon and helium nuclei.

parallactic ellipse The apparent elliptical movement of a nearby star in the plane of the sky relative to distant stars during the course of a year due to parallax.

parallax The quantity that describes the change in direction to a celestial body (relative to a background of far more distant bodies) resulting from a given change in position of the observer perpendicular to the direction of the body. The term parallax is often used to refer specifically to stellar parallax, p, when the change in position of the observer is one astronomical unit. This quantity is important in the determination of the distance of nearby stars.

parsec (pc) The distance to a celestial body that has a stellar parallax of one arc second. 1 pc = 3.09×10^{16} m.

period–luminosity relationship A correlation between period and luminosity; in particular the relationship between period and luminosity of Cepheid variables that enables these stars to be used as standard candles. (Absolute visual magnitude, M_V, is generally used in place of luminosity when using this relationship.)

photodisintegration The process in which a nucleus is split apart by the absorption of a gamma-ray photon. This type of reaction plays an important role in the later stages of stellar nucleosynthesis.

photoemission The process in which a photon is emitted by an atom or molecule.

photoevaporation The process in which an interstellar cloud is eroded by the action of intense ultraviolet radiation from nearby highly luminous stars. The ultraviolet dissociates the hydrogen molecules (H_2) in the cloud into individual hydrogen atoms.

photoexcitation The process in which an atom or molecule is excited by the absorption of a photon.

photometric method A method of determining stellar photospheric temperatures (or the surface temperature of other bodies) by comparing the amount of radiation emitted over at least two different wavelength regions.

photon The particle of electromagnetic radiation in the photon model of light. The photon energy ε is proportional to the frequency f of the associated radiation; $\varepsilon = hf$ where h is the Planck constant.

photon model of light A model in which a ray of visible light, or of any other form of electromagnetic radiation, is treated as a stream of separate particles called photons.

photosphere The 'surface' of the Sun, or of any other star – the thin layer from which comes nearly all of the solar (or stellar) radiation that we observe.

plages Extensive bright regions of the solar chromosphere, seen particularly well in Hα images. They often occur directly above active regions of the photosphere that contain sunspots.

Planck constant (h) The constant that relates the energy ε of a photon to the frequency f of the wave with which it is associated: $\varepsilon = hf$.

Planck curve *See* black-body spectrum.

planetary nebula A shell of material ejected towards the end of its life by a star whose initial mass was less than about $11 M_{\odot}$. The hot central star of the planetary nebula becomes a white dwarf.

plasma A fluid in which there is a high degree of ionization.

polarity The property of the pole of a magnet, or of any other region from which magnetic field lines emanate, that determines the direction of those magnetic field lines. By convention, magnetic field lines are directed away from north magnetic poles (regions of positive polarity) and towards south magnetic poles (regions of negative polarity).

positron (e^+) A kind of positively charged elementary particle which has the same mass, magnitude of charge, and many other properties as an electron.

ppI chain (or cycle) A sequence of nuclear reactions that is responsible for the bulk of the Sun's radiant energy, and that of comparable or lower mass stars. The net effect of the chain is the conversion of four hydrogen nuclei (protons) into one helium nucleus – an example of nuclear fusion.

ppII and ppIII chains Two sequences of nuclear reactions that predominate in the more massive lower main sequence stars. The net effect of each chain is the conversion of four hydrogen nuclei (protons) into one helium nucleus.

pressure broadening The broadening of a spectral line due to the high density of material in certain stellar atmospheres. This in turn is due to the modification of the energy levels of an atom that result from the close proximity of other atoms.

primordial nucleosynthesis The nuclear processes occurring during the first few minutes of the expansion of the Universe that were responsible for forming nuclei of the light elements (such as helium and lithium).

prominence A kind of bright filamentary feature seen beyond the solar limb. Prominences are huge clouds of relatively cool gas held high above the chromosphere by magnetic fields. Also seen as filaments.

proper motion The quantity that describes the angular movement of a star in the plane of the sky, normally expressed in arc seconds per year.

protostar A term used to describe a star during the earliest stage of its life, during which it is gravitationally contracting and before nuclear fusion has been initiated.

pulsar A widely used abbreviation for a pulsating radio star. Pulsars are widely believed to be rapidly rotating neutron stars with strong magnetic fields, which emit beams of radio waves from the vicinity of their two magnetic poles.

pulsating variable A type of intrinsic variable star that changes in luminosity, radius and temperature in a cyclic manner.

quantum theory A wide-ranging theory that describes, amongst other things, the structure and behaviour of atoms and their interaction with electromagnetic radiation. It accounts for the phenomena that are embraced by the photon model of light, and implies the existence of energy levels in atoms.

quark star A hypothetical type of star, even more dense than a neutron star, in which the neutrons lose their individual identities and matter exists as a sea of quarks.

radial velocity The component of a star's velocity in the line of sight of an observer, i.e. in a radial direction towards or away from the observer.

radiation A process of energy transfer in which energy is transported from place to place by the passage of waves or the direct movement of subatomic particles. The term radiation is also used to describe the travelling disturbances responsible for the energy transfer.

radiation pressure A pressure exerted by photons on any object that absorbs or scatters them. Although it is a weak force it is significant for individual atoms and molecules as well as small dust grains.

radiative zone The region in the Sun (or in any other body) in which energy transfer is predominantly by electromagnetic radiation.

radio waves Electromagnetic radiation with the lowest frequencies/longest wavelengths, extending from the lowest frequency/longest wavelength microwaves.

random walk A process in which a particle (such as a photon) encounters other particles, and has an almost equal chance of travelling in any direction after each encounter.

recombination The process in which an electron and an ion combine, i.e. the opposite of ionization. The electron is typically captured into a high-energy orbit and then cascades downward through the atom's energy levels emitting photons as it does so.

red giant branch (RGB) A region on the Hertzsprung–Russell diagram occupied by stars after they leave the main sequence and are undergoing shell hydrogen burning.

red giant A large star with photospheric temperature less than about 6000 K. Main sequence stars with masses of less than about $11M_\odot$ evolve to become red giants.

red-shift The increase in the observed wavelength of radiation relative to the wavelength at which the radiation was emitted (the frequency is correspondingly decreased). A common cause of red-shift is the relative movement of the source away from the observer (an example of the Doppler effect).

regular variable A star whose luminosity varies with time in a regular manner, on a timescale that is very short compared with the star's lifetime (e.g. a Cepheid).

relative spectral flux density The spectral flux density expressed as a fraction of some arbitrarily chosen reference value.

rest energy The energy E of a particle at rest, given by Einstein's equation $E = mc^2$, where m is the mass of the particle, and c is the speed of light in a vacuum.

Roche lobe The pear-shaped surface around one star in a binary star system, inside which the gravitational force of that star dominates. It represents the maximum volume that the star may occupy before it begins to lose mass through the inner Lagrangian point to its companion. Both stars are surrounded by Roche lobes, and in cross-section the surface has a figure-of-eight shape.

Roche lobe overflow The process whereby matter is transferred from one star to another in an interacting binary system. The transfer occurs when the donating star fills its Roche lobe, and takes place through the inner Lagrangian point.

rotational transition A transition in a molecule between two states corresponding to different amounts of molecular rotational energy.

r-process reactions Nuclear reactions that form elements heavier than iron by the rapid absorption of several neutrons. The reactions are thought to occur for a few seconds during supernovae, and are particularly important for building elements from the unstable (radioactive) isotopes that are formed in supernovae.

Russell–Vogt theorem A theorem stating that a certain mass of stellar material of fixed composition can reach only one stable configuration. This stable configuration corresponds to one point on the Hertzsprung–Russell diagram.

scattering The process in which photons bounce off particles in random directions.

Schwarzschild radius (R_S) The radial distance from the centre of a black hole at which the escape speed equals the speed of light. Given by

$$R_\mathrm{S} = \frac{2GM}{c^2}$$

where G is the universal gravitational constant, M is the mass of the black hole, and c is the speed of light in a vacuum.

second law of thermodynamics A law that (among other things) states that heat spontaneously flows from a hotter body to a cooler body but is not spontaneously transferred from a cooler body to a hotter body.

semimajor axis A distance equal to half the longest axis of an ellipse.

shell helium flashes Approximately periodic, explosive releases of energy occurring within the helium-burning shell of material in asymptotic giant branch stars.

shock front The very thin transition zone between an unperturbed medium and a shell of material moving very rapidly through the medium.

sidereal period The period of one revolution of a body in orbit about another, or of one rotation of a body about its own axis, measured relative to the stars.

silicon burning A set of nuclear reactions that involve the fusion of helium nuclei (α-particles) with silicon and the heavier nuclei that are subsequently formed. The α-particles are formed by photodisintegration reactions. Such a process is exothermic up until the point at which elements in the iron group are formed.

SN 1987A The first relatively nearby, unobscured supernova in modern times, occurring about 163 000 light-years away, in the Large Magellanic Cloud, a nearby galaxy. The first radiation from the supernova reached the Earth on 24 February 1987.

solar activity Activity in the Sun, as displayed by a number of phenomena, including sunspots, plages, filaments, prominences and solar flares, all of which are more prevalent at times of higher activity. At such times the corona exhibits streamers jutting out in all directions.

solar activity cycle A roughly 11-year cycle in the level of solar activity (some associated magnetic phenomena have a 22-year cycle).

solar cycle *See* solar activity cycle.

solar flare An energetic event, in which solar radiation at radio, X-ray and other wavelengths increases markedly and suddenly.

solar granulation The pattern of granules on the Sun's photosphere.

solar interior The vast bulk of the Sun that is beyond direct observation, i.e. below the photosphere.

solar limb The edge of the circular disc that the Sun exhibits to an observer.

solar luminosity The rate at which energy is transported away from the Sun by electromagnetic radiation, $L_\odot \approx 3.84 \times 10^{26}\,\text{J}\,\text{s}^{-1}$.

solar models Mathematical models that describe the physical conditions of the solar interior.

solar neutrino problem The apparent discrepancy between the numbers of solar neutrinos measured by detectors on the Earth and those predicted by solar models. This discrepancy has now been explained by the fact that some neutrinos appear to change type between their emission and detection.

solar neutrinos Neutrinos originating in nuclear reactions in the Sun. They provide a direct test of our ideas about solar nuclear reactions.

solar rotation The rotation of the Sun on its axis. The intrinsic (sidereal) period, measured at the photosphere, exhibits differential rotation, varying from about 25 days at the solar equator to about 36 days near the poles.

solar wind A gusty stream of high-speed particles (mainly protons and electrons) that spreads out from the Sun, carrying traces of the Sun's magnetic field with it.

space velocity The velocity of a star through space relative to the Sun. Its components are radial velocity and transverse velocity.

space weather A collective term for the range of effects of solar activity on the Earth's environment.

spectral flux density (F_λ) The quantity that describes the rate at which energy transferred by radiation is received from a source, per unit area facing the source, per unit wavelength range.

spectrometric method A method of determining the photospheric temperature of a star, and its luminosity, by examining the absorption lines in its spectrum.

spectroscopic binaries Binary stars which are identified as such from the Doppler shifts of their spectral lines. Spectroscopic binaries may show two sets of spectral lines that exhibit relative motion as the two stars orbit their common centre of mass or, if one of the stars is very faint, there may be just a single set of observable lines that change their wavelength (or frequency) periodically.

spectroscopic parallax A method for obtaining (approximate) stellar distances based on an observation of a star's flux density and a spectroscopic determination of its approximate luminosity. *See also* luminosity class.

spectroscopy The study of spectra and spectral lines.

spectrum (plural spectra) A representation (usually as a visual image or a graph) of the way in which the strength or intensity of radiation emitted or received from a specified source is distributed with respect to wavelength or frequency.

speed of light in a vacuum (*c*) The speed at which electromagnetic radiation travels through a vacuum, which, to 3 significant figures, is $3.00 \times 10^8\,\text{m}\,\text{s}^{-1}$.

spiral density wave A long-lived, self-consistent pattern of density enhancement that may arise in a disc of stars and gas, thought possibly to account for the pattern of star formation that gives rise to spiral arms in spiral galaxies.

s-process reactions A series of nuclear reactions initiated by the slow absorption of neutrons.

stability The property of a system that enables it to return to its original position or configuration following some (usually small) disturbance or perturbation. In the case of a star, the term stability is often used to refer to the star's ability to counteract any tendency for it to expand or contract. The main sequence lifetime of a star represents an extended demonstration of its stability.

standard candle Any type of object whose luminosity is known from its observable properties, thus allowing its distance to be inferred from the difference between its apparent brightness and its true brightness.

star cluster Any group of stars with more than a few members, in a relatively small volume of space.

Stefan–Boltzmann constant (σ) The constant that relates the power *l* radiated by unit area of a black-body source to its absolute temperature *T*, i.e. $l = \sigma T^4$.

stellar parallax The quantity that describes the change in the direction to a celestial body against a background of far more distant bodies, resulting from a change of one astronomical unit in the observer's position, in a direction perpendicular to the direction to the body.

stellar wind The outflow of material from the surface of a star.

stellar wind accretion A mass transfer process in a binary star system in which some of the strong stellar wind from one star is captured by its companion.

strong shock The effect on a medium of being subjected to the passage of a high-speed flow. The material in the medium is compressed and its temperature raised in the transition zone, called the shock front, between the flow and the unperturbed region. The material is said to have undergone a 'strong shock' or be 'shocked'.

subgiant Any star in the zone of the Hertzsprung–Russell diagram that is between the main sequence and the red giant branch. Most subgiants are on the way to becoming red giants.

sunspots Relatively cool, small patches on the Sun's photosphere. In visible light images they appear as dark patches.

supergiant A star that lies along the top of the Hertzsprung–Russell diagram, i.e. a star with the greatest luminosity. Main sequence stars with masses greater than about $11M_\odot$ evolve to become supergiants. Later, such stars become Type II supernovae.

supergranulation A large-scale pattern of upward and downward motion of the Sun's photosphere, possibly resulting from a deep layer of large convection cells.

supernova An outburst in which a star suddenly increases in brightness by an enormous factor ($\sim 10^6$). Such a star is ending its life in a gigantic explosion resulting from the collapse of its core.

synchrotron radiation Electromagnetic radiation with a continuous spectrum, emitted by electrically charged particles, usually electrons, as they pass through magnetic fields. For appreciable radiation to be emitted, the particles must have very high energies.

T Tauri star A type of irregular variable that exhibits variations in luminosity by factors of two or three over intervals of the order of a few days. They are thought to be very young stars, losing mass before settling on to the main sequence.

termination shock The region in the heliosphere where the solar wind begins to slow down due to the pressure of the interstellar medium.

thermal pulse A rapid release of energy, lasting perhaps a few hundred years, typically caused by a shell helium flash in a star. Thermal pulses are thought to occur almost periodically within stars that have reached an appropriate stage in their evolution, with individual pulses being separated by 10^4 to 10^5 years.

thermal source A source that emits electromagnetic radiation because of its temperature (the higher the temperature, the greater the amount of radiation emitted per unit area of the source).

thermalization The process through which a distribution of photon energies is converted into a black-body spectrum as a result of multiple scattering.

total eclipse of the Sun The natural phenomenon that occurs when the Moon passes between the Earth and the Sun, completely blocking the photosphere from the view of appropriately positioned observers on the Earth's surface.

transition region A thin layer in the Sun's atmosphere, between the chromosphere and the corona.

transverse velocity The component of a star's velocity in the plane of the sky, i.e. in a direction perpendicular to the line of sight of an observer.

trigger mechanism An event or process that initiates some further series of events. In astronomy, it often refers to an event or process that compresses an interstellar cloud and initiates star formation.

trigonometric parallax A method of determining the distances of celestial bodies based on measurements of the parallax of those bodies. *See* parallax.

triple alpha (3α) process The nuclear fusion reactions in which three helium nuclei (alpha particles) combine to form one carbon nucleus. It is the dominant nuclear process in the core of a red giant.

Type II supernova A particular kind of supernova that marks the explosive conclusion of the supergiant phase of the life of stars with masses greater than about $11M_\odot$.

UBV system A set of wavelength ranges commonly used in observations of stars. The wavebands are: U, ultraviolet, central wavelength 360 nm; B, blue, 440 nm; V, visual, 550 nm.

ultraviolet radiation Electromagnetic radiation with frequencies or wavelengths between those of X-rays and visible light.

upper main sequence star A star with a mass greater than about $1.5M_\odot$, occupying the upper part of the main sequence on the Hertzsprung–Russell diagram and in which the CNO cycle dominates the production of energy.

variable star A star whose luminosity varies on a timescale (seconds to years) that is very short compared with the star's lifetime. Variable stars are subdivided into irregular variables and regular variables.

vibrational transition A transition in a molecule between two states corresponding to different amounts of molecular vibrational energy.

visible light Electromagnetic radiation with frequencies or wavelengths between those of ultraviolet radiation and infrared radiation. Our eyes are sensitive to visible light.

visual binary system A binary star in which both stars can be observed as distinct points of light.

warm intercloud medium A low-density component of the interstellar medium characterized by a moderately high temperature; probably derived largely from interstellar clouds, by 'evaporation'.

wavelength The distance over which a periodic wave repeats itself, e.g. the distance from one peak of the wave to the next.

white dwarf The remnant left over when a star sheds a planetary nebula. They lie below the main sequence in the Hertzsprung–Russell diagram.

white light image An image of an object formed using the whole range of visible light wavelengths.

Wien's displacement law A law that, for a black-body source, relates the wavelength λ_{peak} at which the relative spectral flux density is a maximum, to the absolute temperature T of the source. In SI units $(\lambda_{peak}/m) = (2.90 \times 10^{-3})/(T/K)$. It is also known as Wien's law.

Wolf–Rayet star A rare type of hot, massive star which exhibits high mass loss and emission lines, indicating that the outer layers have been ejected and the core revealed.

X-rays Electromagnetic radiation with frequencies or wavelengths between those of gamma-rays and ultraviolet radiation.

X-ray binary An interacting binary system that emits X-rays during the process of accretion. X-ray binaries contain a white dwarf, a neutron star or a black hole and dominate the sky at X-ray wavelengths

X-ray pulsar A neutron star in an accreting binary system which shows regular pulses in its X-ray emission. The origin of the X-ray pulsations is linked to the interaction between the accreting material and the magnetic field of the neutron star.

FURTHER READING

Böhm-Vitense, E. (1989) *Introduction to Stellar Astrophysics, Volume 1 Basic Stellar Observations and Data*, Cambridge University Press, Cambridge.

Charles, P.A. and Seward, F.D. (1995) *Exploring the X-ray Universe*, Cambridge University Press, Cambridge.

Gilmour, I. and Sephton, M.A. (2004) *An Introduction to Astrobiology*, Cambridge University Press, Cambridge.

Jones, M.H. and Lambourne, R.J.A. (2004) *An Introduction to Galaxies and Cosmology*, Cambridge University Press, Cambridge.

Kaler, J.B. (2001) *Extreme Stars – at the Edge of Creation*, Cambridge University Press, Cambridge.

Lang, K.R. (2001) *The Cambridge Encyclopedia of the Sun*, Cambridge University Press, Cambridge.

McBride, N. and Gilmour, I. (2004) *An Introduction to the Solar System*, Cambridge University Press, Cambridge.

Phillips, A.C. (1999) *The Physics of Stars*, 2nd edn, John Wiley and Sons, Chichester.

Phillips, K.J.H. (1992) *Guide to the Sun*, Cambridge University Press, Cambridge.

Prialnik, D. (2000) *An Introduction to the Theory of Stellar Structure and Evolution*, Cambridge University Press, Cambridge.

ACKNOWLEDGEMENTS

The production of this book involved a number of Open University staff, to whom we owe a considerable debt of thanks for their commitment and the high professional standards of their contributions. The production of the Open University course of which this book forms a part was managed by Christopher Edwards, assisted by Valerie Cliff. The text was copy-edited by Rebecca Graham, who also played an invaluable role in steering the project through the production process. The graphic artwork was prepared by Pam Owen with considerable skill and the design and layout was undertaken in an exemplary fashion by Debbie Crouch. The index was prepared by Jane Henley. We are also grateful to Giles Clark (Open University) and Susan Francis (Cambridge University Press) for their support and help with co-publication.

In addition, we wish to thank the following people who commented on earlier versions of the text: Malcolm Longair (University of Cambridge) and Paul Murdin (University of Cambridge) together with anonymous referees appointed by Cambridge University Press. We should also like to acknowledge the contribution made by the following members of the Open University in commenting on early drafts of the text: Andrew Conway, Anthony W. Jones, Andrew Norton and Fiona Vincent. Many other individuals and organizations furnished and/or granted permission for us to use their diagrams or photographs and to them we also express our gratitude.

Grateful acknowledgement is made to the following sources for permission to reproduce material within this book.

Cover

Figures

Observatory; *Figure 1.10* Prof. G. Scharmer, (Royal Swedish Academy of
Sciences); *Figure 1.11* National Optical Astronomy Observatory/Association of
Universities for Research in Astronomy/National Science Foundation; *Figure 1.12*
Copyright: BASS 2000 – Ministere de l'Enseignement Superieur et de la Recherche –
Observatoire de PARIS – Observatoire Midi Pyrenees – INSU/CNRS – France;
Figure 1.15 Science and Society Picture Library; *Figure 1.16* Dr J. Durst,
Schonenberg, Switzerland; *Figure 1.17* George East; *Figures 1.18, 1.20, 1.39c and
1.39d* © Big Bear Solar Observatory/New Jersey Institute of Technology; *Figure
1.19* NASA/Marshall Space Flight Center, Huntsville; *Figure 1.21b* Gordon Garrad/
Science Photo Library; *Figure 1.27* National Solar Observatory/Sacramento Peak;
Figure 1.28 Kitt Peak National Observatory; *Figure 1.29* Royal Astronomical
Society; *Figures 1.33a and 1.33b* copyright 2001 NCAR; *Figure 1.35* Gabriel, A.
(1976) *Philosophical Transactions of the Royal Society of London*, vol. 281, p. 339;
Figure 1.37 and 1.38 Nicholson, I. (1982) *The Sun*, Mitchell Beazley; *Figure 1.39a*
Yohkoh (ISAS); *Figure 1.39e* Nobeyama Radioheliograph, Japan; *Figure 1.40* ©
Stanford University;

Figure 2.4 © Robert Barker/Cornell University; *Figure 2.8* Sky Publishing
Corporation. Reproduced with permission; *Figures 2.9f, 2.20a, 2.20b and 2.30*
National Solar Observatory; *Figure 2.11* © Stanford University; *Figures 2.12 and
2.26* SOHO (ESA & NASA); *Figure 2.13* © Brookhaven National Laboratory;
Figures 2.14a and 2.14b © Sudbury Neutrino Observatory (SNO); *Figure 2.15*
Royal Swedish Academy of Sciences; *Figures 2.21, 2.23, 2.24 and 2.36* © NASA
Headquarters; *Figure 2.35a* © Institut d'Astrophysique, Paris. CNRS; *Figure 2.35b*
Science Applications International Corporation; *Figure 2.38* Lionel F. Stevenson/
Science Photo Library; *Figure 2.39* G. E. Parks (University of Washington) and the
UVI Team, Polar, NASA;

Figure 3.1 Photograph David Malin, Anglo Australian Telescope Board; *Figures 3.4,
3.9 and 3.10* Science Photo Library; *Figures 3.11a, 3.11b, 3.11c and 3.32* Royal
Astronomical Society; *Figure 3.12* © K. A. A. Strand (1973) 'Photographic
measurements of the 6 double stars', Annalen Van De Sterrewacht Te Leiden, Deel
XVIII, Tweede Stuk; *Figure 3.14* Royal Observatory Edinburgh/Anglo Australian
Telescope Board/David Malin; *Figure 3.15* NASA/ESA. Image courtesy of Space
Telescope Science Institution, Baltimore; *Figure 3.17a* P. Warner, MRAO, Cavendish
Laboratories, Cambridge, and William Hershel Telescope, La Palma; *Figure 3.17b*
Courtesy COAST Group, University of Cambridge; *Figure 3.17c* © NASA
Headquarters; *Figure 3.24* Harvard College Observatory/Science Photo Library;
Figure 3.26 Kaufmann, W. J., *Universe*, 5th edition. Copyright © 1985, 1988, 1991
by W. H. Freeman and Company; *Figure 3.30* Abt, H. A. *et al.* (1968) An *Atlas of
Low Dispersion Grating Stellar Spectra* and from a collection at the Royal
Astronomical Society; *Figure 3.37* Prepared with the assistance of M. A. Seeds;

Figure 4.2a © The Royal Astronomical Society; *Figure 4.2b* Science Photo Library;
Figure 4.5 Seeds, M. A. (1984) *Foundations of Astronomy*, Thompson Learning
Global Rights Group; *Figure 4.9* Anglo-Australian Observatory, photograph by David
Malin; *Figure 4.11* Nigel Sharp, Mark Hanna/AURA/NAO/NSF; *Figure 4.13* NASA;
Figure 4.14 © Akira, Fujii, Tokyo;

Figure 5.1 The Royal Astronomical Society; *Figure 5.3* R. Maddalena (NRAO),
M. Morris, J. Moscowitz and P. Thaddeus; *Figure 5.7* Anglo Australian Telescope
Board. Photographs by David Malin Images; *Figure 5.8* N. Walborn (STScI),
R. Barba (La Plata Observatory) and NASA; *Figure 5.11* Ronald Snell, University of

Massachusetts; *Figure 5.13a* C. Burrows (STScI), NASA; *Figure 5.13b* The European Southern Observatory; *Figures 5.14a* and *5.14b* Jeff Hester and Paul Scowen (Arizona State University), NASA; *Figure 5.15a* NASA Headquarters; *Figure 5.15b* Jeff Hester (Arizona State University), NASA; *Figure 5.17* M. J. McCaughrean (MPIA), C. R. O'Dell, Vanderbilt University and NASA; *Figure 5.18* A. Schultz (Computer Sciences Corp.), S. Heap (MASA Goddard Space Flight Center) and NASA; *Figures 5.19a* and *5.19b* Brad Smith (University of Hawaii), Glenn Schneider (University of Arizona), and NASA; *Figure 5.19c* outreach@jach.hawaii.edu;

Figure 6.3 © The Royal Astronomical Society; *Figure 6.11* Atlas Image obtained as part of the Two Micron All Sky Survey (2MASS), a joint project of the University of Massachusetts and the Infrared Processing and Analysis Center/California Institute of Technology, funded by the National Aeronautics and Space Administration and the National Science Foundation; *Figure 6.13a* T. Nakajima (Caltech), S. Durrance (JHU); *Figure 6.13b* S. Kulkarni (Caltech), D.Golimowski (JHU) and NASA; *Figure 6.14a* NASA, K. L. Luhman (Harvard–Smithsonian Center for Astrophysics, Cambridge, Massachusetts) and G. Schneider, E. Young, G. Rieke, A. Cotera, H. Chen, M. Rieke, R. Thompson (Steward Observatory, University of Arizona, Tucson, Arizona); *Figure 6.14b* NASA, C. R. O'Dell and S. K. Wong (Rice University);

Figure 7.12 Yves Grosdidier (University of Montreal and Observatoire de Strasbourg), Anthony Moffat (University of Montreal), Gilles Joncas (Université Laval), Agnes Acker (Observatoire de Strasboug), and NASA;

Figure 8.2 Margarita Karovska (Harvard–Smithsonian Center for Astrophysics) and NASA; *Figure 8.3a* © European Southern Observatory; *Figure 8.3b* © *The Astrophysical Journal*, American Astronomical Society; *Figure 8.4a* The Hubble Heritage Team (AURA/STScI/NASA); *Figure 8.4b* NASA and The Hubble Heritage Team (STScI/AURA); *Figure 8.4c* H. Bond (STScI) and NASA; *Figure 8.4d* Matt Bobrowsky (Orbital Sciences Corporation) and NASA; *Figure 8.4e* NASA, ESA and The Hubble Heritage Team (STScI/AURA); *Figure 8.4f* Bruce Balick (University of Washington), Vincent Icke (Leiden University, The Netherlands), Garrelt Mellema (Stockholm University) and NASA; *Figure 8.5* © Hakon Dahle; *Figure 8.7* 1984–2002, Anglo-Australian Observatory/Royal Observatory, Edinburgh/David Malin; *Figure 8.8* 1989–2002, Anglo-Australian Observatory, photograph by David Malin; *Figure 8.9* © Kamiokande Experiments (http://www.hp.phys.titech.ac.jp/kamioka/ kamioka.html); *Figure 8.11* NASA/STScI; *Figure 8.12* NASA, Peter Challis and Robert Kirshner (Harvard-Smithsonian Center for Astrophysics), Peter Garnavich (University of Notre Dame) and the SINS Collaboration; *Figure 8.13a* VLA http://chandra.harvard.edu/photo; *Figure 8.13b* NASA/CXC/SAO/Rutgers/ J. Hughes; *Figure 8.13c* MDM Observatory http://chandra.harvard.edu/photo; *Figure 8.14a* European Southern Observatory; *Figure 8.14b* W. P. Blair, NASA and The Hubble Heritage Team (STScI/AURA); *Figure 8.15* IRSA at IPAC, California Institute of Technology; *Figure 8.16a* WFAU, Institute for Astronomy, Royal Observatory, Edinburgh; *Figure 8.16b* ROSAT, MPE, NASA;

Figure 9.2 Steve Koppes, University of Chicago News Office; *Figure 9.3(left)* Kitt Peak National Observatory 0.9-meter telescope, National Optical Astronomy Observatories, courtesy of M. Bolte (University of California, Santa Cruz); *Figure 9.3(right)* Harvey Richer (University of British Columbia, Vancouver, Canada) and NASA; *Figure 9.6* © California Institute of Technology; *Figure 9.11* Dr Guy Pooley,

Mullard Radio Astronomy Observatory, Cavendish Laboratory, Cambridge; *Figures 9.12a and 9.12b* Fred Walter (State University of New York at Stony Brook) and NASA; *Figure 9.15* © Rob Hynes, 2000; *Figure 9.16* Michael Muno, MIT; *Figure 9.18* Drawn by F. A. Córdova and H. Papathanassiou using AAVSO data on SS Aurigae; *Figure 9.19* Lick Observatory; *Figures 9.21 and 9.22* Drawn from data provided by Simon Clark, University College, London; *Figure 9.24* European Southern Observatory.

Every effort has been made to contact copyright holders. If any have been inadvertently overlooked the publishers will be pleased to make the necessary arrangements at the first opportunity.

FIGURE REFERENCES

Abt, H.A. (1968) *An Atlas of Low-Dispersion Grating Stellar Spectra*, from a collection at the Royal Astronomical Society.

Ander, E. and Grevesse, N. (1989) *Geochimica et Cosmochimica Acta*, **53**, p.197.

Däppen, W. (2000) Table 3.2 in *Allen's Astrophysical Quantities*, 4th edn, edited by A.N. Cox, Springer-Verlag, New York.

Dayal, A. and Bieging, J.H. (1993) The abundance distribution of C_4H in IRC+10216, *Astrophysical Journal Letters*, **407**, pp.L37–L40.

Fleck, B., Brekke, P., Haugan, S., Sanchez Duarte, L., Domingo, V., Gurman, J.B. and Poland, A.I. (2000) Four years of SOHO discoveries - some highlights, *ESA bulletin*, **102**, pp.68–86.

Foukal, P. (1990) Solar Astrophysics, John Wiley & Sons, Chichester.

Gabriel, A.H. (1976) A magnetic model of the solar transition region, *Philosophical Transactions of the Royal Society of London*, **281**, pp.339–52.

Gosling, J.T. (1999) The Solar Wind, in Weissman, P.R., McFadden, L.A. and Johnson T.V., *Encyclopedia of the Solar System*, Academic Press, London, pp.95–122.

Grevesse, N. and Noels, A. (1993) in *Origin and Evolution of the Elements*, edited by N. Prantzos, E. Vangioni, and M. Casse, Cambridge University Press, Cambridge.

Heyvaerts, J., Priest, E.R. and Rust, D.M. (1977) An emerging flux model for the solar flare phenomenon, *Astrophysical Journal*, **216**, p.123.

Iben, I. (1991), Single and binary star evolution, *Astrophysical Journal Supplement Series*, **76**, pp.55–114.

Kaufmann, W.J. and Freedman, R.A. (1998) *Universe*, W.H. Freeman, New York.

Labs, D. and Neckel, H. (1968), The Radiation of the Solar Photosphere from $2\,000$Å to $100\,\mu$, *Zeitschrift für Astrophysik*, **69**, pp.1–73.

Lang, K.R. (2001) *The Cambridge Encyclopedia of the Sun*, Cambridge University Press, Cambridge.

Mauron, N. and Huggins, P.J. (2000), Multiple shells in IRC+10216: shell properties, *Astronomy and Astrophysics*, **359**, pp.707–15.

Nicholson, I. (1982) *The Sun*, Mitchell Beazley, London.

Pagel, B.E.J. (1997), *Nucleosynthesis and Chemical Evolution of Galaxies*, Cambridge University Press, Cambridge.

Phillips, K.J.H. (1992) *Guide to the Sun*, Cambridge University Press, Cambridge.

Pneuman, G.W. and Kopp, R.A. (1971) Gas-Magnetic Field Interactions in the Solar Corona, *Solar Physics*, **18**, pp.258–70.

Reinecke, M., Hillebrandt, W. and Niemeyer, J.C. (2002) Three-dimensional simulations of type Ia supernovae, *Astronomy and Astrophysics*, **391**, pp.1167–72.

Seeds, M.A. (1984) *Foundations of Astronomy*, Thompson Learning (Global Rights Group).

Strand, K.A.A. (1973) Photographic measurements of the 6 double stars, Annalen Van De Sterrewacht Te Leiden, Deel XVIII, Tweede Stuk.

Sturrock, P.A. (1980) Flare models in Sturrock, P.A., *Solar flares: A monograph from SKYLAB Solar Workshop II*, Colorado Associated University Press, Colorado, pp.411–49.

INDEX

Entries and page numbers in **bold type** refer to key words which are printed in **bold** in the text. Italics indicate items mainly, or wholly, in a figure or table.